T0211512

Disposal of All Forms of Radioactive Waste and Residues

Michael Lersow · Peter Waggitt

Disposal of All Forms of Radioactive Waste and Residues

Long-Term Stable and Safe Storage
in Geotechnical Environmental Structures

 Springer

Michael Lersow
Breitenbrunn, Saxony, Germany

Peter Waggitt
Darwin, NT, Australia

ISBN 978-3-030-32912-9   ISBN 978-3-030-32910-5   (eBook)
https://doi.org/10.1007/978-3-030-32910-5

This Springer imprint is published by the registered company Springer Nature Switzerland AG
The registered company address is: Gewerbestrasse 11, 6330 Cham, Switzerland

Contents

Chapter 1
Introduction

There are 443 nuclear power plants (NPP) currently (2018) in operation in 30 countries for the purpose of generating electrical power [1]. In the future old systems will be decommissioned and new ones will be planned and built. How the share of energy produced by NPPs will develop is not exactly predictable at the time of writing due to economic and political uncertainties. The production however may be slowly decreasing or constant; but it is quite possible that a gentle increase of the share of energy from NPP's could occur in the future. At present, 55 NPPs are under construction against 166 that are now decommissioned and shut down, see [1–3].

What is clear is that the radioactive waste resulting from currently operating plants and residues from the supply and disposal process of NPPs, will need to be disposed of safely. see Fig. 2.6.

Germany is one of the first countries with a plan to discontinue its civil nuclear program by 2022. This followed the events at Fukushima and was further influenced by the planned transition to renewable energy.

This changed the energy policy of the Federal Republic Germany decisively. Germany, as a leading industrialized nation, operated 17 nuclear power plants in 2011, with a grid capacity of 20,490 GWe, which generated approx. 23% of the electricity produced. In 2018, there were still 7 nuclear power plants connect to the grid.

Other industrialized countries, such as Austria and Australia, never used nuclear energy. In 2017 Switzerland decided by referendum for the long-term withdrawal from the nuclear fuel cycle and as a result of the Fukushima accident Italy has already completed such an exit. Other countries such as France, China, USA, Russia, South Korea, Finland, Sweden and Great Britain continue to focus on nuclear energy.

In Germany the policy rejecting the use of nuclear energy remains in place, on the grounds that this is an uncontrollable high-risk technology. However, concerns remain over the risks associated with long term disposal of spent fuel and high level waste from reprocessing. Although developments, especially in Finland, Sweden and France, show encouraging technological disposal solutions, in Germany it remains urgent to close this technical gap in the nuclear fuel cycle with the construction

© Springer Nature Switzerland AG 2020
M. Lersow and P. Waggitt, *Disposal of All Forms of Radioactive Waste and Residues*, https://doi.org/10.1007/978-3-030-32910-5_1

of different geotechnical environmental structures for the disposal of all types of radioactive waste.

Although a less ideologically motivated discussion about the final disposal of radioactive waste and residues was conducted in the USA, the progress in the construction of a repository for Heat-Generating (HG), High Active Waste (HAW) can be compared with that in Germany.

However, a repository for defense-related Transuranic waste has been designed, constructed and put into operation in the USA. The Waste Isolation Pilot Plant (WIPP), see Chap. 7, is an underground facility in bedded salt approximately 658 m (2160 ft.) below the surface in a semi-arid region near Carlsbad, New Mexico, see [4].

The US currently has 99 NPPs in operation with a grid-connected energy of 30,000 GWe. 34 NPPs were permanently shut down, 2 are under construction and 14 NPPs are planned, see [1]. In addition there are the large quantities of radioactive waste and residues from the US military-industrial complex.

With Germany's withdrawal from nuclear power generation, the radioactive and nuclear waste and waste that has accumulated and still accumulates can now be clearly accounted for. This is an essential boundary condition for the construction of repository structures.

For other countries such as the USA, France and Japan, this is not so obviously possible. For these countries, who wish to continue to produce electric energy from nuclear fuel in the future, there is a need to plan and construct the repository for high-level waste (HAW—High Active Waste), and the dimensions of the forecasted volumes and associated inventory must be taken into account.

The annual storage volume and the expected size of the final volume will determine the operational phase of the repository. There may be differences between planning and reality that have to be compensated for during the operating phase. Later the decommissioning phase will take place. For the approved Finnish repository for Heat-generating HAW at Okiluoto, the operating phase is planned to last until about 2112. The Okiluoto repository is to be closed from 2120, see Chap. 7. It is clear with further use of nuclear energy that Finland will need another repository for HAWs, from 2112 approximately, see [2]. The planning, construction, decommissioning and closure of repository structures is a multi-generational task, so that it is expected that the current "generation" of repositories will be further optimized by subsequent generations, incorporating scientific and technological progress.

This book deals with the question of what types of radioactive and nuclear waste and residues have arisen and continue to develop. How can they be disposed of in the long term? Which geotechnical environmental structures already exist or need to be developed and built? The term "long-term security" requires definition not only technically and scientifically, but also legally and ultimately politically. In this context, the remaining residual risk must be clearly stated.

The answer to these questions from an initially complex scientific task does not clearly lead to a general, location-independent solution. Ultimately society is forced to an evaluation of proposed solutions. These must all be subordinated to the goal of designing a final disposal site for radioactive waste and residues in such a way that the harmful effects of the waste on humans and the environment are prevented in the long term. Changes and dispersal of radionuclides from the inventory of the repository in the biosphere can only be tolerated within socially and legally acceptable limits. However, the solution of this task may also be subject to political influence.

The process can be divided into three levels of participation:

– The designer, who develops the task and implements it scientifically and technically,
– The operator who submits the geotechnical environmental structure for approval, according to specification operates and later decommissioned,
– The approver, the appropriate regulating and licensing authority, and possibly also the courts, which ultimately have to make decisions in the case of legal proceedings.

The involvement of the public is an absolute must.

The proper operation of a repository in all the necessary operating phases is an indispensable prerequisite for trouble-free operation and thus also for social acceptance, which is based on trust, transparency and reliability.

What is meant by "proper operation" is generally defined in Germany in the First General Administrative Regulation on the Major Accidents Ordinance—StörfallVwV, of September 20, 1993, [14], under 2.2:

> Operation according to specification is the permissible operation for which a system according its technical purpose is determined, designed and suitable. Operating conditions that do not comply with the granted approval, enforceable subsequent orders or legal provisions do not belong to the operation according to specification.

This is similarly prescribed in other countries. This book is based on the respective German regulations for the long-term safe and stable disposal of radioactive waste. The variations in the individual countries are usually insignificant, because among others the IAEA provides the international oversight. But where significant deviations exist, e.g. in the US, these have been highlighted.

A glossary is attached for clarification of terms. It is based on the terms proposed by the IAEA or the NEA/OECD. But these are not used uniformly in the individual countries. However, the attached sources give the reader the opportunity to identify and follow them up at any time. This also applies to mining terms. In this area the terms follow long mining traditions. The glossary largely balances these out.

In this book, in addition to the general hazard potential of the substances that are put into a repository, the focus is placed on the safe storage of radioactive substances.

Radioactivity is the property of certain nuclides (radionuclides) to transform into a new nuclide (or several new nuclides) without external action. In the conversion of radionuclides, the energy is released in the form of radiation. Radioactivity is thus a value-free physical phenomenon.

Since the discovery of radioactivity by Antoine Henri Becquerel in 1896, the dangers posed by radioactivity and how it directly affects human health and society have been recognised. It is interesting to note that even one of the greatest of the early nuclear physicists, Marie Curie was unaware of the life-threatening properties of ionizing radiation. The acceptable use of artificially generated radioactivity in

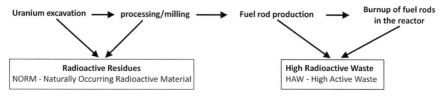

Fig. 1.1 Origination of radioactive waste and residues [2]

particular is politically highly dependent on solving the questions "What types of radioactive materials are generated, where and how, which uses and recycling options for these are possible, which residues, by-products and radioactive waste accumulate and how can they ultimately be safely disposed of".

This book will describe geotechnical environmental structures designed to reduce the risks of ionizing radiation on human populations. Permanent disposal sites depend on the nature and properties of the radioactive material which is placed inside for long term safe and stable disposal. The radioactive waste results primarily from technical processes in which ionizing radiation or nuclear properties of the material used are used, but also includes the residues of processing and milling processes. Reference is made to ICRP[1] Publication 103 [5] and the subsequent European Directive 2013/59/Euratom,[2] [6].

The use of radioactivity first requires the extraction and processing of the appropriate geological raw material. Enrichment of the natural radionuclides and subsequent nuclear reactions produces radioactive material suitable for use in various technical and medical processes, e.g. the manufacture of nuclear medical preparations and fuel assemblies for use in nuclear reactors for power generation. However, radioactive residues and waste are always generated in this process chain, see Fig. 1.1.

By reference to physico-chemical properties, the type of radioactive material will be described by its category of radioactivity. The radioactivity is characterized by the type of radiation, the half-life and the energy of the radiation. The knowledge of these physical quantities is an important prerequisite for disposal.

The radiotoxicity of a radionuclide is a parameter that describes the potential hazard of a radionuclide. Where radiotoxicity is greater, the lower its exemption limit, see German Radiation Protection Ordinance (StrlSchV), Appendix 4, Table 1 columns 2, 3. The exemption limit not only determines the radiotoxicity of a radionuclide, but its biological effect must also be included, see Table 1.1.

According to EURATOM and ICRP [5, 6], radioactive material is any material containing radioactive matter within the meaning of this Directive; that is to say radioactive to the extent that this radioactivity cannot be disregarded. In Germany it is then determined whether the material to be stored in accordance with the Recycling Management Act (KrWG) and Landfill Ordinance (DepV) and can be deposited in

---

[1]ICRP—International Commission on Radiation Protection.

[2]European Atomic Energy Community (EAEC or Euratom).

Table 1.1 According to the ordinance on the protection against damage of ionizing radiation

Radiotoxicity	Radionuclides	Exemptions limit [Bq]; German Radiation Protection Ordinance (StrlSchV), Appendix 4, Table 1
Very high	Sr-90, Y-90, Po-210, Ra-226, Pu-239	1.0E+04; 1.0E+E04; 1.0E+04; 1.0E+04; 1.0E+04
High	Cs-134 Cs-137, U-233	1.0E+04; 1.0E+04; 1.0E+04
Medium	P-32, J-131, Ba-140,	1.0E+05; 1.0E+06; 1.0E+05
Low	H-3, U_{nat}	1.0E+09

German Radiation Protection Ordinance—StrlSchV—Annex 4, Table 1, column 2

a landfill as a material with no regard to the radioactivity; or if it must be stored in a facility according to the Atomic Energy Act (AtG).

In Germany the atomic law forms the legal basis for the establishment of safe systems of managing radioactive materials that pose a significant risk to human health and the environment. According to § 9b of the Atomic Energy Act, for example, the establishment, operation and decommissioning of the federal installations referred to in § 9a (3), as well as the substantial modification of such facilities or their approved operation plan, must include an integrated environmental impact assessment. In cases where the location has been determined by federal law, the planning approval is replaced by a permit. The approval is issued by the Federal Agency for Nuclear Waste Management at the request of the operator.

Planning approval procedures with an integrated environmental impact assessment are also used in the construction of landfills in accordance with the KrWG.

The United States is also now making preparations for the handling and disposal of to the anticipated large volumes of VLLW[3] associated with the decommissioning of nuclear power plants and material sites; this also includes, waste that might be generated by alternative waste streams that may be created by operating reprocessing facilities or a radiological event. The Nuclear Regulatory Commission (NRC) coordinates these tasks. For example, in 2018[4] the NRC commissioned a large scoping study to develop possible options to improve and strengthen the regulatory framework for the disposal of the large volumes of VLLW expected.

However, the process continues to be based on 10 CFR, Part 61—"Licencing of requirements for land disposal of radioactive waste".[5] It contains approval procedures, performance targets and technical requirements for the licensing of radioactive waste (VLLW).

---

[3] VLLW—Very Low Level Waste.

[4] A Notice by the Nuclear Regulatory Commission on 02/14/2018; https://www.federalregister.gov/documents/2018/02/14/2018-03083/very-low-level-radioactive-waste-scoping-study.

[5] https://www.nrc.gov/reading-rm/doc-collections/cfr/part061/.

Deviating from this is the treatment of residues from uranium ore processing[6] is an issue here. To manage this situation and put it in context of the objective of a long-term safe and stable disposal is here considered a necessity.

Geotechnical environmental structures are therefore presented for their characteristics according to specification as landfill sites, as encapsulated uranium tailings ponds and as repository structures for radioactive waste. In the United States, uranium mill tailings ponds are licensed, operated and finally safely deposited on the basis of The Uranium Mill Tailings Radiation Control Act of 1978, see [7].

Although there are differences in the classification of different radioactive materials (VLLW and uranium mill tailings), there are few differences between Germany and the USA in the structure of geotechnical environmental structures for the long-term safe storage of these radio-toxic materials.

An attempt is made to determine which relevant environmental requirements should be used to determine the necessary site-specific environmental structures to be developed. The book will discuss the current state of science and technology as well as which solutions are to be expected on the basis of current research programs. The text will also discuss the legal situation and where knowledge gaps are to be closed. The current state of knowledge and consequent scientific-technical consensus is intended to equip the environmental structures with a multi-barrier system during operation according to the selected specification.

Worldwide, deep geological formations are favored as repository sites for highly radioactive waste. It is the task of the site assessment to develop and fine tune the multi-barrier system best suited to the geological formation chosen to host the final repository.

In Germany, it has not yet been possible to summarize the specifications, regulations and standards for the construction of facilities for long-term safekeeping of radioactive waste and residues in a binding set of rules. However, this is urgently required for such plants, where there is such a high and long-term potential risk for the biosphere.

A set of technical and scientific rules should be based on general and consistent criteria derived from the requirements for long-term safekeeping of the various types of radioactive waste and residues and adapted to the current state of science and technology. The rules should be characterized by high objectivity. The following statements are also intended to contribute to this.

Concentrating on the residue storage facilities from uranium ore milling is significant in that significant concentrations of radioactivity and toxicity remain in the deposits. The text concentrates on the six large uranium tailing ponds of the SDAG, which were designed to accommodate the residues from processing plants in Crossen (Saxony) and Seelingstädt (Thuringia).

---

[6]Concentrating on the residue reservoirs from uranium ore processing is significant in that significant concentrations of radioactivity and toxicity been deposited in the basins. The concentration on the 6 large uranium tailing ponds of the SDAG, which were assigned to the processing plants in Crossen (Saxony) and Seelingstädt (Thuringia), sufficiently concentrate the long-term safe, long-term stable closure of the residues requiring monitoring.

These sites are responsible for the long-term safe and stable custody of the residues and require monitoring. Strictly speaking, the large stockpile systems, opencast mines, underground mining, processing plants, etc., could of course, also be used to accommodate the accumulated waste residues of the legacy of uranium mining and processing.

The responsible authorities in Germany as well as in the USA, Canada, UK, etc. also have an obligation to analyze and to make readily useable their international experience and the resulting recommendations for a repository concept for highly radioactive waste. Following the decision in several countries, such as Finland and Sweden, to decide on the construction of repositories based on approved long-term safety proofs and legally binding decisions, it is now possible to better target the discussion regarding a repository concept. The explanations here do not only describe concept basics, but also provide solution-oriented stimulations.

The radionuclide vectors of the inventories of repository sites of for High Active Waste (fuel assemblies, vitrified high-level radioactive waste) are compared. These have many similarities. The radiotoxicity, indicated by a dimensionless radiotoxicity index A_i/F_i (A_i activity of the radionuclide amount, F_i exemption limit of the radionuclide), of all the repository sites listed here is approximately 1E+16, regardless of host rock type. This suggests that a site-specific long-term safety case always requires a multi-barrier system tailored to the host rock type. The barriers have to be well coordinated, so that their mechanical, hydraulic, chemical effects complement each other optimally and ultimately the hydrochemical mobility of the radionuclides in the waste matrices is reduced very much, ideally to the point of immobilization. Some international waste container developments should also be considered for example for a German and also in a Canadian repository concept for Heat-generating HAW, see Chap. 7.

The discussion here deals extensively with the storage of all types of radioactive waste and residues in Germany. Based on the dimensionless radiotoxicity indices of the respective inventories in the respective geotechnical environmental structures, the reader is given a well-founded overview of the degree of unequal treatment in the safety requirements and thus the necessary transparency is established, [13].

In order to make repository sites comparable with each other, this task requires that one first refers to a legal and regulatory framework, as this is the only way to make geotechnical environmental structures comparable. Even if the authors refer in the following to the German regulations as a matter of priority, it is in any case necessary to make a comparison with the legal and regulatory framework of the USA. The authors are keen to emphasise this because, on the one hand, the USA has the world's greatest experience in dealing with radioactivity and, on the other hand, it is there that by far the largest quantities of radioactive waste and residues are generated annually, as a result of the peaceful and military uses of nuclear energy. Thirdly, because the USA uses a different classification of radioactive waste to Germany, for example. This is due to tradition and because in the USA the need for classification arose earlier, before the IAEA developed and promulgated a classification.

According to the World Nuclear Association (WNA), 12,000 tons of high-level radioactive waste are produced worldwide every year. As a result of Germany's

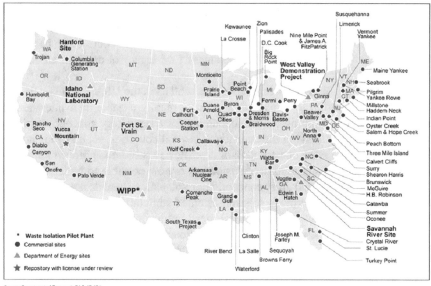

Fig. 1.2 Current storage sites for HLW and SNF and repository with license under review (*Source* U.S. Government Accountability Office and in addition WIPP inserted by the authors)

announced withdrawal from the peaceful use of nuclear energy, the amount of high-level radioactive waste is known quite precisely and after 2022 only extremely small amounts will be added each year. According to the US Department of Energy (US DOE), approximately 90,000 tons of highly radioactive, heat-generating waste are currently stored at 80 sites in 35 states in the USA, 14,000 tons of which come from the US government's nuclear weapons program. Approximately 3000 tons are added annually. However, this corresponds to less than 0.5% of total radioactive waste and residues, see Fig. 1.2.

In order to achieve the goal of "best possible final disposal" in Germany, the law on the search and selection of a site for a repository for heat-generating radioactive waste (Standortauswahlgesetz[7]—StandAG) of 23 July 2013 was adopted together with the further development of 23 March 2017. This created an important prerequisite for the "path to the repository". In Germany, in 2016, the legislature also enacted a reorganization of the organizational structure in the area of radioactive waste disposal, which should better integrate the described levels of participation but still allow them to be able act independently of each other, see [8]. These are similar to structures in comparable industrialized countries or have been adapted from such structures. With the reorganization of the organizational structure in the field of final disposal of radioactive waste, a pluralistic National Accompanying Board (NBG) was set up by law, for the benefit of the public; the Board also has the task of overseeing the implementation of public participation. However, it must be assumed that the

---

[7]Site Selection Act.

organizational structure in the field of final disposal should be further optimized. On the one hand the levels of participation need to be clearly differentiated from each other and on the other hand to be able to develop the best possible repository system for Germany from the various task-solving possibilities.

In order to facilitate a functional design and construction of the repositories a considerable number of geological, hydrogeological, geochemical, geotechnical and technical issues have to be worked out. The social, political and legal problems should also not be forgotten. It is expected that legislation will be required to create updates to the StandAG to facilitate the disposal of highly reactive, radioactive, heat-Generating waste, see also [12].

So part of the question to be answered is, which independent monitoring options must be available, to demonstrate the safety of the storage of radioactive material in the various stages of final disposal. This activity must take place from construction to closure and beyond, and be able to identify early any security-relevant changes and thus to be able to respond with countermeasures.

Earlier repository concepts for highly radioactive, Heat-generating waste were planned on the basis that the decommissioning and sealing of the repository would take place as soon as possible after operation, [9] BMU (2010). Today the necessity of the timely the installation of a requirement-oriented monitoring concept will be discussed intensively. This should be an integral part of the approval process.

A disadvantage of disposal in deep geological formations is that—measured by the longevity of the waste—only a limited observation of the repository process is possible, e.g. the geochemical interactions. The input values in the sensitivity analysis of a long-term safety record, to which the result is most sensitive, and the enclosed uncertainty analysis, are only tools to narrow the solution range. You cannot completely rule out observational errors. Significant disturbances which may present a risk can only be detected at a very large distance after the occurrence of any malfunction or interference. Even a wrong location decision would then no longer be correctable and repair measures in the repository itself would be virtually impossible. The provisions for reversibility, accessibility, recoverability and retrievability are immense and it is not certain that these will be viable in the very long time. The reversibility of the highly radioactive waste stored in the repository building should be ensured over 500 years according to [10] StandAG (2017).

When deciding on a disposal in deep geological strata, there are high demands on the site selection procedure, namely the proof of suitability for the repository and, in particular, the procedure for creating the long-term safety record. It should be noted that the essential foundations for the long-term safety proof and the achievable fore-cast reliability are already established within the location definition. The procedures must therefore be methodologically correct and coherent, meet the legal and social framework conditions in an appropriate and binding manner, and be comprehensible to non-technical readers. This includes the teaching of the scientific and technical fundamentals and the international state of scientific and technical knowledge, as described here.

It is a requirement that the safety of the personnel must be considered during the operation according to current health and safety regulation. The safety of personnel is not only important in the storage and closure phase of repository structures but also in scenarios of retrieval, recovery, excavation etc. and therefore deserves very high attention, see Chaps. 2 and 6.

Related to these matters are also questions of the integration of interim storage facilities and the establishment of transitional camps on the basis of a participation procedure and which tasks and functions they could take over in a repository concept, see Chap. 7.

A set of rules must also take account of social conditions. There is a necessity that the rules will be updated frequently to reflect increases in scientific and technical knowledge. Due to the long-term risk potential for the biosphere, this task has not only a technical-scientific dimension but will require transparent information to be given to the public via appropriate authorities. Although the main focus here is on the technical-scientific dimension of this task, one cannot completely hide the fact that the implementation of a "best possible" disposal concept, especially for highly radioactive waste, requires credible confidence-building and communication by the responsible bodies and organizations. A few remarks on this are included in this article, while ethical-moral aspects of this task have not been specifically included in this text.

It also becomes clear that the proof of safety in the post-closure phase cannot be carried out in a strictly scientific sense, since the potential consequences of implementing the repository are beyond the scope of meteorological verification due to the long observation period (detection period) of 1 million years. Until the long-term safe and stable decommissioning, this not only requires a continuous optimization process with periodic safety checks, but it is pointed out that there are possibilities to reduce or split the observation period. One in which the long-term safety monitoring and recording time windows, taking into consideration different uncertainties in the probability of occurrence of events and risks is facilitated, without having to reduce the security requirements.

The main objective of the book is to present the technical-scientific dimension of the construction of environmental structures for the disposal of any radioactive and nuclear waste and to inform a broad social public in this task. This is especially important as the development, construction, operation and closure of the respective repositories is a cross-generational task.

Acknowledgements We would also like to thank SKB AB (Sweden), Posiva Oy (Finland), US Department of Energy (DOE), Bundesgesellschaft für Endlagerung mbH[8] and the Thuringian State Institute for the Environment and Geology for the provision of graphics and data, with which complex connections can be presented more clearly.
Special thanks are due to Prof. Dr. rer. nat. Bruno Thomauske, former Director at Research Centre Jülich; Prof. Dr. rer. nat. habil. Gert Bernhard, former Director of the Institute for Resource Ecology (Radiochemistry) of the Helmholtz-Zentrum Dresden-Rossendorf and Prof. Dr. rer. nat. habil. Broder J. Merkel, former Director of the Institute of Geology of the TU Bergakademie Freiberg. Not only did they provide support in the preparation of the German text, but they also encouraged us to undertake the project.

---

[8]Federal company for radioactive waste disposal.

The authors are particularly grateful to Springer nature for encouraging this work. The basis for this presentation here is a German version [11] on the same topic. However, the internationality of the topic was reflected in this English language edition.

References

1. World Nuclear Power Reactors & Uranium Requirements; http://www.world-nuclear.org; July 2019.
2. Lersow, M.: Energy Source Uranium – Resources, Production and Adequacy; Glueckauf Mining Reporter; Verlag der Bergbau-Verwaltungsgesellschaft mbH, Shamrockring 1, 44623 Herne; 153(3) p. 178–194; 6/2017.
3. Power Reactor Information System (PRIS); the International Atomic Energy Agency (IAEA); https://pris.iaea.org/PRIS/WorldStatistics/OperationalReactorsByCountry.aspx.
4. U.S. Department of Energy - Waste Isolation Pilot Plant, Annual Site, Environmental Report for 2007; DOE/WIPP-08-2225; September 2008.
5. Published by ICRP – International Commission on Radiation Protection Annals of the ICRP: ICRP Publication 103 – The 2007 Recommendations of the International Commission on Radiological Protection. Elsevier, Amsterdam 2007, ISBN 978-0-7020-3048-2.
6. COUNCIL DIRECTIVE 2013/59/EURATOM of 5 December 2013: Laying down basic safety standards for protection against the dangers arising from exposure to ionising radiation, and repealing Directives 89/618/Euratom, 90/641/Euratom, 96/29/Euratom, 97/43/Euratom and 2003/122/Euratom.
7. The Uranium Mill Tailings Radiation Control Act (UMTRCA) of 1978, Department of Energy, Public law 95-604; 09. November 1978.
8. Responsibility for the future a fair and transparent procedure for the selection of a national repository site, Commission for the Storage of Highly Radioactive Waste in accordance with § 3 of the Site Selection Act, StandAG, K-Drs. 268, May 2016, German only.
9. The safety requirements governing the final disposal of heat-generating radioactive waste of 30.09.2010 by the Federal Ministry for the Environment, Nature Conservation and Nuclear Safety (BMU).
10. Law on Search and Selection of a Site for a Repository for Heat-Generating Radioactive Waste (Site Selection Act - StandAG); of 23 July 2013 and its development of 23 March 2017.
11. Lersow, M.: Endlagerung aller Arten von radioaktiven Abfällen und Rückständen; Springer Spectrum; ISBN 978-3-662-57821-6, 1. Aufl. 2018, 448 pages.
12. Abschlussbericht der Kommission Lagerung hochradioaktiver Abfallstoffe "Verantwortung für die Zukunft", Deutscher Bundestag (2016); English translation: Report of the German Commission on the Storage of High Level Radioactive Waste; July 2016; https://www.gruene-bundestag.de/fileadmin/media/gruenebundestag_de/themen_az/Gorleben_PUA/Report-German-Commission-Storage-High-Level-Radioactive-Waste.pdf.
13. Lersow, M.; Gellermann, R (2015).; Langzeitstabile, langzeitsichere Verwahrung von Rückständen und radioaktiven Abfällen – Sachstand und Beitrag zur Diskussion um Lagerung (Endlagerung); Ernst & Sohn Verlag für Architektur und technische Wissenschaften GmbH & Co. KG, Berlin geotechnik 38 (2015), Heft 3, S. 173–192.
14. 1.StörfallVwV - Erste Allgemeine Verwaltungsvorschrift zur Störfall-Verordnung; Vom 20. September 1993, (GMBl. S. 582, berichtigt GMBl. 1994 S. 820); https://www.umwelt-online.de/recht/luft/bimschg/vo/12v1v_gs.htm

Chapter 2
Radioactivity in Waste and Residues

Currently, public discussion on the disposal of radioactive substances in Germany and in other parts of the world is focused on the search for a suitable location for the repository for heat-generating radioactive (highly radioactive) waste. In order to begin the discussion on technical and scientific aspects relating to the long-term stable and safe disposal of radioactive waste, it is necessary to clearly define the initial parameters which are scientifically credible and socially acceptable.

It is necessary to achieve consensus to achieve the following aim:

- identification of which radioactive substances (quantity, quality) must be disposed of,
- where and how are these created and accumulated,
- what are the basic potential dangers to the biosphere arising from this waste,
- which technical storage solutions are possible, in principle,
- what are the selection criteria for location (host formation, etc.),
- definition of the safety case,
- establishing the criteria and use of data (databases), models and programs for the long-term safety analysis,
- on which legal basis the safekeeping or storage (disposal) will be based: this ideally will remain unchanged in the future,
- which parameters must be adhered to,
- which criteria will be selected to prove compliance with the parameters over long periods of time.

2.1 Classification of Residues and Radioactive Waste

A first difficulty in this description results from the fact that, although radioactivity is considered a value-free physical phenomenon, radionuclides of natural origin were considered differently in the legal framework compared to radionuclides produced using technology or used for the production of nuclear energy.

© Springer Nature Switzerland AG 2020
M. Lersow and P. Waggitt, *Disposal of All Forms of Radioactive Waste and Residues*, https://doi.org/10.1007/978-3-030-32910-5_2

Table 2.1 Occurrence of radionuclides

Naturally occurring radionuclides	Artificial radionuclides
K-40 (contained in the natural isotope mixture K_{nat} with 0.0117%, $T_{1/2} = 1.277E+09$ a	**Artificial radionuclides** are those that result from nuclear reactions induced by humans
Radionuclides of the natural decay chains (e.g. radionuclides occurring in the natural isotope mixture U_{nat}—U-235 and U-238)	**Activation products**: from nuclides which have become radioactive due to neutron radiation in a nuclear reactor outside the fuel, e.g. in the fuel rod cladding tubes, in the coolant, etc.
C-14, cosmogenic radionuclide, $T_{1/2} = 5.730$ $\pm$ 40 a	**Fission products**: nuclides formed by nuclear fission

While Directive 96/29/Euratom [1], based on ICRP Publication No 60, has planned a separate regulatory regime for naturally occurring radionuclides and radioactive substances and their ionizing radiation or nuclear properties, the new European Directive 2013/59/Euratom [2], based on the ICRP publication 103 [3], intends to repeal as far as possible this conceptual separation (Table 2.1).

The central definition in radiation protection, which remains unchanged in its core, is that of "radioactive substance". This term describes a substance whose activity cannot be ignored under radiation protection requirements due to the activity concentration. To clarify this, the definitions given in Directive 96/29/Euratom and Directive 2013/59/Euratom are compared:

96/29/Euratom: *""Radioactive substance": any substance that contains one or more radionuclides the activity or concentration of which cannot be disregarded as far as radiation protection is concerned."*

And

2013/59/Euratom: *""Radioactive substance" means any substance that contains one or more radionuclides the activity or activity concentration of which cannot be disregarded from a radiation protection point of view."*

The German Radiation Protection Act (StrlSchG) of 27 June 2017 transposes Directive 2013/59/Euratom into national law. It contains an authorization to recast the Radiation Protection Ordinance (StrlSchV).

However, the reorganisation of the legal framework for natural and closure situations has led to some issues, e.g. consideration of the uranium mill tailings ponds of the SDAG Wismut which were constructed according to the old regulations, including the safety definitions. A subsequent change to the design is impossible.

Since in reality radioactive substances are normally mixed with other chemical substances, Directive 2013/59/Euratom introduces the term of radioactive material. Radioactive material is therefore any material which contains radioactive substances within the meaning of the Directive. to the extent that this radioactivity cannot be disregarded—"radioactive material" therefore means material incorporating radioactive substances.

The property "radioactive" is thus bound by these definitions to a legal assessment ("cannot be ignored") and thus not attached to the scientific property of radioactivity. Waste materials, i.e. substances that an owner wants to discard and for which no further use is foreseen (see EU Waste Directive), which always contain radionuclides in the physical sense, are not radioactive substances, unless they are based on a legal assessment due to their particular radiological properties or are a quantity sufficient to be declared as radioactive. These substances are dealt with in Chap. 3 with reference to the Closed Substance Cycle and Waste Management Act (KrWG) and the Landfill Ordinance. Substantial amounts of such material will be encountered when dismantling the NPPs.

The USA is now making preparations on how to deal with the disposal of anticipated large volumes of Very Low Level Waste (VLLW) associated with the decommissioning of nuclear power plants and material sites, as well as waste that might be generated from alternative waste streams that may be created by operating reprocessing facilities or a radiological event. The Nuclear Regulatory Commission (NRC) coordinates these tasks. For example, in 2018 the NRC commissioned a large scoping study to develop possible options to improve and strengthen the regulatory framework for the disposal of the expected large volumes of VLLW. However, it continues to be based on 10 CFR, Part 61—"Licensing of requirements for land disposal of radioactive waste".[1] This document contains approval procedures, performance targets and technical requirements for the licensing of disposal of radioactive waste (VLLW).

Against this background, the depiction of the disposal of all the wastes, which are monitored by radiation protection law due to their radioactivity, is considered here comprehensively. Two categories of waste are to be distinguished in the current radiation protection system in Germany:

- Radioactive waste which is radioactive material within the meaning of § 2 Abs.1 of the Atomic Energy Act (AtG), and must be disposed of according to § 9a AtG. Such materials originate (as a rule) from a radiation-protection or nuclear-licensing situations in which the radioactivity or the radiation emanating from it was used purposefully and which are regulated under Part 2 of the Radiation Protection Ordinance.
- Residues which are materials that are produced in the industrial and mining processes listed in Annex 12 Part A of the German Radiation Protection Ordinance and whose recovery or disposal is regulated in Part 3 of the Radiation Protection Ordinance. These materials are substances that contain naturally occurring radionuclides or are contaminated with such substances.

In contrast to Closed Substance Cycle and Waste Management Act (KrWG), in which the concept of residual material was consistently replaced by the concept of waste, the Atomic Energy Act (AtG) continues to contain this term in § 9a. Radioactive waste generated in a plant regulated under nuclear law or radiation-protection law can then be disposed of freely or shall be disposed of in an orderly

---

[1] https://www.nrc.gov/reading-rm/doc-collections/cfr/part061/.

manner as radioactive waste (direct disposal). Thus, the term "radioactive waste" in the narrower sense refers only to radioactive materials that have to be disposed of in special environmental structures (e.g. repositories) under radiation protection supervision during specified normal operational conditions. With regard to the long-term safekeeping of waste that is state-monitored due to its radioactivity, four areas of origin can be distinguished:

(1) The technical application of radioactive materials in medicine, industry and research, which leads to radioactive contaminated residues for which no further use is envisaged.

(2) The use of nuclear energy for power generation (abroad, in some cases, the development and production of nuclear weapons),

    (a) their operation (for example during plant maintenance) produces radioactive contaminated residues that cannot be reused

    (b) dismantling of nuclear power plants after the withdrawal of Germany from the peaceful use of nuclear energy will result in an increased amount of radioactively contaminated waste.

(3) Mining or industrial processes resulting in residues of naturally occurring radionuclides ("NORM"), which are governed by Directive 2013/59/Euratom.

(4) Mining for the extraction of uranium (and/or thorium) and the processing of ores obtained as a special part of the nuclear energy programme is given a special role.

Legend to Fig. 2.1**: Classification according to IAEA (2009):**

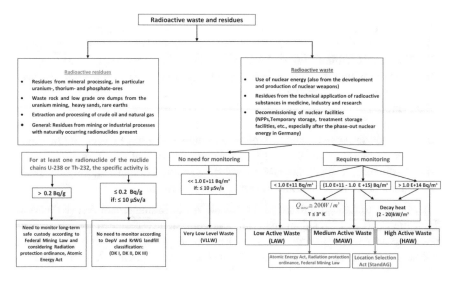

Fig. 2.1 Classification of radioactive waste and residues proposed for use in Germany [4]

EW   Exempt waste, Waste exempted or excluded from the fuel cycle by regula-
             tory control and not declared as waste, e.g. demolition materials from the
             dismantling of NPPs for potential recycling as building material,
VLLW   Very Low Level Waste,
LLW   Low Level Waste (Low Active Waste—**LAW**),
ILW   Intermediate Level Waste (Medium Active Waste—**MAW**). ($10^{10}-10^{15}$)
             Bq/m^3 Activity range, according to IAEA classification,
HLW   High Level Waste (High Active Waste—**HAW**). A $> 10^{14}$ Bq/m^3 Activity
             range according to IAEA classification,
HGW   Heat Generating Waste,
             Active Waste with Negligible Heat Generation ($\Delta T \leq 3$ K)—Temperature
             range for repository "Shaft Konrad"—(NHGW)

These four areas are characterized by (very) different legal framework conditions. Consequently, the current conditions of the various long-term safekeeping methods (final disposal) for these wastes are very different. Looking at the already proven concepts e.g. as the safekeeping of mine structures and dumps of tailings from uranium mining or tailings ponds from uranium ore processing but also the existing facilities for the disposal of low and medium radioactive waste (in Germany, the final disposal for radioactive waste Morsleben, ERAM, and the planned, but not in operation final disposal "Shaft Konrad"), see Chap. 6. It can be seen that in solving the disposal problem of radioactive waste and residues, a wide range of technical options can be used, but these must be tailored to the respective location and the waste to be disposed. Figure 2.1 shows the classification of radioactive wastes and residues currently being established in Germany and the classification of radioactive waste according to the IAEA; see also Table 2.2.

2.2   Description of Residues and Radioactive Waste

In all four of the abovementioned areas of origin, there are subsets of waste which, because of the low radioactive hazard potential, do not require radiation protection monitoring and those which require special requirements on disposal. The German Commission on Radiological Protection (SSK) has issued a number of recommendations on the release of slightly radioactive substances, which have been summarized in a general recommendation [5]. These recommendations are based on the *de minimis* concept[2] of the IAEA [6], which provides for the population to receive a dose limit of 10 μSv per calendar year for substances that are no longer subject to surveillance. These are called substances with very low radioactivity (VLLW) and are subject to Closed Substance Cycle and Waste Management Act (KrWG). This has resulted in

[2]The de minimis concept defines a dose at which potential risks are so low that these are outside of the need for regulation.

Table 2.2 Disposal (recycling) of radioactive waste and residues according

Type of waste	EW (released from the fuel cycle)	VLLW	Residues not requiring special supervision	LAW (LLW) and MAW (ILW) as NHGW	Residues requiring special supervision	HAW (HLW) and MAW as HGW
Waste management	Released for recycling	According to KrWG/DepV[a] and StrlSchV[b] on landfills of DK II and III	According to KrWG/DepV and StrlSchV on landfills of DK II and III	In plants according to AtG § 9a	According to Federal Mining Act (BBergG) and radiation protection application for authorisation according to "Regulation for ensuring on nuclear safety and radiation protection" (VOAS) and according to "Order for ensuring of radiation protection in dumps and industrial tailings ponds and in use of therein deposited materials" (HaldAO)	In plants according to AtG § 9a
Example	E.g. concrete demolition during dismantling of NPP's	E.g. solid/liquid material from the control area of a NPP with release according to § 29 StrlSchV	E.g. red mud/sludges from oil and gas production	Operation and dismantling of NPP's as well as in medicine, industry and research	For example uranium tailings from the uranium ore milling	Spent fuel elements and waste from reprocessing
In this book	Chapter 4	Chapter 4	Chapter 4	Chapter 6	Chapter 5	Chapter 7

To the Classification of radioactive waste—General Safety guide, IAEA [4]; Assignment to the geotechnical environmental structures Chaps. 4–7

[a]KrWG—Closed Substance Cycle and Waste Management Act; DepV—Landfill Ordinance

[b]Radiation protection ordinance

certain requirements for continued and future landfills (DepV) and disposal facilities as well as the associated work processes that influence the dose calculation (in particular soil sealing, cover, pre-treatment of waste).

The Commission for Radiological Protection was commissioned by the Federal Ministry for the Environment, Nature Conservation and Nuclear Safety[3] to draw up a recommendation for the determination of the clearance levels for disposal. In addition to the above-mentioned boundary conditions, the size (annual capacity) of the disposal plants to be established was also re-assessed. In addition, in the case of the annual approved mass of released waste, which may be fed to a single disposal plant, it was considered that in the future increased demolition projects with huge waste streams would become relevant and should be included in the modeling.

Examples of producers of large quantities of very low radioactivity waste (VLLW):

- Aluminum industry: red mud
- Lignite and coal-fired power stations—power station ashes; flue gas cleaning residues, fluidized bed ashes
- Waste incineration ash: residues
- Sewage treatment plants; sewage sludge
- Oil and gas industry: sludge, oil residues
- Decommissioning of nuclear power plants: especially demolition materials.

The annual amount of waste with a low level of radioactivity throughout Germany, is currently forecast to be approximately 10,000 m^3, with an upward trend in the coming years.

For comparison the comparable volume produced in the USA in 2017[4] was 5,053,885 cubic feet $\approx$ 143,000 m^3 with an activity of 57,179 Ci $\approx$ 2.116 PBq. Most of this was disposed at the Clive/Utah repository, see Chap. 6.

The low-level radioactive waste (LAW) and medium-level radioactive waste (MAW) come from nuclear facilities as well as research facilities and industrial or medical applications of radioactive materials. Typical radionuclides in the inventories[5] are H-3, C-14, Co-60, Ni-63, Sr-90, Cs-137, Ra-226 but also actinides (uranium, plutonium, americium). In the course of the approval process for the retrofitting of the "Shaft Konrad" to the repository for LAW and MAW, it was crucial to consider the decay heat of the radioactive materials to be deposited and its possible effect on the host rock in the deposition areas. The establishment of a repository for radioactive waste was approved with the criterion that the temperature increase of the host rock caused by the heat of disintegration must not exceed on average $\Delta T \leq 3$ K. Finally, a radionuclide vector [7] was approved, which guarantees compliance with this criterion, see Chaps. 6 and 7 (Table 2.3).

The HAW primarily includes spent fuel elements from German nuclear power plants as well as waste from the reprocessing of fuel elements from German nuclear

---

[3]The Federal Ministry was renamed several times after task relocation. The term BMU subsumes this.

[4]According to U.S. NRC—Nuclear Regulatory Commission.

[5]A repository inventory is an exact inventory of all radionuclides with furthermore characterising, repository-relevant information.

Table 2.3 Radionuclides of spent fuel elements, which significantly determine the thermal output (Q) during the decay period up to 20,000 years

Isotope	$T_{1/2}$ in a	Q in keV	Decay
Short-lived			
Sr-90	28.90	2826	β
Cs-137	30.23	1176	βγ
Minor actinides			
Am-241	433	5486	α
Am-243	7.39E+03	5438	α
Cf-249	351	6295	α
Cf-250	13.08	6128	α
Cf-251	900	6176	α
Cf-252	2.645	6217	α
Cu-242	162.8 d	6216	α
Cu-243	29.1 a	6169	α
Cu-244	18.1 a	5902	α
Cu-245	8.50E+03	5623	α
Cu-246	4.76E+03	5475	α
Cu-247	1.56E+07	5353	α
Cu-248	3.48E+05	5162	α
Np-237	2.14E+06	4959	α

power plants. Typical radionuclides in the inventories are C-14, Cl-36, Sr-90, Cs-137, I-129, Nb-94, Pu-238, Pu-239, Pu-240, Pu-241, Am-241, U-234, U-238. In the final disposal of HAW, in addition to the activity, the decay heat of radionuclides plays a significant role. The spectrum of decay heat is described as (2–20) kW/m^3. One speaks of heat-producing HAW, if its heat production causes it to have a significant influence on the host rock or the design of the repository, see Chap. 7. The wastes are referred to as HGW.[6] The effects on the host rock are significant and may be a determining factor in site selection, see also [8] TIMODAZ—Thermal Impact on the Damaged Zone Around a Radioactive Waste Disposal in Clay Host Rocks (2008). This can be mitigated if the disposal provides measures which counteract the heat concentration. These include the positioning of the containers or the number of containers per storage chamber, etc.

The thermal performance of spent fuel elements and vitrified waste during the first 200 years is mainly determined by the short-lived fission products strontium-90 and cesium-137.

In the period up to 100,000 years, it is mainly the minor actinides that contribute to thermal output. The radioactive nuclides converted from U-238 are called "transurans" (minor actinides). The major isotopes of minor actinides in spent nuclear fuel are Neptunium-237, Americium-241, Americium-243, Curium-242 to -248, and Californium-249 to -252.

---

[6]HGW—Heat Generating Waste.

Residues to be considered according to Annex 12 Part A German Radiation Protection Ordinance arise from many industrial processes:

– In the case of oil and gas extraction as sludges, scales, and deposits,
– When processing raw phosphate (phosphorite) as dusts, scales and slags,
– In the extraction and treatment of bauxite, columbite, pyrochlore, micro-polythene, euxenite, copper shale, tin, rare-metal and uranium ore as by-products, sludges, sands, slags and dusts,
– For flue gas cleaning in the primary smelting of iron ores, non-ferrous ores as dusts and sludges,
– As filter ashes in lignite and hard coal-fired power plants,
– etc.

In Germany, however, the largest amounts of residues, arose from the uranium ore mining and processing operations of the SDAG Wismut in Thuringia and Saxony between 1946 and 1991; see Chap. 5 and [9].

2.3   Change in Radioactivity Over Time in Waste and Residues

In most cases, radioactive nuclei (radionuclides) are transformed into other atomic nuclei by emission of α-particles (He nuclei), electrons (β^--particles), positrons (β^+-particles) or electron capture. The number of nuclear transformations per unit time is the activity, their unit of measure in the SI-system is the Becquerel (Bq). A Becquerel corresponds to one nuclear transformation per second (1 Bq = 1 decay/s). Radioactive decay decreases the activity of radionuclides. In addition to the level of activity or activity concentration, the half-life of the radioactive decay is a major criterion for assessing radioactive materials. The half-life of a radionuclide is the time after which the radioactivity in the substance has decayed to half the initial level. The radionuclides found in residues or radioactive waste have extremely different half-lives—from less than an hour to millions of years. A repository-relevant selection is compiled in Table 2.4.

For many residues arising from nuclear medicine or in laboratories, due to the short half-life of their radionuclides, their radiological relevance decreases very quickly and the radioactivity loses its ecological relevance after a certain time. Therefore, considering the issue of long-term stable and safe storage of radioactive waste, the only relevant radionuclides, are those which have such a long half-life for which a long-term technical decay storage is necessary. Such wastes are considered below and consideration is given as to whether the decay energy can lead to heat generation that would be relevant to disposal, see Table 2.3.

Nuclides, which occur in the natural decay series, play an important role in the extraction of energy from nuclear fission. A decay series is the transformation of natural uranium and thorium radionuclides to individual radioactive decay products. The four decay series are shown in Fig. 2.2. The Neptunium series is no longer

Table 2.4 Classification of radionuclides according to their half-lives

Ranges of half-lives ($T_{1/2}$)	Radionuclide (kind of transformation)	Remarks
less as 1 a	I-125(β), I-131(β) Rn-222(α), Po-210(α)	Nuclear medicine nat. decay series
1 a up to 100 a	H-3(β); Co-60(β); Sr-90(β); Cs-137(β) Ra-228(β); Pb-210(β)	
100 a up to 10.000 a	C-14(β); Ra-226(α) Pu-240(α); Am-241(α) Am-243 (α)	
10.000 a up to 1 Mio. a	Cl-36(β); Kr-81(β) Tc-99(β); Th-230(α) Pu-239(α); Pu-242(α)	
More than 1 Mio. a	K-40(β); Cs-135(β) I-129(β); U-235(α) U-238(α); Th-232(α) Pu-244 (α)	

observable today. It is therefore also called a "prehistoric" series, since the Np-237 present in the formation of the earth (see half-life) has already completely decayed. Artificially, it is relevant again because radionuclides such as Pu-241, Np-237, and U-237 are being bred in nuclear reactors.

The nuclides of the decay series must also be taken into account in the storage of residues and radioactive waste.

The general decay law describes how the number of not yet decayed atomic nuclei (activity A_0) of a radioactive substance decreases over time [activity A(t)], see Eq. 2.1.

$$A(t) = A_0 \cdot 2^{-\frac{t}{T_{1/2}}} = A_0 \cdot e^{-\lambda \cdot t} \tag{2.1}$$

t—is the time when the activity of a radioactive substance has decreased from A_0 to A (t)
λ—is the decay constant of the radioactive substance measured in 1/s.

Consider one 1 kg U-235 with a $T_{1/2}$ = 703.8 million years and an activity A_0 = 80.0 MBq, from spent fuel as an individual process; after 10,000 years the activity would be reduced to A(10,000 a) = 79.9992 MBq. U-235 is a very long-lived radionuclide. It would take 703.8 million years, before the activity was reduced to 40 MBq.

The decay behavior of an isotopic mixture in radioactive waste does not behave in this ideal way. This is because, although the chain reaction in the reactor ceases, no energy is released by the controlled nuclear fission, but spontaneous decomposition of the uranium and fission products in the fuel rods of the reactor core continues.

<table>
<tr><th></th><th colspan="9">U-238 - decay series (4n+2)</th><th colspan="7">Th-232 - decay series (4n)</th><th colspan="7">U-235 - decay series (4n+3)</th><th colspan="6">Np-237 - decay series (4n+1)</th></tr>
<tr><td>Cf</td><td></td><td></td><td>250</td><td></td><td></td><td></td><td></td><td></td><td></td><td>252</td><td></td><td></td><td></td><td></td><td></td><td></td><td>251</td><td></td><td></td><td></td><td></td><td></td><td></td><td>249</td><td></td><td></td><td></td><td></td><td></td></tr>
<tr><td>Bk</td><td></td><td></td><td>|</td><td></td><td></td><td></td><td></td><td></td><td></td><td>|</td><td></td><td></td><td></td><td></td><td></td><td></td><td>|</td><td>247</td><td></td><td></td><td></td><td></td><td></td><td>|</td><td></td><td></td><td></td><td></td><td></td></tr>
<tr><td>Cm</td><td>250</td><td></td><td>246</td><td></td><td></td><td></td><td></td><td></td><td></td><td>248</td><td></td><td>244</td><td></td><td></td><td></td><td></td><td>247</td><td>|</td><td>243</td><td></td><td></td><td></td><td></td><td>245</td><td></td><td></td><td></td><td></td><td></td></tr>
<tr><td>Am</td><td>|</td><td>246</td><td>|</td><td></td><td></td><td></td><td></td><td></td><td></td><td>|</td><td></td><td>|</td><td></td><td></td><td></td><td></td><td>|</td><td>243</td><td>|</td><td></td><td></td><td></td><td></td><td>|</td><td>241</td><td></td><td></td><td></td><td></td></tr>
<tr><td>Pu</td><td>246</td><td></td><td>242</td><td></td><td>238</td><td></td><td></td><td></td><td></td><td>244</td><td></td><td>240</td><td></td><td>236</td><td></td><td></td><td>243</td><td>|</td><td>239</td><td></td><td></td><td></td><td></td><td>241</td><td>|</td><td></td><td></td><td></td><td></td></tr>
<tr><td>Np</td><td></td><td></td><td>|</td><td></td><td>|</td><td></td><td></td><td></td><td></td><td>|</td><td>240</td><td>|</td><td></td><td>|</td><td></td><td></td><td></td><td>239</td><td>|</td><td></td><td></td><td></td><td></td><td></td><td>237</td><td></td><td></td><td></td><td></td></tr>
<tr><td>U</td><td></td><td></td><td>238</td><td></td><td>234</td><td></td><td></td><td></td><td></td><td>240</td><td></td><td>236</td><td></td><td>232</td><td></td><td></td><td></td><td></td><td>235</td><td></td><td></td><td></td><td></td><td></td><td>|</td><td>233</td><td></td><td></td><td></td></tr>
<tr><td>Pa</td><td></td><td></td><td>|</td><td>234</td><td>|</td><td></td><td></td><td></td><td></td><td></td><td></td><td>|</td><td></td><td>|</td><td></td><td></td><td></td><td></td><td>|</td><td>231</td><td></td><td></td><td></td><td></td><td>233</td><td>|</td><td></td><td></td><td></td></tr>
<tr><td>Th</td><td></td><td></td><td>234</td><td></td><td>230</td><td></td><td></td><td></td><td></td><td></td><td></td><td>232</td><td></td><td>228</td><td></td><td></td><td></td><td></td><td>231</td><td>|</td><td>227</td><td></td><td></td><td></td><td></td><td>229</td><td></td><td></td><td></td></tr>
<tr><td>Ac</td><td></td><td></td><td></td><td></td><td>|</td><td></td><td></td><td></td><td></td><td></td><td></td><td>|</td><td>228</td><td>|</td><td></td><td></td><td></td><td></td><td></td><td>227</td><td>|</td><td></td><td></td><td></td><td></td><td>|</td><td>225</td><td></td><td></td></tr>
<tr><td>Ra</td><td></td><td></td><td></td><td></td><td>226</td><td></td><td></td><td></td><td></td><td></td><td></td><td>228</td><td></td><td>224</td><td></td><td></td><td></td><td></td><td></td><td>|</td><td>223</td><td></td><td></td><td></td><td></td><td>225</td><td>|</td><td></td><td></td></tr>
<tr><td>Fr</td><td></td><td></td><td></td><td></td><td>|</td><td></td><td></td><td></td><td></td><td></td><td></td><td></td><td></td><td>|</td><td></td><td></td><td></td><td></td><td></td><td>(223</td><td>|</td><td></td><td></td><td></td><td></td><td></td><td>221</td><td></td><td></td></tr>
<tr><td>Rn</td><td></td><td></td><td></td><td></td><td>222</td><td></td><td></td><td></td><td></td><td></td><td></td><td></td><td></td><td>220</td><td></td><td></td><td></td><td></td><td></td><td></td><td>219</td><td></td><td></td><td></td><td></td><td></td><td>|</td><td></td><td></td></tr>
<tr><td>At</td><td></td><td></td><td></td><td></td><td>|</td><td></td><td></td><td></td><td></td><td></td><td></td><td></td><td></td><td>|</td><td></td><td></td><td></td><td></td><td></td><td></td><td>|</td><td></td><td></td><td></td><td></td><td></td><td>217</td><td></td><td></td></tr>
<tr><td>Po</td><td></td><td></td><td></td><td></td><td>218</td><td></td><td>214</td><td></td><td>210</td><td></td><td></td><td></td><td></td><td>216</td><td></td><td>212</td><td></td><td></td><td></td><td></td><td>215</td><td></td><td>(211)</td><td></td><td></td><td></td><td>|</td><td>213</td><td></td></tr>
<tr><td>Bi</td><td></td><td></td><td></td><td></td><td>|</td><td>214</td><td>|</td><td>210</td><td>|</td><td></td><td></td><td></td><td></td><td>|</td><td>212</td><td>|</td><td></td><td></td><td></td><td></td><td>|</td><td>211</td><td>|</td><td></td><td></td><td></td><td>213</td><td>|</td><td>209</td></tr>
<tr><td>Pb</td><td></td><td></td><td></td><td></td><td>214</td><td></td><td>210</td><td></td><td>206</td><td></td><td></td><td></td><td></td><td>212</td><td>|</td><td>208</td><td></td><td></td><td></td><td></td><td>211</td><td>|</td><td>207</td><td></td><td></td><td></td><td></td><td>209</td><td>|</td></tr>
<tr><td>Tl</td><td></td><td></td><td></td><td></td><td></td><td></td><td></td><td></td><td></td><td></td><td></td><td></td><td></td><td></td><td>208</td><td></td><td></td><td></td><td></td><td></td><td></td><td>207</td><td></td><td></td><td></td><td></td><td></td><td></td><td>(205)</td></tr>
</table>

Fig. 2.2 Radioactive decay series shown as mass numbers of the atomic nuclei of the chemical elements shown in the column on the left. Artificial radionuclides (yellow), naturally occurring radionuclides (green) and stable end products (gray). Vertical transitions—α-emitters; Diagonal transitions β-emitter. Shown are only those chains in which the initial members have half lives of more than 1 year

This releases energy. This released energy is called decay heat, see also Table 2.3. As each nuclear fission produces radioactive fission products, see also decay series, the composition of the radioactive inventory in the fuel rods during the burn up, and thus in the radioactive waste, gradually changes (Fig. 2.3).

As uranium-235 decreases as a result of nuclear fission, the amount of radioactive fission products increases. The decay behavior and evolution of the radioactive waste inventory of spent fuel assemblies are shown in Fig. 2.4.

For waste with low heat generation and residues (for example, residues from uranium ore processing), the decay processes are subject to the same physical phenomena, but the resulting potential dangers are different. The parent radionuclide, U-238, of the decay series of the same name has a half-life of 4.468 billion years. The decay results in Ra-226 and from it the radon isotope Rn-222, a noble gas with a half-life of 3.825 days. Due to the long half-life of the parent radionuclide U-238, Rn-222 is produced permanently. The low half-life of Ra-222 is therefore not the determining factor in this decay process but that of U-238. Radon-222 breaks down in turn into the nuclides shown in Fig. 2.3. This is the simplified last part of the uranium 238 series. When Rn-222 enters the lungs through the air in large concentrations, the stable nuclides Pb-206, which are responsible for lung carcinomas, are deposited there following radio decomposition. Within the thorium-232 decay series Rn-220. arises.

The parent radionuclide Th-232 has a half-life 14,050 billion years. Rn-220 produces the stable nuclide Pb-208 (simplified last part of the Thorium-232 series), which can also cause carcinomas in the lungs. It can be seen that in the former mining areas of the SDAG Wismut there is a focus on excluding radon exhalations as far as possible or allowing them only within socially and legally acceptable limits.

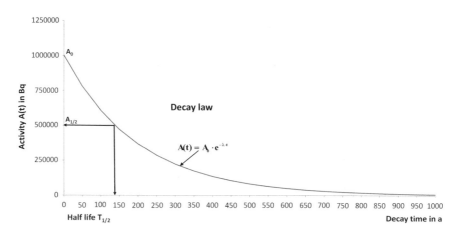

Fig. 2.3 General decay law for a radioactive substance, Eq. 2.1

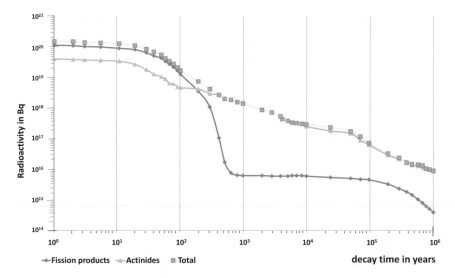

Fig. 2.4 Radioactivity inventory [Bq] of spent fuel as a function of decay time (principle—double logarithmic), see also [10] acatech 2014

2.4   Origin of Radioactive Waste and Residues

Considering the origin of radioactive waste and residues, the resulting amounts are important to develop an overall understanding of the task. Although each type of waste is referred to in the individual chapters, as shown in Fig. 1.1 or here in Fig. 2.1, nevertheless a summarised presentation is included. With regard to the long-term safekeeping of radioactive waste and residues, one can refer to three very different areas of origin.

(a)   Origin in the generation of energy from nuclear fuels

The following subprocesses are called supply, see Fig. 2.5.

First, the nuclear fuel must be produced. This happens mostly from uranium or thorium[7] ores. Usually, the focus is on the uranium ores. After mining uranium ores, the uranium-containing material is separated from the rest of the rock, crushed and ground. Thereafter, the uranium is extracted from a rock compound in a chemical treatment and cleaned. The result is a yellow powder that consists of about 90% of a uranium-oxygen compound such as Triuranium octoxide (U_3O_8), in addition to ammonium or magnesium diuranate. Because of the yellow appearance this is called "yellow cake". The International Atomic Energy Agency (IAEA) monitors the extraction of uranium and thorium and records in the "Red Book" the annual production of U_3O_8. An interesting compilation of the issue can be found in [11].

---

[7]Thorium occurs in the earth's crust at a frequency of 7–13 mg/kg; it is twice to three times as abundant as uranium.

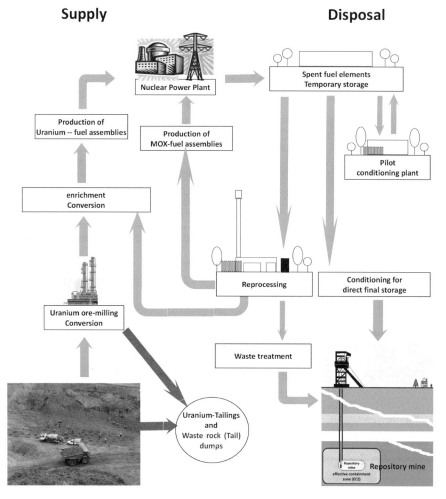

Fig. 2.5 Radioactive waste and residues in the supply and disposal process of NPPs; without dismantling

The volume of (U3O8) produced in 2016 was 73,148 t worldwide, compared to 71,343 t in 2015, see Lersow [12].

In this sub-process, tailings are generated during the mining and processing of the ores. The residues from the wet chemical treatment are usually stored in tailings ponds. The long-term management of uranium tailings ponds is described in Chap. 5.

Thereafter, the production of the nuclear fuel by enrichment and manufacturing of fuel elements, which are then burned up in the reactor. This happens in the following way. The natural uranium and "yellow cake" consist mainly of U-238 and only 0.7% of U-235. This is not enough to maintain a chain reaction in light water reactors. The natural uranium must therefore be enriched, i.e. the proportion of U-235 is increased until the product contains 4–5% U-235. This product will be used further. Before enrichment, however, the "yellow cake" must be brought into a form suitable for further processing. This step is called conversion. The uranium oxide is converted into uranium hexafluoride (UF6), a white, salt-like compound.

After the enrichment process, depleted uranium remains in the enrichment facilities (called Tails or Depleted Uranium). It contains only a little uranium-235 and is stored. However, depleted uranium is not waste, but potential nuclear fuel for enrichment and/or use in fast breeders and may have other potential uses.

The enriched uranium is transported to the fuel element factory and, after several processing steps, used to manufacture fuel rods. These rods, depending on the reactor type, are bundled into fuel elements of different sizes. The facilities for enrichment and for fuel element manufacture are under the control of the Nuclear Regulatory Authority of the respective country and usually monitored by the IAEA.

The following sub-processes are included in the term disposal, see Fig. 2.5.

For disposal, all residues from the mining and processing of the uranium ore, as described above, are involved. Disposal is described in Chap. 5.

Following nuclear fission reaction in the fuel elements of the reactor, uranium-235 levels are depleted to less than 0.5%. The spent fuel elements are highly radioactive and emit a large amount of heat. They are stored in the cooling pond of the nuclear power plant until they have cooled to the point where they may be transported safely. Afterwards, they will be taken to an interim storage facility of the respective nuclear power plant. There they remain, according to the current disposal philosophy in Germany, until the repository for high-level radioactive, heat-generating waste is ready to commence operations.

A spent fuel element consists of only about four percent radioactive waste. The remaining 96% continues to be potentially usable as nuclear fuel. In reprocessing plants, the reusable nuclear fuel is separated from the radioactive waste. The material is thus available again for power generation. The recycled U-235 and the plutonium dioxide produced in the spent fuel elements can be reused in existing nuclear power plants together with fresh uranium dioxide as so-called mixed oxide (MOX) fuel. The high-level radioactive waste left over during reprocessing is melted into a glass (glass blocks), packed in steel containers and transported to an intermediate storage facility in Castoren. In other countries, such as France and the UK, reprocessing is an important part of the fuel cycle.

However, the total amount of radioactive wastes incurred and not reused will increase during reprocessing. Such reprocessing technology has been prohibited by law in Germany since 2005.

During the operation of a nuclear power plant, however, not only high-level radioactive, heat-producing waste accumulates, but also a considerable amount of low and medium-level waste, which does not come from the reactor; this includes items such as pipeline material, water and consumables. These are also collected in the federal state collection depots or in interim storage facilities at the site and will be stored until the completion of the repository "Shaft Konrad".

Due to Germany's exit from nuclear energy production, the nuclear power plants, which have been taken out of service, must be decommissioned and dismantled. In addition to the former reactor core a lot of other waste types will accumulate, which are either low or medium radioactive or waste with radioactivity (quantities of non-hazardous waste), which can be disposed of according to the Closed Substance Cycle and Waste Management Act (KrWG) and Landfill Ordinance (DepV). The recording of the quantities is presented in Chap. 4, Fig. 4.1. Most of the waste generated can be disposed of in class II and III landfills.

In the USA, there are similar issues, especially with regard to managing the large volumes of VLLW related to the decommissioning of nuclear power plants as well as additional waste streams arising from the reprocessing of fuels and new types of nuclear installations or from a radiological event. To this end, provision will also be made in the USA to develop possible options for improving and strengthening the regulatory framework for the disposal of these expected large volumes of VLVW.

The radioactive waste to be expected from nuclear energy production is shown in Figs. 2.5 and 2.6.

(b)   Origin from isotope production and application of the products

In the meantime, there is a wide range of applications in which radioactive isotopes are used. In nature, isotopes are generally found everywhere in equal mixing ratios and it is now the to separate or enrich the isotopes by suitable methods so that they can be used. In terms of quantity, the most important isotope separation is the uranium enrichment, increasing the U-235 content, for the production of nuclear fuel for nuclear power plants as described under (a).

The basis of isotope production is research reactors and installed accelerators. Especially in nuclear medicine, but also in technical procedures and in research, radioactive isotopes are needed. This application also leads to production of radioactive contaminated residues, which cannot be further used and must be disposed of. However, the resulting amounts of radioactive contaminated residues are small in relation to those from the supply and disposal processes of nuclear power plants.

Most of the radioactive contaminated waste is low and medium level radioactive waste, which in Germany is first collected at the federal state collection facilities or in suitable interim storage facilities. Research reactors also produce highly radioactive, heat-generating wastes that are subject to the same disposal criteria as those described in Chap. 7.

The economy of a separation method depends on: the properties of the element, the desired degree of enrichment, the amount needed and the specific energy consumption. Small amounts of different isotopes can be separated very precisely with the aid of mass spectrometers (mass spectrograph). Gas diffusion and gas centrifuge methods are now used on a large scale for uranium enrichment.

In recent years, dynamic separation techniques have been developed, such as laser isotope separation and the nuclear spin method. Using laser isotope separation, one expects separation factors of 15–20 at a low specific energy consumption; Their development is therefore being driven forward, especially in the USA. Isotope production is subject to the same monitoring as the above-described uranium enrichment, the handling of the isotope preparations, probes, radiation sources, etc. and their disposal is monitored by the state including storage in a suitable repository.

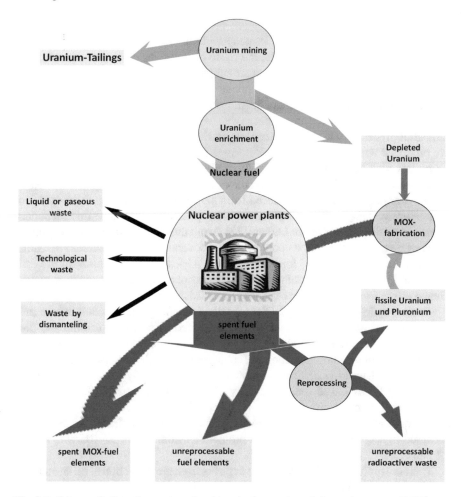

Fig. 2.6 Scheme: Radioactive waste and residues in the supply and disposal process of NPP's

(c) Mining or industrial processes leading to residues of naturally occurring radionuclides ("NORM")

Also described under (a) was the extraction and treatment of uranium and thorium
ores, which lead to radioactive residues requiring monitoring. However, there are a
whole series of other industrial processes in which residues with naturally occurring
radionuclides also arise. In many cases, these residues are not subject to monitoring
requirements and can be disposed of in landfills in Germany, mostly landfill classes
II and III, in accordance with the law of Closed Substance Cycle and Waste Management Act and Landfill Ordinance. These industrial processes include the production
of aluminum, during which so-called red mud is accumulated; and lignite and hard
coal power plants, where large amounts of filter ash can be accumulated.

The release of radioactive substances (according to § 29 radiation protection ordinance) can be carried out in Germany without restrictions or restrictions for a certain disposal route (for example, the reuse of buildings or disposal at landfills). After release, the materials are no longer subject to any further radiation protection monitoring and are waste within the meaning of the Closed Substance Cycle and Waste Management Act, see Chap. 3. An overview of this is also given in Fig. 2.1.

Rare earth extraction and treatment as well as phosphate production are two further industrial processes where considerable quantities of radioactive residues (tailings) are produced which may require monitoring. The largest rare earth occurrences are found in China (in Inner Mongolia), Australia (Western Australia), Greenland and Canada. China is by far the largest producer of rare earths, currently about 100,000 t/year. Uranium may also be produced from phosphate deposits in a secondary process. However, production costs and uranium prices usually make this secondary process uneconomic at present Lersow [12]. More than 80% of the phosphate minerals currently considered to be economically recoverable are concentrated in China, Morocco, the United States, Jordan and South Africa. The world reserves are estimated at 71 billion tons. Another source of phosphorus is bird droppings, guano. Guano arises from the pasty excrement of seabirds.

Another industry producing NORM waste is oil and gas production, where millions of tonnes of radioactive residues are produced each year as drilling and production sludge and scales. Sludge and waste water pumped to the surface in the context of hydrocarbon extraction contain Technologically Enhanced Natural Occurring Radioactive Material (TENORM) substances; i.e. the highly toxic and long lived Ra-226 and Po-210. The specific activity of the waste is between 0.1 and 15,000 becquerels (Bq) per gram. In Germany, where about 1000–2000 tons of dry matter are produced each year by this industry, the residues requiring monitoring are recorded in accordance with radiation protection ordinance §§ 97-102 with Annex 12 Part A.

Therein are also included sludge and deposits from the extraction of oil and natural gas. The prescribed rules for disposal must be observed.

The determination of the specific activity of the sludge is carried out in a laboratory and serves for the classification of radioactive substances with regard to the supervision by the authorities, their disposal and transport. For this purpose, the long-lived radionuclides present in the residues and their specific activity must be identified in the declaration analyses so that the nuclide composition of the U-238 and the Th-232 decay series can be stated with suitable reliability. For this purpose, the radionuclides Pb-210, Ra-226, Ra-228 and Th-228 are usually determined. Determination of U-238 and Th-232 can be omitted for exploration and production (E & P Industry) residues, as these nuclides are not enriched in the deposits and sludges. Disposal has to be oriented to the criteria for release of residues requiring monitoring and transporting sludges in mandatory steel IBCs ("Intermediate Bulk Containers") with "inner packaging" (e.g. plastic drums). A simplified proof of release of residues from radiation protection monitoring is the compliance with the direct dose of 1 mSv/a. The recovery or disposal route of radioactive residues is specified in an approved waste management plan from the exploration and production company, see Chap. 4.

The E & P industry uses a significant amount of radioactive sources (for example from geophysical borehole logging measures). For the transport of radioactive sources both radiation protection and dangerous goods standards apply. The provisions of the Radiation Protection Ordinance, the Law on the Carriage of Dangerous Goods and, e.g. the dangerous goods regulations for road and/or railway. The transport of radioactive materials is subject to a license under the radiation protection ordinance. Only radioactive sources of low activity are exempt. The license holder is usually the operator of the radioactive sources, in most cases the geophysical borehole logging company.

(d) Radioactive waste and residues resulting from the development, production and selection of nuclear weapons, radioactive munitions

This type of waste does not accrue in Germany because Germany has no nuclear weapons. A repository for radioactive waste from the military application of nuclear energy is described in Chap. 7, WIPP. The Waste Isolation Pilot Plant, or WIPP, is a repository for the permanent disposal of transuranic radioactive waste in deep geological salt horizons. It is located approximately 26 miles (42 km) east of Carlsbad, New Mexico, see Chap. 7.

One fact should be mentioned here nevertheless. On the one hand, plutonium-239 is generated during the operation of nuclear power plants powered by uranium. The Pu-239, which is suitable for the construction of a bomb, is obtained by chemical means from the burnup of nuclear power plant's fuel, see Fig. 2.7. The operation of nuclear power plants thus increases the risk of nuclear proliferation. To minimize this, various international treaties have been concluded. The most important of these contracts is the Non-Proliferation Treaty. In nuclear weapons producing countries such as Russia, USA, India nuclear reactors may be operated only for this purpose.

Residues from the nuclear chain also include depleted uranium which may be used for ammunition production. The uranium ammunition or depleted uranium (DU) projectiles contain depleted uranium. The depleted uranium contains a smaller proportion of fissile uranium isotopes U-234 and U-235 than natural uranium and thus consists largely of material not capable of chain reaction, i.e. isotope U-238 Due to the high density of the depleted uranium, the projectiles have a very great penetration impact (armor-piercing Ammunition). This ammunition has been used quite often in military conflicts. There is a considerable worldwide call for the prohibition of such depleted uranium ammunition.

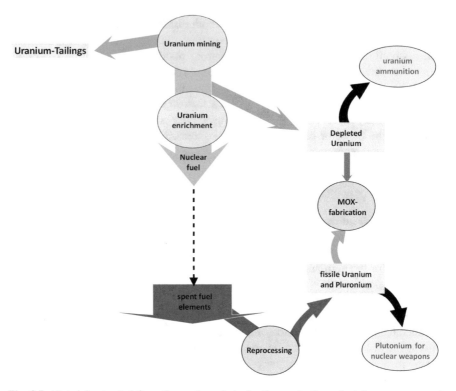

Fig. 2.7 Material extracted from the nuclear chain for the production of nuclear weapons and uranium ammunition

References

1. COUNCIL DIRECTIVE 96/29/EURATOM of 13 May 1996: Laying down basic safety standards for the protection of the health of workers and the general public against the dangers arising from ionizing radiation.
2. COUNCIL DIRECTIVE 2013/59/EURATOM of 5 December 2013: Laying down basic safety standards for protection against the dangers arising from exposure to ionising radiation, and repealing Directives 89/618/Euratom, 90/641/Euratom, 96/29/Euratom, 97/43/Euratom and 2003/122/Euratom.
3. Published by ICRP – International Commission on Radiation Protection *Annals of the ICRP: ICRP Publication 103 – The 2007 Recommendations of the International Commission on Radiological Protection.* Elsevier, Amsterdam 2007, ISBN 978-0-7020-3048-2.
4. Classification of radioactive waste - General Safety guide - No. GSG-1, IAEA, Vienna 2009.
5. Recommendation of the Commission on Radiological Protection: "Release of materials, buildings and soil surfaces with low radioactivity from handling subject to notification or approval"; Adopted at the 151st meeting of the Commission on Radiological Protection on 12 February 1998; Reports of the Commission on Radiological Protection, Volume 11, in German language.
6. International Atomic Energy Agency (IAEA): Principles for the Exemption of Radiation Sources and Practices from Regulatory Control. IAEA safety standards, 23 Seiten, Safety Series No. 89, ISBN 9201238886: Vienna, 1988.

7. Brennecke, P.; Steyer, St.: Repository "Shaft Konrad" - Balancing provision for radionu-clides/radionuclide groups and non-radioactive, noxious substances; BfS; SE-IB-33/09-REV-1; Stand: 07. December 2010, in German language.

8. TIMODAZ - Thermal Impact on the Damaged Zone Around a Radioactive Waste Disposal in Clay Host Rocks, (Contract Number: FI6 W-CT-2006-036449), Date of issue of this report: 10/12/08.

9. Status and Results of the WISMUT Environmental Remediation Project; Presentation at the Second Joint Convention Review Meeting; Vienna, May 17th, 2006.

10. Partitioning and Transmutation, acatech STUDY, Research - Development - Social implica-tions; Center for Interdisciplinary Risk and Innovation Studies (ZIRIUS) at the University of Stuttgart, publisher: Prof. Dr. med. Dr. hc Ortwin Renn, acatech - GERMAN ACADEMY OF TECHNOLOGICAL SCIENCES, 2014.

11. Rübel, A., Müller-Lyda, I., Storck, R.: The classification of radioactive waste with regard to disposal; Gesellschaft für Anlagen- und Reaktorsicherheit (GRS) gGmbH; GRS - 203 ISBN 3-931995-70-4; Cologne, December 2004, in German language.

12. Lersow, M.: Energy Source Uranium – Resources, Production and Adequacy; Glueckauf Mining Reporter; Verlag der Bergbau-Verwaltungsgesellschaft mbH, Shamrockring 1, 44623 Herne; 153(3) p. 178–194; 6/2017.

13. Gangulay, C., Slezak, j., Waggitt, P., Hanly, A. et. al.: Uranium Raw Material for the Nuclear Fuel Cycle: Exploration, Mining, Production, Supply and Demand, Economics and Environ-mental Issues (URAM-2009) Proceedings of an International Symposium Organized by the International Atomic Energy Agency in Cooperation with the OECD Nuclear Energy Agency, the Nuclear Energy Institute and the World Nuclear Association and held in Vienna, Austria, 22–26 June 2009; Technical Report May 2014; Report number: IAEA- TECDOC-1739.

Chapter 3
Fundamentals of Final Disposal of Radioactive Waste and Residues in Geotechnical Environmental Structures

3.1 Task

The task and goal of final disposal is to simultaneously safeguard radioactive material subject to supervision in a Geotechnical Environmental Structure and permanently prevent the dispersal of any radionuclides released from the repository into the biosphere. This means that the geotechnical environmental structure and the storage of the radioactive material must be carried out so that this goal is achieved with great certainty. Internationally, limits are set for the release of radio-toxic nuclides into the biosphere, see Euratom [1] and ICRP [2], but these may need to be supplemented with national standards. Geotechnical environmental structures that are used for the final disposal of radioactive waste undergo various phases of development, which work together to achieve the long-term safety of the geotechnical environmental structure:

- The phase of planning and construction of a repository to the standards required by the Atomic Energy Act for such wastes must be taken into account considering state-of-the-art science and technology.
- The operating phase (with an approved operational safety and operating permit) in which the radioactive waste is accepted and placed in the storage areas permanently (disposal).
- The post-operational phase (closure phase), in which the geotechnical environmental structure is shut down, closed and monitored.
- The post-closure phase, during which the geotechnical environmental structure will be monitored to ensure the structure performs as designed and remains safe in the long-term.

This process must allow optimization of the final closure of the repository in such a way that the target is reached with the highest possible level of safety, including any necessary recovery of any of the inventory contents.

The geotechnical environmental structure is built as a multi-barrier system. It consists of geological, technical and geotechnical barriers, which together form the

© Springer Nature Switzerland AG 2020
M. Lersow and P. Waggitt, *Disposal of All Forms of Radioactive Waste and Residues*, https://doi.org/10.1007/978-3-030-32910-5_3

effective containment zone and which, in their interaction, should ensure the isolation of the waste for the required period. For high-level radioactive waste (HAW), the multi-barrier system, according to current regulations in Germany, must be able to guarantee an isolation period of the order of one million years. The fact that such a period under review (detection period) does not lead to more security seems to be undisputed. Thus, it seems more appropriate to divide the entire period under review into periods with different probability of failure; this is because the probability of failure does not remain constant over a period of 1 million years. This suggestion should be put up for discussion.

Achieving the objective of ensuring compliance with the protection criteria requires both a permanent monitoring of the environment of the repository site and crisis management in the case of any exceedance of limit values (failure scenarios).

Chapter 8 is devoted almost exclusively to the long-term safety of radioactive waste repositories, with a focus on heat-generating HAW. The long-term safety for the final disposal of residues from uranium ore treatment is dealt with in Chap. 5. In order not to confuse matters and repeat material, a summary of this is also included in Chap. 8.

In Chap. 5 the example of the uranium-tailings-ponds of the SDAG Wismut is used to demonstrate that long-term safety evidence for these residues cannot be achieved because a base sealing is missing in the repositories. As a consequence radionuclides are released continuously from the inventory into the biosphere; via the aquatic pathway at a potentially damaging level such discharges have to be collected. Both the collection of the contaminated waters and their treatment are described in detail in Chap. 5. This ensures that the final disposal of the uranium tailings in the closed tailings ponds could be assessed as safe for the long-term.

The objective is not to present an algorithm for the provision of a long-term safety record for a license application for the safe disposal of radioactive waste and residues. The purpose of this presentation is to show the essential requirements and relationships for a long-term safety assessment; thus the dimension of the task to be undertaken for the approval of a radiotoxic waste and residue disposal system is made clear. A prerequisite for a permit is to demonstrate that the repository system can operate with a probability bordering on certainty to prevent the spread of radionuclides into the biosphere permanently. Using different probable and less probable scenarios, it should be shown that compliance with the protection criteria may be ensured through permanent monitoring of the environment at the repository site. Furthermore, a plan for crisis management is prepared so that when limit values are exceeded (accident/incident) the system remains permanently functional. Only in this way can a permit for the construction of a geotechnical environmental structure as a repository structure be expected to be approved.

The statements may also give some suggestions on the design of a long-term safety assessment for a future repository for HAW.

3.2 Demands

The international and national legal and subordinate regulations define the essential conditions for the final disposal of radioactive waste. In Germany these include the Atomic Energy Act (AtG), the Radiation Protection Ordinance (StrlSchV) and the Federal Mining Act (BBergG) with the associated Federal Mining Ordinance (ABBergV). To this end, the relevant international recommendations of the ICRP, the IAEA and the OECD NEA must be taken into account, especially if they contain additions or exceptions to the national regulations. The presentation here will be based on the existing legal standards and other applicable documents in Germany or internationally for the existing and/or planned or constructed final disposal facilities and, for future repository concepts for high-level radioactive, heat-generating waste. These should reflect not only the current legal requirements, but also the scientific findings on the subject, which decisively determine the safety philosophy. These are:

[3] The safety requirements governing the final disposal of heat-generating radioactive waste of 30.09.2010 by the Federal Ministry for the Environment, Nature Conservation and Nuclear Safety (BMU;

[4] Law on Search and Selection of a Site for a Repository for Heat-Generating Radioactive Waste (Site Selection Act—StandAG); of 23 July 2013 and its development of 23 March 2017;

[5] Abschlussbericht der Kommission Lagerung hochradioaktiver Abfallstoffe „Verantwortung für die Zukunft", Deutscher Bundestag (2016); English translation: Report of the German Commission on the Storage of High Level Radioactive Waste; July 2016; https://www.gruene-bundestag.de/fileadmin/media/gruenebundestag_ de/themen_az/Gorleben_PUA/Report-German-Commission-Storage-High-Level-Radioactive-Waste.pdf;

[6] Plan approval decision and approval decision for the construction and operation of the Konrad mine in Salzgitter as a facility for the disposal of solid or solidified radioactive waste with negligible heat generation of 22 May 2002; Lower Saxony Ministry of the Environment (2002);

[7] Decommissioning of the industrial tailing plants (IAA) of the SDAG Wismut - Federal Mining Act (BBergG), Regulation for ensuring on nuclear safety and radiation protection (VOAS), Order for ensuring of radiation protection in dumps and industrial tailings ponds and in use of therein deposited materials (HaldAO), Radiation Protection Ordinance (StrlSchV), Atomic Energy Act (AtG) etc. [8]

[9] The Safety Case and Safety Assessment for the Disposal of Radioactive Waste for protecting people and the environment; No. SSG-23 Specific Safety Guide; International Atomic Energy Agency (IAEA), Vienna, 2012;

[10] IAEA Safety Standards—Geological Disposal Facilities for Radioactive Waste Specific Safety Guide No. SSG-14; published 2011;

[11] Horizon 2020: European Joint Programme in Radioactive Waste Management— EC Policy change, and Joint Programming: the issue and aim; JOPRAD programme document workshop, London, 04 April 2017;

[12] U.S. NRC: PART 20—STANDARDS FOR PROTECTION AGAINST RADI-ATION; https://www.nrc.gov/.

In the following chapters it will be shown that the respective assessments for the different radioactive waste or residues are individual solutions for the various different geotechnical environmental structures, each of which has a high, site specific, dependency. Similarities in the basic requirements for the long-term safety of the geotechnical environmental structures are emphasised and the differences in the requirements for a structure are presented. The fact that these cannot be justified with the respective radioactive-toxic inventory, however, shows that no uniform requirement concept for all types of radioactive waste and residues has been developed in Germany and the different safety standards and designs of geotechnical environmental structures show a very varied picture; see also Chaps. 5, 6 and 7.

For example, in the USA all types of radioactive material are described in U.S. NRC publications and other documentation is subordinate. This leads to nationally uniform treatment and supervision. However, there are always difficulties and variations in the implementation of the concept.

3.3 Criteria and Conditions for the Repository Site

As the name implies, waste and residues are stored permanently at suitable, prepared locations. It is not intended to return them to the economic cycle. Suitable locations for the storage of radioactive waste are dry sites. Depending on the category of waste, the waste needs to be sufficiently far away from aquifers and streams, and sites with low fissure and crack systems, low porosity and permeability in the immediate environment of the waste (dry storage concept). However, extensive investigations must take place before the location is determined and before a site can be shortlisted, see [4] StandAG and [10]. The greater the amount of activity to be stored or its radiotoxicity, the heat production and the volume of the inventory, the more extensive the site investigations have to be.

The preparation of the site for deposition includes **storage on surface**: the creation of hollow forms such as trenches, pits, valley impoundments, former open pits, etc. and the introduction of a base sealing and preparation of the containment area.

Storage in an underground facility shafts, drifts, galleries, caverns, chambers, tunnels and transport routes etc. and their construction in dry and highly stable and dense strata.

There must also be consideration of the waste category: Very Low Activity Radioactive Waste (VLLW), these are landfill sites, mostly located at the surface, depositories (DK) II and III (DepV), see Chap. 4.

For uranium mining residues, these are mostly heaps, dumps and tailings ponds, see Chap. 5.

For the waste categories: LAW and MAW with low heat generation, see Chap. 6, very different solutions are known, such as:

- **Final disposal on the surface, examples**:

  - in the Slovak Republic (Mochovce) in two double-row concrete boxes [13]
  - American surface disposal waste for civil use, in trenches, e.g. Savannah River Site; Aiken, South Carolina [14].

- **Final disposal close to the surface in caverns, examples**:

  - Hostím (Czech: Úložiště Hostím) Srbsko/Central Bohemia/Czech Republic; Cavern (limestone), ground-level access, closed/sealed,
  - Himdalen/25 km east of Oslo/Norway; Cavern (basalt), ground-level access, currently a final disposal operation.

- **Final disposal in deep geological formations, examples**:

  - Radioactive waste repository Morsleben (ERAM), Morsleben/Saxony-Anhalt/Germany; geological formation: salt, facility currently in decommissioning, see Chap. 6 and [15]
  - Repository "Shaft Konrad", Salzgitter/Lower Saxony/Germany, geological formation: iron ore, under construction, see Chap. 6.

For disposal of the category: heat-generating HAW, repository mines predominate, with very different location solutions, depending on the legal and social situation in the respective countries. Various site and country specific solutions are described, see Chap. 7.

Search for and selection of a site

(a) Geotechnical environmental structure (in accordance with the Landfill Ordinance) for the long-term safe and stable storage of waste and residues not requiring monitoring

The site requirements are described in the Landfill Ordinance (DepV) Annex 1. It states in the introduction: "The suitability of the site for a landfill is a necessary prerequisite for ensuring that the public good is not impaired by the landfill in accordance with § 15 paragraph 2 of the Closed Substance Cycle Waste Management Act". The general protection objective is the public good. However, this protection objective can be determined by overriding interests and thus by overriding laws. For example, a strategic environmental impact assessment may lead to a decision that the location becomes excluded as a landfill site [(e.g. regional planning), in accordance with the Environmental Impact Assessment Act (UVPG), see also EC Directive (85/337/EEC) and/or the post-use conditions of the Federal Mining Act,]. DepV Annex 1 lists which parameters of the site are to be considered in principle. These include the geological and hydrogeological conditions of the area, underground requirements such as permeability and bearing capacity as well as the expected upper free groundwater level up to the top of the base sealing system, etc.

(b) Geotechnical environmental structure (Tailings pond) for the long-term safe, long-term stable storage of uranium processing residues (Uranium mill tailings)

The site search is initially determined by the fact that the processing tailings are to be stored in the vicinity of the processing site, since transport elsewhere cannot be designed effectively and safely. Suitable landforms are abandoned open pits and terrain gaps. e.g. river valleys etc., which can be closed with a dam structure. Ideal are sites with a geological barrier, especially under the contact area of the tailings deposit. If these are not present, a geotechnical barrier [geosynthetic seals, clay seals (bentonites) etc.] should be introduced to achieve the isolation of the radioactive residues, and in particular to offer groundwater protection to prevent highly contaminated pore water from the tailings body leaving the site.

(c) Geotechnical environmental structure for the long-term safe, long-term stable storage of radioactive waste

The construction of a geotechnical environmental structure (repository building) begins with the search for a suitable location. The location determines the existing geological (natural) barrier. There are several possibilities, all of which guarantee the principle of concentrating and isolating the disposal of radioactive waste without the intention of retrieval: construction of a repository in a repository mine/EBW/in deep geological strata, see BMU [3]; Deposition of radioactive waste in deep boreholes/TBL/, see Natilus [16]; Final disposal in caverns or bunkers/KuB/(close to the surface), see EPA [17] etc. The different types of final repository must comply with the scheme in Fig. 3.2.

In Germany, three suitable formations of host rocks are found: clay, salt rock and crystalline rocks (magmatic formations, e.g., granites). In other countries, such host rock diversity may not be available, see IAEA [10]. The different host rocks have different isolation potentials in the period under review. Any geotechnical environmental structure used as a repository, especially for heat-generating, high-level radioactive waste, must be permanently characterized by integrity, reliability and robustness. These requirements are inextricably linked. Robustness is defined as the insensitivity of the safety functions of a repository system and its barriers to internal and external influences and disturbances as well as the insensitivity of the results of the safety analysis to deviations from the underlying assumptions, see BMUB [3].

The preservation of the properties of the containment capacity of the effective containment zone of a repository is called integrity. The choice of site (including the host rock) thus determines how a repository system consisting of the site conditions and the technical and geotechnical components of the repository is developed, how the required level of safety can be guaranteed and what specific security features its components and subsystems must fulfil. The different host rocks have different isolation (retention) and integrity properties both with respect to the detection period and in particular for heat generating waste, with respect to temperature, see BfS [18]. However, disadvantages may be offset or overcome by technical and/or geotechnical barriers.

Where there is a choice to be made between alternative repository locations, a comparative evaluation is applied. First of all, it is necessary to develop a target system that includes all the targets for a repository system before selecting a site. When considering a site as a component of the repository selection, only some of the formulated objectives are queried. Consequently, the goal of the "best possible" safety of a repository system cannot be achieved through the site selection process alone. The target system is therefore the yardstick for the transparent selection of a preferred repository system with regard to radionuclide confinement when comparing two repository systems for approval. The level of safety is determined (in the technical sense) when the specified protection goals can be met with a high degree of certainty with the geotechnical environmental structure (repository system) to be used in accordance with a long-term assessment of safety. "Best-safe" is initially not a category in this assessment. The attributes "best-safe" and "best-possible" are not suitable for describing the security of a complex repository system. In addition, long-term safety achievement does not necessarily entail authorisation of the repository, if, for example, acceptance in the region is not achievable. In [4] StandAG Part B is to § 1 (Purpose of the law) noted: "The site selection process should be self-interrogative and learning designed. Central to a successful overall learning process that ultimately leads to a disposal with the best possible safety is the claim of all persons and institutions involved in the site selection process to constantly question themselves and each other throughout the entire process of final disposal and to engage and to practice continuously in systematic self-critical analysis of the achieved status." That provides a definition with which the site selection procedures can be started. Since the site selection process in Germany is going to take a very long time and its real-time untimed duration cannot be estimated, the passage should consider: "… constantly questioning ourselves and each other and practicing systematically and continuously the self-critical analysis of the achieved status.", whilst also referring to this law and thus to the definition of "best-safe" and "best-possible".

A geotechnical environmental structure is built with the aim of both preserving the environment and protecting it sustainably from existing harmful influences. Since a geotechnical environmental structure cannot be built without intervening in the environment, it must be proven that the goal to serve and protect, justifies an intervention. A repository must meet this requirement. The radioactive waste and residues as such are not part of the geotechnical environmental structure, but initially part of the environment itself.

The location and geotechnical environmental structure for radioactive waste and residues should only be designed and constructed in a receiving community (municipality) if the regional authority (municipal executive, district) agrees to this project. In order to be able to make the decision in a qualified manner, it must be scientifically accurate and otherwise credible so the operator is able to prove that the goal will achieved, i.e. that the radio-toxic inventory is isolated over the period under consideration (detection period), will remain stable in the long-term and isolate material from the biosphere. Radioactivity is only allowed to escape within socially and legally accepted limits. The evidence should also be supported by long-term environmental monitoring.

The isolation of radioactive waste and residues pre-supposes that they are encapsulated in such a way that, on the one hand, the personnel and the environment (e.g. transport) suffer no damage and, on the other hand, radioactive substances will not escape into the biosphere, or only within socially and legally accepted limits. For this purpose, the repository will be equipped with an optimally matched, multiple barrier system (multi-barrier system).

3.4 Multi-barrier Concept

Regardless of the quantity and quality of the radioactive waste to be stored, it is advantageous to prevent the escape of radioactivity by means of a multi-barrier system. A multi-barrier system will usually consist of several optimally coordinated technical, geotechnical and geological barriers.

(a) Multi-barrier concept for a geotechnical environmental structure for the long-term safe and stable storage of waste with very low radioactivity and with no need for monitoring the waste

The deposition of waste with very low levels of radioactivity not requiring monitoring of the residues is carried out according to the waste code and the European Waste Catalog (Waste Catalog Ordinance—AVV) on surface disposal sites of DK II or III. According to the Landfill Ordinance (DepV), 4 landfill classes (DK) are to be distinguished, see Table 3.1.

Annex 1 of the Landfill Ordinance describes site requirements, the geological barrier, basic and surface sealing systems: classes 0, I, II and III landfills. In addition to the 3 barriers described in the DepV, 3 additional barriers can optionally supplement a general multi-barrier concept of a site-specific landfill project:

– A possible waste treatment, this must be submitted with an application for approval;
– Measures on the nature of the landfill body, these are also subject to approval, and
– The follow-up and repair measures, these are prescribed by the approval authority, where appropriate.

Further details see Chap. 4.

Table 3.1 Landfill classes according to landfill ordinance

Classes	DK 0	DK I	DK II	DK III	DK IV
Arrangement	Surface deposited	Surface deposited	Surface deposited	Surface deposited	Underground deposited
Typical waste	Unpolluted soil—inert waste	Building rubble, soil, slag (mineral waste)—municipal waste/commercial waste	Commercial waste	Hazardous waste	Hazardous waste

- **Barrier location**: When choosing a location, particular consideration must be given to:

  - geological and hydrogeological conditions of the region;
  - specially protected or reserved areas;
  - adequate safety distance to sensitive areas;
  - risk of earthquakes, floods and other natural phenomena;
  - drainage of collected leachate.

- **Barrier Basic sealing**: if a basic sealing in the subsoil is available, it must be ensured that:

  - in the presence of a geological barrier, that it is guaranteed that the distance of the top edge of the geological barrier to the highest expected free groundwater level is at least 1 m at all times.
  - For the improvement of the geological barrier and technical measures as a substitute for the geological barrier—as well as of the sealing system: materials, components or systems may only be used if they comply with the state of the art and that this has been demonstrated by the competent authority. Materials used are: geosynthetics, polymers, etc.
  - Structure of the geological barrier and the base waterproofing system according to landfill classes, see Chap. 4.

- **Barrier Surface sealing**: In the decommissioning phase (also stepwise in the operating phase), the operator of a class 0, I, II or III landfill must immediately take all necessary measures to set up the surface sealing system. The surface sealing system consists of: mineral sealing component; protection mat/protective layer; drainage layer; recultivation layer; technical functional layer. For details of the structure of the surface sealing system according to landfill classes, see Chap. 4.

(b) Multi-barrier concept for a geotechnical environmental structure for the long-term safe and stable storage of uranium processing residues (residues requiring special supervision)

The uranium tailings are first flushed from the milling plant into the available disposal site, usually the sedimentation pond, via a pipeline system. The tailings storage management (TSM), is the chain of: generation of uranium tailings → Transport through piping system (pipe transportation) → storage in the prepared land form → Operating phase of the uranium tailing pond. The most important functional elements, the multi-barrier system of a geotechnical environmental structure for long-term stable safe and stable in situ storage of tailings from uranium ore processing (see Fig. 3.1) consists of:

- Structural barrier (tailings dam)
- Surface coverage of the tailings (geotechnical)
- Basic sealing (geological and/or geotechnical)[1]

---

[1]Geotechnically can also be mineral, i.e. a clay seal.

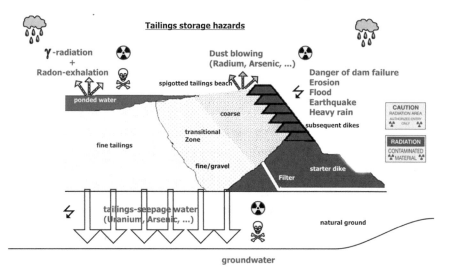

Fig. 3.1 Schematic figure of a uranium tailing pond with potential discharge hazards. *Source* Lersow [19]

- Water management system (Collection of surface and seepage water, wells for groundwater lowering, water treatment plant, groundwater wells and groundwater measuring points, data collection, logging, analysis and intervention plan)
- Long-term environmental monitoring system.
  Developed from basic monitoring and accompanying monitoring during time of long-term safe closure works of the tailings pond. This chain, which describes the operating phase of the uranium tailings pond, is not included in the descriptions below, as there is no longer any impact on the uranium tailings ponds in Germany (SDAG Wismut).

The site-specific geotechnical environmental structure has to meet the requirements for safe, long-term stable in situ storage of residues from uranium ore processing.
  The requirements are:

- Increase in and guarantee of stability
- Protection against infiltration of the surface cover
- Radiation protection
- Emission protection
- Groundwater protection
- Erosion protection, revegetation and
- Long-term environmental monitoring.

(c) Multi-barrier concept for a geotechnical environmental structure for the long-term safe and stable storage of radioactive waste

The repository system, see Fig. 3.2, results from the detailed requirements for safe, long-term stable storage and its protective function. This is to guarantee that the

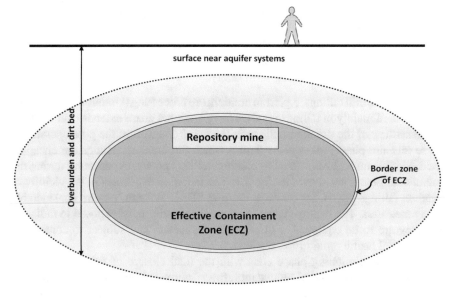

Fig. 3.2 Schematic of a repository system in deep geological formations, see also [20]

discharge of radio-toxic waste from an effective containment zone (ECZ) will be limited and then only within the socially and legally accepted exemption limits.

Humans and the environment thus remain permanently protected from the effects of ionizing radiation and its toxic effects. "In deep geological formations an effective containment zone in the host rock is defined according to the definition of the BMU safety requirements, see Baltes et al. [21], as the part of the repository system which ensures the containment of waste, in interaction with the technical closures (repository containers, shaft closures, chamber closure structures, dam structures, backfilling, etc.),". Host rock refers to the geological formation in which the effective containment zone is placed. The host rock is thus part of the multi-barrier system of the repository.

3.5 Basic Terms of Long-Term Safety

(a) The task or goal of the final storage of radioactive material requiring special monitoring in a Geotechnical Environmental Structure is to separate the stored material from material cycles of the biosphere for the required period of time. This means that the geotechnical environmental structure and the storage of the radioactive material must be equipped and arranged in such a way that this goal is very likely to be achieved. Geotechnical environmental structures that receive radioactive waste for final disposal have an operating phase (with a specified operational safety and operating license), in which the radioactive waste is accepted and stored permanently in the storage areas, a closure phase in which the operating is finished and the Geotechnical environmental structure is finally closed, and a post-closure phase (post-operational phase), for which the geotechnical environmental structure will be designed for long-term safety. Chapter 8 is devoted to long-term safety.

Radioactive mill tailings, e.g. from uranium ore processing (Uranium mill tailings) are deposited mainly in tailings ponds. The operational phase extends over the time of the settling of the uranium tailings in the pond. Thereafter, the post-operational phase (closure phase) takes place during which the uranium tailings are encapsulated. Significant hazards related to uranium tailings ponds, in particular during the operational phase, are shown in Fig. 3.1. The radiation exposures in the vicinity of uranium tailings ponds, for example those at SDAG Wismut, require considerable safety measures. The deployed personnel are the persons most exposed to radiation and they are to be equipped with personal dosimeters and subject to continuous occupational health monitoring, see Sect. 3.7.2.

The decommissioning phase ends with the final closure of the tailings ponds and the decommissioning of redundant infrastructure. This is followed by the post-operational phase, in which the stored uranium tailings must not pose any, or only socially accepted hazards, to the biosphere's material cycles. The required period for the post-operational phase is set by the responsible mining authority; namely when the mining authority releases the geotechnical environmental structure from the mining operator's supervision. For further information see Sect. 3.7.1.

The post-closure phase is accompanied by a comprehensive, demand-oriented long-term monitoring programme, which can be adapted to the respective situation and which is associated with an approved emergency plan.

(b) Both radioactive waste with negligible heat generation and heat-generating waste are stored in Germany in a repository mine for the long term, see Fig. 3.2. The operating phase must be accompanied by an operating license, which also includes the acceptance and protection criteria for the personnel. In the operating phase, therefore, the radioactive waste is accepted and deposited in the prepared storage areas or storage field in a prescribed manner. The delivery takes place in final disposal containers, with prescribed shielding and, in the case of HGW containers, have a high insulation capacity and are designed for a long service life. The waste containers for HGW are designed as a technical barrier and thus form part of the multi-barrier system. Here, too, the personnel employed are persons exposed to radiation who are required to be equipped with personal dosimeters and who are subject to continuous occupational medical surveillance, see Sect. 3.7.2. When operated as intended, however, the risks posed by radioactive waste are many times lower than with uranium tailings. This is followed by the decommissioning phase in which the emplacement fields and areas are filled with suitable backfill or insulation material or the repository mine is filled as required and containments are closed out using geotechnical structures, so that the radioactive material is permanently isolated for disposal. The decommissioning phase ends with the final closure of the shafts and the dismantling of the surface facilities. The required period for the post-operational and monitoring phase is determined in the approval of the final disposal.

The geotechnical environmental structures consist of geological (natural), technical and geotechnical barriers, which differ according to the type of radioactive material but perform similar functions. In their interaction they should ensure the isolation of the radioactive material for the required isolation period. Such a multi-barrier system consists of various system components, each of which has its specific role in the isolation of the waste. No element of the multi-barrier system is absolute and forever impermeable. It is important that the barriers are coordinated so that their mechanical, hydraulic and chemical effects complement each other optimally and reduce the hydrochemical mobility of the radionuclides in the waste matrices until immobilization, thus creating an overall insulation capacity. With heat-generating HAW this function takes over the effective containment area (ECZ) for as long as possible before radionuclides are released and could reach the biosphere on a propagation path. Whether it happens at all, and in what quantity and concentration, and which radiological effects may be expected in the biosphere, is determined by the safety analysis and the long-term safety proof.

For high-level radioactive waste (HLW) in Germany, the barriers, waste matrices, waste containers, chamber and shaft closure structures, the effective containment zone (ECZ) and the surrounding or overlying (geological) strata (see Fig. 3.3) are currently required to guarantee an isolation period of the order of a million years. It is also important to understand that the barriers listed above are passive safety barriers.

Within the history of humanity, 1 million years is a very long period. Internationally there are diverging views on the duration of the monitoring or observation period for the final disposal of HGW-HLW. This is explained by the fact that the risks associated with disposal, even with very long observation times, cannot be completely

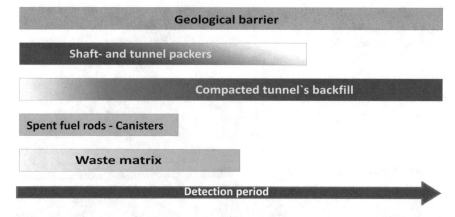

Fig. 3.3 Scheme of the temporal effect of various barriers in the repository system, in the detection period of one million years (not to scale)

eliminated (the remaining residual risk). The remaining residual risk and the partial risks are rated differently internationally, so that different periods under review (monitoring periods) are based on such data as is available. This in turn influences the long-term safety assessment. Ultimately, the national approval authority will decide on the application for the construction and operation of a repository structure.

Of course, the consideration of the long-term safety assessments of geotechnical environmental structures and repository structures for the several types of radioactive material should not obscure the fact that there are not only considerable differences in the potential hazards, but also in the safety precautions. A long-term safety assessment means the proof of the permanent safe, stable isolation of radioactive waste and residues in a geotechnical environmental structure. The robustness, integrity and reliability of the repository system must be proven over the detection period, see Chap. 8.

- For landfill structures according to the landfill and long-term storage ordinance (DepV), long-term safety assessments are generally not done. The wastes are trapped between the surface- and base sealing. When deposited in hollow forms, the embankment dams (slopes) are part of the sealing system. At the request of the operator, the waste authority decides whether the landfill site will be exempt from (groundwater) monitoring. However, the landfill site remains registered in the regional land register for an indefinite period as an old deposit. As a rule, the construction of the landfill requires a plan approval procedure according to the Federal Immission Control Act—BimSchG, with Environmental Impact Assessment (UVP) for DK (landfill class) II and III, see Chap. 4.
- Uranium tailings ponds are subject to the Federal Mining Act-BBergG. The long-term safe and stable custody of a tailing pond is carried out worldwide on the basis of a site redevelopment concept [Conceptual Site Model (CSM)], WIS [22] and NAC [23], see Fig. 3.4.

Fig. 3.4 Schematic of the conceptual site model (CSM)

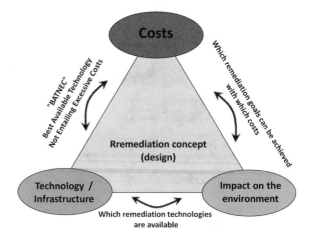

The long-term safe and stable closure of uranium tailing ponds is carried out in Germany according to a closure plan approved by the competent mining office. The closure plan is based on a Conceptual Site Model (CSM). Site-specific, the encapsulation of the radioactive residues is guaranteed by the following functional elements: Uranium tailings cover (Multifunctional Cover System), the shaped and stabilised dam structures and a geological and/or geotechnical sealing of the base. The closure technology used is usually unique as it has to be site specific. Therefore, the optimization criteria and the long-term safety investigations and proofs of each existing storage method must always be geared to the achievement of site-specific objectives. However, there are also objective criteria to follow, IAEA [24].

It is acknowledged worldwide that the long-term safety assessment of mechanical, hydromechanical, chemical and radiological safety for site-specific environmental structures is to be done for a period of 1000 years (in exceptional cases for only 200 years). This period is also referred to here as the period under review (detection or monitoring period). However, a detection period of 1000 years is not scientifically proven. For radiation protection, limit values of the radiation protection ordinance must be observed [25, 26], for example:

In the U.S. a radon exhalation rate[2] of RER $\leq$ 20 pCi m^{-2}s^{-1}

In Germany a radiation exposure of ODL $\leq$ 0.3 μSv/h

The geotechnical environmental structures to be operated must ensure that no (or only within the socially accepted limits) radioactive and toxic substances enter the biosphere, in particular the prevention of contamination of groundwater. Limit values are specified for the individual substances: in USA by US Code of Federal Regulations (CFR), EPA [27], in Germany by the recommendations of the radiation protection commission SSK (98) [28].

Construction, operation and safekeeping must be authorized by the competent mining authority in separate procedures. Post closure phase: If proof is provided that all relevant environmental data since closure are within the specified limits, the site of the uranium tailings (tailing pond) may be released from the supervision of the mining authority at the request of the operator. However, the file with the final documentation of the location with all relevant data remains permanently in the files of the responsible mining authority. This also provides some protection against human intervention (consciously or unconsciously) into the closure structure (Human Intrusion Scenario).

- For the final disposal of highly radioactive, heat-generating radioactive waste, a site selection procedure must be carried out first. Site selection compares long-term safety analyses for different locations with possibly different host rocks and thus also with different technical and geotechnical barriers tailored to the respective host rock. This first condition requires the development of benchmarks, methods and tools as well as the objectives, possibly with weighting of results of individual considerations, in order to be able to compare the long-term safety analyses of

---

[2]Output radon flux by earthen uranium mill tailings cover.

different repository systems in different host rock types. The aim of this procedure is to develop the scientific and technical basis for a safety-oriented, methodically systematic and transparent comparison between sites that have already been identified as safe; and thus to eventually propose the best repository site for HGW-HAW. Naturally, acceptance for that site should also be sought from relevant stakeholders. The repository model that can be approved is therefore also a Conceptual Site Model (CSM).

In order to assess the possible types of final disposal in different types of host rock (repository mine/deep boreholes/caverns, etc.), the safety-relevant functional elements must be put together for consideration in the safety assessment. According to BMU [3] these are the following: (see also Glossary).

- **Barrier** (natural, technical or geotechnical)—A barrier is a natural or technical component of the final disposal (repository) system, which completely or partially prevents or delays the transport of radionuclides or other substances into the biosphere. A distinction is made between technical, geological and geotechnical barriers. Such barriers are, for example, the waste matrices, the waste containers (canisters), the chamber and shaft gate structures, the effective containment zone (ECZ) and surrounding or overlying geological strata;
- **Possibilities of planned and unplanned retrieval (emergency planning) of radioactive waste**—The unplanned retrieval is called recovery (salvage). During the operation of the repository, the retrievability of the radioactive waste must be possible at any time. In an emergency, the repository containers must be able to be salvaged 500 years after closure of the repository;
- **Overburden**—This refers to the geological strata above the effective containment zone (ECZ);
- **Containment**—Radioactive waste is trapped in a defined containment zone in such a way that it essentially remains at the location of storage and at most low amounts of material leave this zone.
- **Effective Containment Zone (ECZ)**: Part of the repository system which, in cooperation with the technical closure elements (shaft closures, chamber closure structures, dam structures, backfill, …), ensures the containment of waste—the integrity of this containment zone must be across the period of review ensured of one million year;
- **Repository mine**—A repository mine consists of various components such as shafts, drifts and chambers with the waste packages stored therein, filling and seal elements, impermeable walls;
- **Repository system (Geotechnical environmental structure)**—A repository system consists of the repository mine, the effective containment zone (ECZ) and the surrounding or overlying geological strata up to the earth's surface, insofar as they are important in terms of safety.

In the case of deep geological disposal, a sufficient volume of the host rock (Thickness, extent) increases the likelihood that sufficiently large, undisturbed areas will be found that fulfill the requirements of the effective containment zone and in which the waste can be stored. Naturally, uniform rock formations and comparatively simple geological conditions can be characterised more reliably and their properties better predicted. According to current philosophy, a sufficient depth is required, since the geological stability, especially with respect to possible time depends on changes in the environment at the earth's surface. However, such disturbances generally increase with depth also. These are examples of parameters which differ greatly in the various host rock types (clay rock, salt rock and crystalline rocks). Various waste containers and waste matrices have been developed and matched to the particular host rock, in order to compensate for the individual shortcomings of particular host rocks. The contention that geological formations have proven their stability over very long periods of time and that technical or technogenic material have not, does not take into account that the technical and technological materials did not yet have the chance to be fully proven, and the aging of geological material has not been studied over such long periods of time either. The earth was not created for the purpose of a repository and thus is not in an ideal condition for that purpose. The authors are of the opinion that it is also not proven that only deep geological repositories are all suitable. Perhaps future generations will have very different options for action and come to other decisions. We have to permit future generations to have such opportunities.

From the current point of view, it can be considered as a proposal that an engineering-technical, geotechnical structure repository close to the surface with only relatively thin capping could also be suitable. It is possible that such a geotechnical environmental structure would pass a safety assessment successfully. Furthermore, if the period of recoverability of 500 years is legally agreed in Germany, a repository must fulfil this requirement completely. If international cooperation in the research and design of repositories is further strengthened, future generations so will be able to benefit greatly from the repository developments of the present generation.

With a long-term safety proof for the final disposal of heat generating HAW it can be scientifically demonstrated which radiological effects on the biosphere can emanate from the repository during the next one million years and to what extent that may happen. For this purpose, the possible future states, events and processes such as the thermal influence of the host rock, erosion, gas formation, human intrusion (human intrusion scenarios), the development of an ice age are combined into scenarios. Using numerical simulations, the consequences of these scenarios are determined in order to prove whether, or how, the entry of fluids from the overburden into the ECZ is possible; and when and how the escape of fluids and gases from the ECZ could be expected and how likely these conditions are to occur. The numerical solutions are complemented by observations of nature or natural analogues and by safety indicators as instruments for building confidence in, and credibility of, the results. This is because over such a long period the behavior of the repository and its elements cannot be predicted with certainty.

Therefore an important element of a long-term safety assessment for a repository for radioactive waste is the long-term safety analysis. It examines the future behavior of the repository and determines the possible future radiation exposure at the repository site. Final disposal must ensure that releases of radioactive substances from the repository will in the long term only marginally increase the risks arising from natural radiation exposure. For less likely developments in the post-closure phase, it must be demonstrated that the additional effective dose caused by the release of radionuclides originating from the stored radioactive waste does not exceed 0.1 millisievert per year for the people affected.

The author is aware that the proof of safety in the post-closure phase of the repository cannot be conducted in a strictly scientific sense, since the potential consequences from the operation of the repository excludes a metrological testing or verification due to the long period of review. In this respect, the proof of safety represents a probable prognosis, whereby its probability of occurrence cannot be determined exactly.

Achieving the goal of ensuring compliance with the protection goals requires both permanent monitoring of the repository site and its environment, as well as appropriate emergency management in the event of intervention thresholds being exceeded, see Chaps. 8 and 3.8.4.

3.6   Crisis Management/Emergency Plan

In the case of final disposal, particularly for radioactive material subject to monitoring, extraordinary conditions may occur which were not predicted during assessment and which may lead to considerable, irreparable damage in the biosphere, e.g. leakage of considerable amounts of radioactivity from the repository in the event of loss of integrity. To resolve such a crisis situation requires appropriate management usually involving an emergency plan. For crisis management in the final disposal of HGW-HAW[3] the terms to be considered as part of the safety assessment include reversibility/retrievability/recoverability.

The term reversibility is defined as follows IAEA [9]:

"The more general concept of reversibility denotes the possibility of reversing one or a series of steps in the planning or development of the disposal facility. This implies the review and, if necessary, re-evaluation of earlier decisions, as well as availability of the means (technical, financial, etc.) to reverse a step."

Thus, the general concept of reversibility means the ability to reverse one or more of a series of steps in the design or development of the disposal facility. This implies the review and, where appropriate, the reassessment of previous decisions and the availability of resources (technical, financial, etc.) to reverse a step.

Usually previous repository concepts have been designed to shut down and to seal the repository as soon as possible after operation; this was justified by the fact that the problem of final disposal of radioactive waste needed to be imposed promptly. The discussion that has now got under way about the task to be solved here has meant that the so-called "generational justice" which is linked to this task, is not endangered, because:

---

[3]That applies similar to radioactive waste with negligible heat generation.

(a) Due to the longevity of radionuclides subsequent generations are faced with providing repositories for radioactive waste, whether they like it or not. In the process, subsequent generations must also be given the opportunity, at least in principle, to intervene and to contribute their ideas, to revise decisions and possibly to reverse them (dissolution of the repository, salvage of the inventory).

(b) The principle of rapid decommissioning does not give future generations the opportunity to alter decisions of previous generations or to install new technologies and procedures.

(c) It is often overlooked that future generations will have to build repositories. In many countries the peaceful use of nuclear energy is likely to be maintained in the long term and repositories will be needed for increasing amounts of radioactive waste. It may be useful to be able to draw on results and knowledge in the disposal of radioactive waste and to combine these with our own findings in such a way that the safety level can be further increased. It can thus be demonstrated that the robustness and integrity of repositories with a probability bordering on safety may persist for very long periods of time.

(d) No reason exists for great haste in developing final storages for HG-HLW, since sufficient interim storage capacities are currently available, at least in Germany. However, currently many operating licenses have only been granted for a period of 40 years. It could become part of a German repository concept to design and build pilot facilities, into which the heat-generating HAW can be taken over and from there transferred to a final disposal, see Chap. 7—Modular Repository Concept.

(e) The most important question for future generations is whether they have to pay for the radioactive waste of today's generation. Chapter 7 addresses this question amongst other issues, including whether the Nuclear Waste Disposal Fund of about EUR 24 billion will be sufficient to cover the tasks to be financed, KFK [29]. It also depends on how it is planned to develop, and effectively implement, a repository concept for HG-HLW in Germany.

In the case of radioactive residues, especially uranium processing residues, closure models for the tailings storage facilities are developed on the basis of Conceptional Site Models (CSM), see also Fig. 3.1, into which so-called decision points are inserted, at which time the model is checked again and a decision is made as to which path is to be taken from there or whether another jump back in the procedure with one or more steps is necessary, see also Benchmarks. CSM's are used internationally to plan the long-term safe closure of uranium mill tailings ponds. In the USA CSM is also used for low- and intermediate-level radioactive waste disposal planning, see Conceptual Site Model for Disposal of Depleted Uranium at the Clive[4] Facility [23]. In Germany, the closure of uranium mill tailings ponds is undertaken using closure plans in accordance with the Federal Mining Act—These are another form of conceptional site model.

---

[4]Clive is one of the three licensed, commercial repositories for low and intermediate level radioactive waste in the US. These repositories are located in Barnwell (South Carolina), Richland (Washington) and Clive (Utah).

Especially when optimizing a concept from planning to the closure of a repository, CSM are important. This applies both to the formulation of the optimization goals and to their achievement. Optimization goals are:

– Radiation protection in the operating phase
– Long-term safety
– Operational safety of the repository
– Reliability and quality of long-term containment (isolation) of waste
– Safety management
– Technical and financial feasibility.

Closely related to reversibility are the terms retrievability/recoverability. These are planned possibilities during the operational phase (retrieval) of a repository and in the subsequent post-operational phase/recovery (salvage)—currently estimated in Germany at 500 years—to bring the repository containers back to the surface, see BMU [3]: "*Measures taken to ensure the possibility of retrieval or recovery must not affect the passive safety barriers and thus the long-term safety*".

Retrieval and recovery may be accompanied by increased risks of radiation exposure for the staff. This should be noted in the consideration of scenarios of retrievability/recoverability, as well as the selection of technical equipment including the use of robots and the storage possibilities of the tailings material in such processes.

The above considerations exclude the operating phase of the repository and waste management from the point of origin to storage in a repository. The chain—place of origin → temporary storage → conditioning plant → transport → acceptance conditions of radioactive waste in temporary storage—is in the view of the authors merely part of a repository concept. The requirements for conditioning, transport, acceptance conditions, etc. are largely excluded here, see Sect. 3.7.

3.7 Sources of Hazards and Protective Measures at Geotechnical Environmental Structures for Radioactive Waste and Residues Disposal

In what follows, the risks to the biosphere shall be compiled so that measures could be taken to avoid or limit these to a non-damaging level. This includes dangers to individuals and groups emanating from radioactive waste and residues and which remain in existence during the entire service life of the geotechnical environmental structures receiving them. It is important to determine during the operational phase and decommissioning phase of the uranium tailings storage what risks are posed and the dangers that are different from those that can be expected in the post closure phase of the encapsulated radioactive waste or a final closed repository, see Fig. 3.1. The basic safety standards for the protection of the health of workers and the general public against the dangers of ionizing radiation are defined and compiled in [30] Directive 96/29/Euratom and transposed into national law. In the U.S., this is defined in [31]: "*Criteria Relating to the Operation of Uranium Mills and the Disposition of Tailings or Wastes Produced by the Extraction or Concentration of Source Material from Ores Processed Primarily for Their Source Material Content.*"

3.7.1 Uranium Mill Tailings Storages (Tailings Ponds)

Uranium mill tailings storages, inventories, long-term safety etc. are described in detail in Chap. 5. In order to describe the dangers emanating from uranium tailings storages, it is first of all important to know the sources of danger. Thus, in the processing of ores large volumes of rock are crushed, ground and then subjected to a chemical and/or mechanical separation process. The resulting process residues are, for example, flushed into surface residue storages as a suspension or slurry and stored, see Fig. 5.6. As a result, the solid components of the slurry settle down into the pond and thus form a water-saturated solid layer, the thickness of which increases over time.

Depending on the climatic zone, a liquid layer of process water and accumulated surface water forms above the water-saturated solid layer. The tailings storage is generally perceived as a "pond storage", from which the term "tailings pond"—residue storage—is derived. Water or liquid is initially a good protection against radioactive emanations. Reference is made here to Canadian uranium tailings ponds, which are usually covered with a clear water lamella. This protection can be significantly reduced in the operating phase of the uranium tailings ponds, in particular by evaporation. This leads to uncovering, especially of the beaches (tailings beach). Radon emanation, radioactive dust blow and gamma radiation may all contribute to significant radiation exposure in the vicinity of uranium tailings ponds; so also in case of SDAG Wismut, see Fig. 3.1. The damaging effect of the radiation exposure becomes clear: Dust adhered to clothing, was inhaled or ingested, passed into the groundwater, in the soil and was distributed with the wind in the environment, deposited there, so that radiotoxicity also enters the food chain, see Fig. 3.10 with well known consequences.

The staff of SDAG Wismut was also affected, as can be seen from the statistics of recorded cancer cases at the German Employer's Liability Insurance Association for the Raw Materials and Chemical Industries (RCI[5]). The damage to the population has not been measured, but could be determined from the cancer registry in the regions of Saxony and Thuringia.

Following the German reunification and the beginning of the remediation of the uranium tailings ponds of SDAG Wismut, the radiation exposure did not disappear immediately. This was followed by a lengthy remediation process of the so-called Wismut region in Saxony and Thuringia, which continues to this day, see Chap. 5 for further details.

---

[5]The Employer's Liability Insurance Association for mining (BBG) was merged into RCI.

3.7.2 Damage of Individuals by Direct Contact and Radiation—Radiation Protection

The radioactive contamination of substances is expressed as the activity of a radionuclide per mass (specific activity, unit Bq/kg) or as activity per volume (activity concentration, unit Bq/l or Bq/m^3). In some cases it is also useful to relate the activity to the area, e.g. when radionuclides are deposited on the ground. In order to be able to make statements about the possible health hazard to humans, the measured activity (per mass or volume) of a radioactive substance must be converted into a dose (organ dose or effective dose, unit Sv). The Radiation Protection Ordinance and the X-Ray Ordinance lay down dose limits for occupationally exposed persons and for the general population. In general, any use of ionising radiation must be justified and radiation exposure must be kept as low as reasonably achievable even below the limit values. For the final disposal of radioactive waste to be discussed here, both the occupationally exposed persons—in the phases of operation, storage and possibly recoverability/retrievability—and the general population in the post-closure phase are considered.

§ 54 in the German Radiation Protection Ordinance defines as occupationally exposed persons, those persons who are exposed to occupational radiation which, in a calendar year, may result in an effective dose of more than 1 mSv or an organ dose higher than 15 mSv for the eye lens or an organ dose higher than 50 mSv for the skin, hands, forearms, feet or ankles externally, to fall into Category A (equivalent to Category B).

Pursuant to § 54 of the Radiation Protection Ordinance, occupationally exposed persons who are exposed to occupational radiation exposure by activities according to § 2 (1) No. 1 are classified in category A and B.

Category A: Persons who get exposed to an occupational (external) radiation exposure that can lead during a calendar year to an effective dose of more than 6 mSv or more than 45 mSv for the lens of the eyes or more than 150 mSv for the skin, the hands, the forearms, the feet or ankles.

The effective dose for occupationally exposed persons may not exceed 20 mSv per calendar year in accordance with §55 of the Radiation Protection Ordinance. In individual cases, the competent authority may authorize 50 mSv for a single year, but for five consecutive years, 100 mSv may not be exceeded. According to §56 of the Radiation Protection Ordinance, the occupational life-dose may not exceed 400 mSv.

According to §46 Radiation Protection Ordinance, the limit value for the effective dose for the protection of individuals of the population is 1 mSv per calendar year. This value refers to all radiation exposure from nuclear and other facilities for the generation of ionizing radiation and the handling of radioactive materials. This means that the limit applies to the sum of the radiation exposures from direct radiation and the radiation exposures from discharges (emissions) from nuclear installations.

According to § 47 Radiation Protection Ordinance, the radiation exposure from a single installation via the exposure pathway wastewater and exhaust air may not exceed the value of 0.3 mSv per year.

In the case of radioactive substances acting on humans from the outside (external radiation exposure), the amount of dose besides the type of radionuclide and its activity authoritative also the distribution in the environment (e.g. in the soil, in building materials) such as the location and duration of stay of the persons must be considered.

When radioactive substances enter the human body (internal radiation exposure), the dose level is determined by the type of radionuclide, the activity absorbed, the pathway of intake (respiratory or ingestion) and the chemical form of the radionuclide.

The adsorption of radioactive material, here radioactive waste or residues; by contamination, is the worst case for incorporation. In the medical sense, incorporation is the deliberate or unintentional intake of substances. When radioactive substances enter the human body (internal radiation exposure), the dose level is determined by:

– the type of radionuclide,
– the nature and amount of radioactivity taken in,
– the intake pathway (respiration or ingestion) and
– the chemical form of the radionuclide.

The methods of incorporation that may occur include:

– Ingestion, radioactive material is absorbed through the mouth, usually in solid or liquid form,
– Aspiration, the entry of radioactive material into the respiratory tract,
– Inhalation, Inhalation of radioactive material,
– Transdermal absorption of radioactive material via the skin or mucous membrane.

Not only the type of radionuclide and its activity is important for the potential health hazard of humans, but also whether the radionuclide is on the outside parts of the human body or enters the human body. So for example, an alpha emitter outside the body is completely harmless, since it is already completely shielded by a few inches of air, regardless of how high its activity. If, however, this alpha emitter is disturbed and a larger amount (higher activity) enters the human body by breathing, this can lead to health damage.

In the US, the dose limits for external and internal radiation exposure are set and summarized in [12, 32]. In [32] "*Summary of dose limits & target populations*" a maximum equivalent dose to the skin of an occupational workers for emergency life-saving efforts of 500 rem (5 Sv) is stipulated.

If this equivalent dose should be achieved (measured by dosimeter), the worker is taken out of the emergency team.

And the "*Annual Exposure Limit for Occupational Workers (NRC, DOE & States)*" is defined in the same document as 5 rem (50 mSv). The Standards for Protection against Radiation follow similar principles as are laid down in the German Radiation Protection Act and in the Radiation Protection Ordinance.

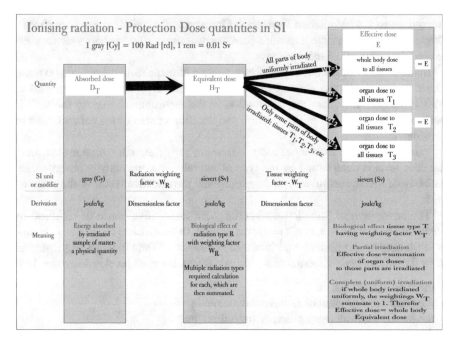

Fig. 3.5   Relationships between different dose types of ionizing radiation

The relationship between absorbed dose, equivalent dose and effective dose in the context of external and internal exposure (ionizing radiation) is shown in Fig. 3.5.

Dose limits do not serve as a dividing line between hazardous and non-hazardous radiation exposure. Rather, the exceeding of a limit value means that the likelihood of the occurrence of health consequences (in particular of cancers) is above a value that has been established as socially acceptable. The limit values are specified in the Radiation Protection Ordinance and in the X-ray Ordinance.

From the perspective of medicine it has been shown: If the radiation exceeds one Sievert (Sv) (300 times the natural radiation in the year), it leads to symptoms such as skin redness, hair loss and burns and acute radiation sickness. This damages the hematopoietic system in the body's bone marrow. At higher doses (above 10 Sv) the gastrointestinal tract and the cardiovascular organs are attacked. The risk for cancer is increased. Released radioactive iodine, e.g. Iodine-131 or Iodine-133, can be inhaled or swallowed; it accumulates, especially in children and young adults, in the thyroid gland and increases the risk of cancer there. With Radon exhalation of Rn-222,[6] this can reach the lungs via the respiratory tract. There, the stable but toxic

---

[6]Radon also occurs in nature and is part of environmental radioactivity. It is emitted by natural radionuclides in soils, rocks and air. Increased in areas with high uranium and thorium content (U-238 and U-235 as well as Th-232) and potassium isotope K-40 in soils and rocks. These are mainly the low mountain ranges of granite rock, in Germany primarily the Bavarian Forest, the Fichtelgebirge, the Black Forest and the Erzgebirge. The dose rate of radiation there is up to 1.3 mSv/a.

decay products (lead, mercury) are deposited and these can lead to lung carcinomas, Bay [33].

Radiation damage to the extent described above is to be expected, in particular if persons are exposed directly to high- and medium-level radioactive waste (HAW and MAW). With fuel elements, a relatively short exposure time is sufficient to produce radiation damage. In extreme cases, this can lead to death, even after a relatively short time.

Three considerations lead to the following presentations:

- Radioactive waste and residues accumulate in various technological processes or from material flows; they are hazardous goods, from which the biosphere has to be reliably protected, not only humans. Protection must be guaranteed at all stages of these processes: at the point of origin and in the technological process; in conditioning and in packaging; during transport of radioactive waste and residues; in the handling of uranium tailings or in the interim storage facility and for uranium tailings management of the tailings storage or in the storage areas of the repository. The protection of personnel and the population has the highest priority in the company. Considering permanent storage in a repository, radioactivity must be prevented from being released to the biosphere, especially at levels above the socially accepted limits. This protection may be guaranteed in a variety of different ways.
- In the case of radioactive residues, in particular uranium mill tailings, radioactive material can spread directly and enter the material cycle in the operation and decommissioning phase, since the uranium tailings are not covered and only the water lamella forms a (low) protective function, see Fig. 3.1. The general population was banned from accessing the areas, but this protection was weak and ineffective, especially for animals, see Fig. 3.6.
- Radioactive waste and residues accumulate in many ways and one will not be able to avoid them in the future either. On the one hand, the danger arising therefrom must not be unrecognized, underestimated or even played down, but on the other hand there should be no exaggeration of the kind that ultimately gives the impression that a long-term safe and stable final disposal is not possible, or that the entire process is uncontrollable.

The radiation protection supervisor is responsible for the radiation protection of the operator of the various facilities. He has to ensure that the protective measures, the three elements, are implemented in the handling of radioactive substances throughout the entire security area (control area):

- Duration of stay (only as long as necessary),
- Distance (intensity of the radiation decreases with the square of the distance from the source),
- Shielding.

This includes exposure monitoring, the wearing of protective clothing, control of inlet and outlet sluices, and the wearing of personal dosimeters (officially approved dosimeters), their readout and prescribed assessment (usually monthly or quarterly).

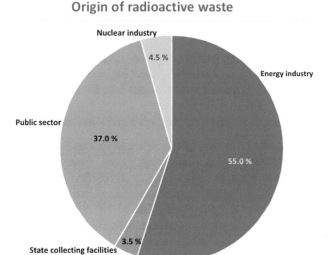

Fig. 3.6 Forecast: origin of waste to be stored in Schacht Konrad according to information BfS [34]

Although the legal frameworks for the different types of radioactive waste and residues are very different, it is useful to compare the radioactive inventories with the waste categories described above to illustrate the problems. However, such a comparison encounters difficulties, as very different radionuclides dominate in waste categories, see Chaps. 5, 6 and 7. Since the physical variable "activity" is easily measurable but not very meaningful in terms of the potentially associated hazards, different parameters are introduced for the comparison of inventories. For example, the calculated doses can be determined, which result from a complete consumption of the waste ("ingestion").

The doses thus obtained are completely fictitious and can be manipulatively mis-used[7] by conversion to "cancer deaths". Therefore, a different indicator is used here. By referring the activity inventories Ai for each radionuclide i to the exemption limit of the total activity FGi of the respective radionuclide according to StrlSchV (Annex 4, Table 1, column 2), a dimensionless ratio Ai/Fi results. The sum of these ratios across all radionuclides yields a dimensionless activity index capable of characterizing the overall inventory of activity inventories of different composition. The radiological hazard (radiotoxicity) of individual nuclides is taken into account by way of comparison of the respective exemption limits of the individual nuclides, see Table 3.2.

As shown in Table 3.2, the radionuclide iodine-131 is very important with a HWZ of only 8 days. It has the highest toxicity of the radionuclides listed in Table 3.2.

---

[7]An effective method, but not absolute protection, is the uninterrupted medical supervision of staff and complex environmental monitoring.

Table 3.2 Selected radionuclides and their radiotoxicity

Radionuclide	Specific activity Ai* in Bq/kg	Exemption limit Fi* in Bq/kg StrlSchV (Annex 4, Table 1, column 2)	Radiotoxicity Ai*/Fi* [−]	$T_{1/2}$ in a
I-131	4.60E+18	1.0E+05	4.60E+13	8.0 days
Pu-239	2.30E+15	1.0E+04	2.30E+11	2.4E+04
Th-232	4.06E+06	1.0E+03	4.06E+03	1.4E+10
U-234	2.30E+11	1.0E+04	2.30E+07	2.5E+05
U-235	8.00E+07	1.0E+04	8.00E+03	7.0E+08
U-238	1.20E+07	1.0E+04	1.20E+03	4.4E+09

3.7.3   Final Disposal of Radioactive Waste

3.7.3.1   Shielding, Transport and Temporary Storage of Radioactive Waste

The disposal of radioactive waste is characterized using various successive process steps such as gathering, sorting, registering, conditioning, packaging, interim storage and transfer to the final disposal. The declared LAW and MAW waste is first delivered and collected in an interim storage facility or a state collecting facility in approved intermediate storage containers. The respective land collection center and the interim storage facility cover the approved transport containers with a corresponding shielding, which allow safe transport and safe handling, see Fig. 3.8. Upon delivery to the interim storage facility, the waste is accepted and inventoried. From there it is transferred to the final storage. However, the waste would first have to be repackaged into final disposal waste packages, see Fig. 3.6. Thus the waste will be embedded in a waste matrix and surrounded by a container.

For heat-generating radioactive waste (fuel rods) it should be noted that these are first stored after use in the reactor in a water storage (cooling pond). As already stated they remain there until their heat development has subsided to such an extent that transport with air-cooled castor containers to the interim storage facility is permitted, see Fig. 3.7. Interim storage, as currently interpreted, is the storage of conditioned or partially conditioned waste for interim disposal. From there the material can be transferred to a final disposal, if the final disposal is constructed and was officially released for operation. HG-HLW, however, has to be conditioned and packaged before being considered to be in a suitable condition for final storage. Conditioning is the production of waste packages by treatment and/or packaging of radioactive waste. A waste treatment for final disposal is the processing of radioactive waste into waste products, which may include the following sub-steps: pressing, drying, cementing.

On the basis of the "Act on the Reorganisation of Responsibility in Nuclear Waste Management" [36], the federal BGZ Company for Interim Storage mbH mbH, located in Essen, was founded. From 2019, BGZ mbH will be responsible for the 12

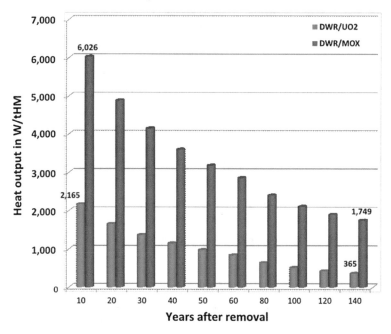

Fig. 3.7 Heat output of spent fuel after removal from the reactor, from data from GRS [35]

interim storage facilities at the sites of the German nuclear power plants; and from 2020 also for the twelve interim storage facilities with low and intermediate level radioactive waste from the operation and decommissioning of the NPPs. In Germany, the Federal Government is thus responsible for the interim storage of all radioactive waste subject to monitoring. NPP operators are left with only decommissioning and dismantling.

There are major differences between the waste containers for temporary storage which only have to fulfill the functions of shielding, and those used for containment of radioactive material in the container and discharge of heat over the entire interim storage period. Of course, these containers must be stable for the entire time of storage, must not corrode etc. This is also permanently monitored.

But in the post-closure phase of the repository the waste containers for final disposal have in addition a containment function (or insulation function) for the radionuclides of the inventory and thus constitute an important geotechnical barrier[8] in the repository concept.

---

[8] Applies only to repositories for HG-HLW.

Fig. 3.8 Packaging: 200-l-sheet steel drum; on the test bench for activity measurement. *Source* www. vgb.org/abfallmanagement. html

For the defined waste product groups,[9] the permitted activity limits must be observed in order to ensure the requirements of the repository operation as well as subcriticality (Fig. 3.8).

3.7.3.2 Waste Packages—Waste Matrix and Container as Geotechnical Barrier

The requirements for waste products and their chemical/physical form of the waste matrix arise from their behavior in the specified normal operation of interim and final storage facilities, as well as in the case of potential incidents and the final disposal of heat-generating HLW. In addition, the waste matrix also takes over functions for the containment of nuclides, including the interim storage building and the waste bin.

Special attention must be paid to the waste containers for the disposal of HGW-HLW, since (only for these wastes) the waste containers are also assigned a function as a technical barrier.

The vitrified, reprocessed waste and the spent fuel make up the largest part of the heat-generating waste occurring in Germany. In the vitrification of the radioactive substances extracted in the reprocessing of irradiated fuel elements are fused with a glass substance to become HLW-glass.[10] The glass matrix is relatively homogeneous and contains only a few phase precipitates. These phase precipitates mainly contain precious metals. The corrosion resistance of the vitrified material must be considered for the long-term safety considerations in the repository.

---

[9]Waste product—processed radioactive waste without packaging.

[10]High Level Waste (HLW); Borosilicate and phosphate glasses are considered for the glass matrix. The glass of German waste, the so-called Cogema glass, is a borosilicate glass with a silicon oxide content of about 50%.

There is thus a distinction to be established between conditioned radioactive waste for intermediate storage or for final disposal. For disposal, the prerequisite is that the waste matrix meets the criteria of long-term stability. In addition, of course, the waste containers must meet all requirements of operational safety, especially radiation protection (shielding) and transport and thus guarantee the safety at work for personnel when handling the dangerous goods containers.

A conditioning procedure for HLW must also take into account that a limitation of the activity in the waste product or in the packaging must be completed in accordance with the terms of delivery to a repository. This activity limit determines the extent of potential release of radioactive material from the waste package. In addition, sub-criticality is to be ensured. Subcriticality means that a self-sustaining chain reaction cannot occur.

The longest possible retention of the radionuclides in the waste matrix and in the waste container will have a generally favourable effect in terms of safety. It can be stated that the waste packages represent a technical barrier for the final disposal of heat-generating HLW, because the contribution of the waste matrix and the waste containers to the isolation of the radionuclides does exist. The release of the radionuclides from the waste matrix and from the waste container into the effective containment zone must be limited; or, if possible, not be allowed at all. An important finding is that the repository containers for heat-generating HLW also have to be adapted to the host rock. For example, different repository containers will have to be developed for salt rock than for granites, since underground cavities (storage fields) converge in the salt rock. The containers surrounding the waste matrix may be destroyed in the post-closure phase, with the salt rock then completely enclosing the radioactive waste. Cavities (emplacement fields) in the granite remain stable. However, since water ingress in the post-closure phase is likely, the containers are surrounded with bentonite (clay), which results in a high sealing and low corrosion capacity and thus a high immobilization capacity (retention capacity), see Fig. 3.9. Claystone in turn has a different long-term behavior.

The effective containment zone (ECZ) is defined according to the definition of the BMU safety requirements, see Baltes et al. [21], as that part of the repository system which,

Fig. 3.9 Technical barrier developed for the disposal of HGW-HLW in crystalline host rocks (granites). *source* SKB [37]

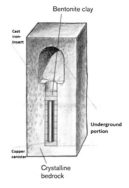

in conjunction with the technical closures (repository containers, shaft packers, chamber closure structures, dam structures, backfilling, etc.), ensures the containment of waste.

In Germany, apart from the requirements of retrievability and recoverability, there are currently no regulatory provisions or other specifications for requirements for the long-term behavior of waste containers and the waste matrix of heat-generating waste. However, these are necessary if the repository concept should assign a safety function for a duration of the period of review for the waste matrix and containers as a component of the repository system, see also Fig. 3.3.

Serious progress in the field of container development has been made internationally, see also Chaps. 7 and 8.

3.8 Dispersal Possibilities of Radionuclides from Geotechnical Environmental Structures

The dispersal of radionuclides from any repository structures is mostly possible only via water as the transport medium. However, release via the gas phase (radon release, for example) must also be considered. If water comes into contact with the stored waste, it may result in a variety of interaction processes depending on the specific geochemical environment. This significantly determines the extent of release and retention of the radionuclides. The overall assessment of the potential release pathways must also take into account the other substances introduced to the disposal including the materials used as technical and geotechnical barriers. In addition to the hydro-geological conditions, the chemical binding form of the radionuclide (speciation) has a significant effect on the dispersal. Due to the different reactions of the radionuclide, such as redox reactions, hydrolysis, complexation,[11] sorption, desorption, remineralization or colloid formation, its speciation along the migration path may change. The knowledge of the thermodynamic data for the relevant binding form forms of the corresponding radionuclides in possible scenarios in the different environmental structures is a prerequisite for the assessment of the long-term safety of these environmental structures. The discovery of the uncharged Calcium uranyl carbonate complex both in seepage waters of uranium ore mining and in pore waters of clays, as well as in saline waters, was a milestone in the proper description of the environmental chemistry of uranium and provides a starting point to reconsider the complexation of uranium in environmentally relevant waters, see Bernhard et al. [38].

The creation of necessary databases with validated data such as the "Thermodynamic Reference Database THEREDA (www.thereda.de)" or the "Sorption Database RES³T (www.hzdr.de/res3t)" are listed here as examples.

Nuclide transport through the geosphere is of crucial importance, as the geological medium is the last barrier to the biosphere, see Fig. 3.10. In the risk analyses for

---

[11]Complexation: Encapsulation of ions with other neutral or polar substances by forming of complexes.

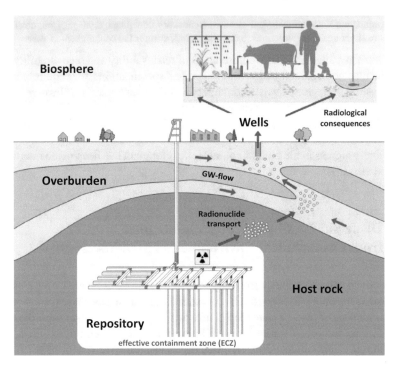

Fig. 3.10 Release path of radionuclides into the biosphere and radiological consequences, see also [20]

repositories, the nuclide migration in the geosphere is calculated by means of transport models. The development and further development of transport models and the corresponding programs, which comprehensively reflect both the hydrogeological, geochemical and biological parameters, is an indispensable requirement.

For critical migration routes, the concentration reduction along the migration path has hitherto been determined, as far as this can be determined. In essence, this concerns the plausible assumption of the relationship between the nuclide concentrations in the liquid and solid phases and the instantaneous equilibrium. In the first step, a linear relationship is assumed.

Of essential influence for each special task is the given source term and the boundary conditions under which the radionuclide transport should take place. An essential constraint is whether the geosphere can be modeled as a layered, porous geological body, or whether the transport of nuclides occurs in naturally existing fracture systems (crystalline host rock) that are much more difficult to detect. This also applies to the transport equation, in which the radioactive decay is incorporated with consideration of decay chains. Added to this is the influence of other important geological parameters such as porosity or groundwater flow velocity.

It becomes clear that there cannot be a general solution for transport models adapted to locations. The site conditions and the multi-barrier systems used are too different for this. The contamination transport models developed as site specific models therefore require a coordinated monitoring program that accompanies the development process in order to have enough calibration points (supporting points) permanently available with which the programme can be verified, see also Chap. 5.

Another observation is that through diffusion/dispersion processes radionuclides can also get into layers in which they are not expected to from the results of a one-dimensional view. Transversal diffusion/dispersion process means that the radionuclides move perpendicular to the flow direction of an aquifer, towards the next higher one. The transversal diffusion/dispersion process can thus represent an additional risk factor, in particular if the migration path is thereby shortened and/or the flow velocity of the "new" aquifer is significantly higher than the original one.

With low dispersivity, a dilution effect only becomes noticeable at a certain distance from the storage area; for large dispersivity values, this effect is already present at the storage location itself. For an example with a transversal rather than longitudinal dispersivity value that is three times smaller, the extension of the "nuclide cloud" perpendicular to the groundwater flow after a distance of 1 km from the repository was already several hundred meters, see Schmocker [39]. In heterogeneous layer structures, this transversal expansion of the "nuclide cloud" can be limited by practically water-impermeable layers or by media with very small transversal dispersion; the diffusion in such layers is only a few meters. When selecting a repository site, not only the host rock and the effective containment zone must be investigated, but also the adjacent geological formations are to be checked for their geological structure and their chemical properties.

Special protection against the discharge of radio-toxic contaminants, in particular into groundwater, is provided in all geotechnical environmental structures.

Since radioactivity can escape from the repository to the biosphere, the investigation of the transport of nuclides through the geosphere is always followed by an exposure analysis in the biosphere when investigating the long-term behavior of a repository system. For example, the final disposal must ensure that releases of radioactive substances from the repository will only marginally increase the risks resulting from natural radiation exposure in the long term. For less likely developments in the post-closure phase, it should be demonstrated that the additional effective dose caused by the possible release of radionuclides from the stored radioactive waste cannot exceed 0.1 mSv per year, for the affected people see Chap. 8.

3.8.1 Measures for Permanently Retention of Radionuclides from on Landfills Stored Waste with Very Low Radioactivity (VLLW)

The disposal of very low radioactivity waste (VLLW) varies by country. The U.S. Radioactive Waste Classification System is defined in detail in various Regulations of the Code of Federal Regulations [40]. There are two main categories of radioactive waste in the US. Low Level Waste (LLW) and High Level Waste (HLW). In addition, there are some special categories. There are 4 waste classes at LLW, with class A coming closest to the wastes with very low radioactivity, see Table 3.3.

The LLW class A is deposited near the surface, mostly in containers. One disposal location is at Clive, located in Utah's West Desert, approximately 120 miles west of Salt Lake City, see Chaps. 5 and 6.

The following statements are in turn based on the legislation in Germany. The reader should compare these conditions to those in their own country.

The disposal of waste with very low radioactivity (VLLW) takes place in Germany in landfills. KrWG, usually dry, but not always. However, during the operational phase of the landfill, surface waters, e.g. precipitation, penetrate unhindered into the landfill body. The spread of dust clouds from the landfill site is not uncommon during the operational phase prior to application of the surface sealing.

As a rule, landfills of DK II and III are planned for these wastes. At the decommissioning of the landfill the body is covered with a surface sealing primarily as a seal against groundwater; in the DK II and III sites base sealings or geological barriers will be installed.

Suitable base sealings are natural (geological), geotechnical base sealings. The base sealing also contains a seepage water collection system which is connected with a water treatment system. In the case of the DK III, the geological barrier must be at least five meters thick and additionally equipped with a leak detection system. When choosing a mineral base sealing, the mineral sealant material in a combination-base-seal system must be selected such that among others the following conditions are met, see also Chap. 4:

– Tightness (permeability) $k_f \leq 5 \times 10^{-10}$ m/s,
– Minimal susceptibility to cracking,

Table 3.3 US commercial radioactive waste classification [40]

Waste class	Description
Class A LLW	The physical form and characteristics must meet the minimum requirements in 10 CFR 61.56
Class B LLW	Waste that must meet more rigorous requirements on waste form than Class A waste to ensure stability. The physical form and characteristics of Class B waste must meet both the minimum and stability requirements in 10 CFR 61.56
Class C LLW	Waste that not only must meet more rigorous requirements on waste form than Class B waste to ensure stability but also requires additional measures at the disposal facility to protect against inadvertent intrusion. The physical form and characteristics of Class C waste must meet both the minimum and stability requirements in 10 CFR 61.56
Greater than Class C LLW	Waste that exceeds the concentration limits of radionuclides established for Class C waste in 10 CFR 61.55. GTCC waste is not generally acceptable for near surface disposal

- Resistance to hydraulic effects (suffusion and erosion),
- Sufficient deformation capacity,
- Resistance to chemical and biological influences as well as weather influences and adverse material changes caused by ageing.

In the presence of a geological barrier, these criteria are transferable.

Based on the geological and hydrogeological conditions of the site, a permanent distance of at least one metre is expected between the edge of the geological barrier or geotechnical expected barrier [such as Plastic geomembranes: clay and polyethylene: High Density Polyethylene (HDPE)[12]] and the highest free groundwater level The control and maintenance system also includes environmental monitoring, which also monitors the chemical-physical composition of the groundwater in the inflow and outflow to the landfill in relation to the permitted limit values.

3.8.2 Measures for Permanently Retaining Radionuclides from Uranium Mill Tailings Ponds (Uranium Mill Tailings Requiring Monitoring)

The limits for residues requiring monitoring are shown in Fig. 2.1. Uranium mill tailings ponds of SDAG Wismut contain a considerable radioactive inventory, see Chap. 5. The restructuring of the uranium mill tailings ponds are subject to special provisions from the German unification treaty. It is the aim of the remediation measures that are still ongoing today to reduce this mining-related radiation exposure as far as possible. The guideline is an additional effective dose of 1 millisievert per year (mSv/a) according to the "Regulation for ensuring on nuclear safety and radiation protection" (VOAS) and its implementing regulations, see Table 2.2 [8].

In the U.S.A., Uranium mill tailings are under the supervision of U.S. NRC. As defined in Title 10, Part 40 of the Code of Federal Regulations (10 CFR Part 40), the tailings or wastes produced by the extraction or concentration of uranium or thorium is a byproduct material. Most of the regulations that the U.S. Nuclear Regulatory Commission (NRC) has established are for this type of byproduct material see 10 CFR Part 40, Appendix A [31].

It is common in the world that tailings from uranium ore processing, are deposited in pits of abandoned opencast mines. It is usual today to provide a base seal before the uranium tailings are placed into the pond and to cover this after completion of the operating phase with a surface sealing. However, this was not always the case. For both the abandoned uranium mine sites in the USA and those of SDAG Wismut the pits have been prepared without base sealing. For the legacy of uranium ore processing, the base sealing cannot be retrofitted. Where there are suitable replacement sites, the uranium tailings can be extracted and transported to the replacement site, which is equipped with a base seal. In Germany, this method is not possible due to

---

[12]In Europe called PEHD.

the very high population density. Here the closure takes place exclusively in situ. In the USA, there are examples of relocating uranium tailings, see Chap. 5.

With regard to the uranium tailings ponds of SDAG Wismut, it can be stated, that here may be unimpeded draining of radio-toxic pore waters over the contact surface (base) of the completely water-saturated uranium tailings body (sediment body) in the post-closure phase because of the missing base sealing, t. In order to prevent uncontrolled spread over the groundwater path, significant expenditure has to be made over an extremely long period of time in order to keep contamination spread within the limits of the permit. Here, the contaminated waters, surface, seepage and pore water, are collected over an extremely long period of time and fed to a water treatment plant, cleaned to the required standard and discharged into the receiving waters.

In particular, the seepage and pore waters emitted are diffuse and it is difficult to prevent spread of a contaminant plume. It may be helpful in this case for a source-term to be formulated and mass transport modeling can be carried out, Metschies [22].

In the context of the tailings ponds of SDAG Wismut,, this requires the installation of a complex long-term monitoring programme and aftercare system and maintenance thereof over the entire monitoring period. It is also necessary to develop and update all site models to describe the mass transport and the spread of contamination. These processes must be integrated into the programmes of site monitoring and water collection and treatment, see Chap. 5.

3.8.3 Measures for Permanently Containment of Radionuclides from Inventory of Final Disposal of Radioactive Waste in Deep Geological Formations

In Germany radioactive wastes LAW/MAW with negligible heat generation and the heat-generating, highly radioactive wastes are stored in disposal containers in deep geological dry strata in the storage fields according to a storage scheme. Part of the repository concept is that the repository areas remain essentially dry for as long as possible. A multi-barrier system should ensure that no, or only minor amounts, of radioactive substances (radiotoxicity) can escape from the effective containment zone (ECZ) through interaction with aqueous solutions.

In the U.S.A, waste is divided into two main groups: LLW and HLW waste. The LLW classes are in turn divided [40], see Table 3.3. High-level radioactive wastes are the highly radioactive materials produced as a byproduct of the reactions that occur inside nuclear reactors. High-level wastes take one of two forms:

– Spent (used) reactor fuel when it is accepted for disposal,
– Waste materials remaining after spent fuel is reprocessed.

The delimitation from the LLW is defined as "more than ILW". It is also stipulated that near-surface disposal is planned for the LLW. Limits are formulated for the individual emitters, above which an assignment to the HLW is made. Waste category LLW has following limits:

– α-emitters with a maximum activity of 3700 Bq/g (100 nCi/g) with a half-life above 20 years.

Furthermore, LLW activity limits are prescribed for each single nuclide and the sum of the specified β-emitter:

– For the sum of short-lived nuclides Ni-63, Sr-90 und Cs-137 an upper limit of the specific activity of 8.9×10^7 Bq/g (2.4 mCi/g) applies.
– For the sum of long-lived nuclides C-14, Ni-59, Nb-94, Tc-99 und I-129 an upper limit of the specific activity of 9.0×10^5 Bq/g (24.3 μCi/g) applies.

These limits are derived from conversions, because the data in the Code of Federal Regulations is in Curie per cubic meter of waste and the standard in Europe is in Becquerel per gram. For this, a density of the waste of 2 g/cm^3 was assumed. The limit values can also be considered as the waste of class C.

This also makes it clear why the U.S. classification suggests a near-surface storage of LLW and makes it acceptable. In Germany, the LAW/MAW with negligible heat generating capacity are permanently stored in deep geological repositories, see Chap. 6.

The following section will address storage of heat generating HAW in Germany. These statements also apply in principle to LAW/MAW with negligible heat generating capacity.

In the final storage of heat-generating waste in rock salt, the contact of waste with an aqueous solution and thus a mobilization of the pollutants is possible only when assuming a solution access scenario. The safety concept stipulates that such a contact is to be prevented for as long as possible in this case. In the safety concepts for repositories in the other host rocks, on the other hand, it must be taken into account that the waste, even in undisturbed development, comes into contact with the natural moisture in the rock. By choosing the final disposal container material, the corrosion rate and thus the service life of the container can be influenced. The chemical composition of the existing solution must be taken into account. Depending on the weight placed on the safety concept on the barrier effect of the repository containers, long-term corrosion-resistant materials can also be selected. Outer and internal sheathing of the waste container with bentonite considerably hampers water access to the waste matrix and retards it for a long time due to low corrosion rates, see Fig. 3.9. Container lifetimes of a few hundred thousand years seem attainable.

Radiolysis can cause gas formation inside and outside the stored waste packages. Investigations so far show that internal gas formation for MAW containers can be significant because of their higher water content and the decomposition of existing organics and that radiolysis outside the containers can be caused by radiation from the container or by leaked radionuclides after a container failure. Thus, radiolysis can be of crucial importance for corrosion of the containers and the waste matrix.

The extent of gas formation depends on the amount of reacting substances present and, in particular, the availability of water. The gases can impair the effectiveness of barriers, affect the chemical environment and thus the mobility of radionuclides and act as a driving force for the spread of radionuclides and aqueous solutions.

Due to the large shielding by coverings with deep geological strata, very large wall thicknesses of caverns and bunkers or water depths in wet storage, etc., radioactive radiation from the effective containment zone cannot enter the biosphere.

The danger therefore arises from the release of radionuclides from the waste matrix and the waste container into the effective containment zone and thus defines with what probability radionuclides can reach the biosphere via pathways from the effective containment zone, see Fig. 3.10. Such a release of radioactive substances can only occur if the radionuclides are mobilized from the waste containers and transported into the biosphere in a suitable transport medium (solution- or gas phase), GRS [41]. By adding immobilizing agents into the fixative, the release protection of the packages can be increased.

A driving force for the formation of flow processes are the gas formation rates coming from, in particular, corrosion processes or to a lesser extent by radiolysis. Previous knowledge allows the statement that the hydrogen formation rates are dependent on the external environment of the respective chemical environment and corrosion of containers. By influencing the chemical environment in the ECZ, these gas formation rates can be kept low. In salt rock, the rock convergence [7], which compresses the mine, is an essential driver for the formation of flow processes. The development of waste packages, which are adapted to the host rock in the repository horizon and to the long-term safety of the repository, represents an important research goal, see also Frei [42].

Water as a transport medium also plays an outstanding role in answering the question of how it can be possible that radionuclides can enter the biosphere against gravity from the effective containment zone. In order to calculate the likelihood that the released radionuclides find their way through the pathways at all, the responsible radiological, chemical and hydraulic processes must be included.

Possibly the milieu of the waters does not allow any transport, because it practically causes a perfect immobilization. There are hardly any complex examinations or even results. However, it has been proven that only a fraction of the radionuclides in the inventory can be mobilized and from this another part is immobilized. The source-term formation cannot be determined exactly and is somewhere between complete immobilization and 100% mobilization.

Two factors that negatively influence the probability of occurrence:

- By choosing extremely dry effective containment zones, and
- Long-term stable waste matrices and final disposal containers will be developed with service lives $>10^5$ a, see Fig. 3.3. Of course, the dryness of the effective containment zone in turn also directly affects the service lives of the waste matrix and of the container.

The formation of pathways, in addition to those which exist in the geological formations, is influenced by the heat generated by the HAW and is therefore another effect. But this can be reduced by:

- Long cooling times in wet storage,
- Length of stay in transitional storages (based long-term interim storage) [43],
- Design of the storage areas and arrangement of the waste packages in it,
- Thermal protection (cooling) between wall of storage areas and containers for a time.

In the case of rock salt as the host rock, it should be noted that this behaves plastically under pressure, so that the occurrence of contiguous joints and crevices is largely prevented. If there are fissures, they "heal" quickly. To some degree, heat can favour this process.

Release of radionuclides into the biosphere with a significant hazard to individuals and groups of persons may also be caused by unintentional opening of the effective containment zone from outside, e.g. by sinking of a well or exploration drilling (human Intrusion). In order to reduce the likelihood of such an occurrence to such an extent that it becomes unlikely, repositories should be placed in non-prospective areas (formations) according to best available knowledge. Thus, in the opinion of the authors, any kind of known mineral deposits should be excluded from consideration as a repository site and thus also the likelihood that exploratory drillings, of whatever kind, will be sunk into the effective containment zone or into the overburden is minimised.

This lists the key security parameters (barrier criteria) of a repository concept:

- Dry effective containment zone,
- Long-term stable waste matrix and containers with service lives $>10^5$ a,
- Search criterion for a repository site in non-prospective areas (formations).

The fact that a radioactive waste and residue disposal concept must permanently prevent radioactive contaminants from entering the biosphere, or only within socially and legally acceptable limits, is due to the fact that stable nuclei remain after the radionuclides have decayed. The repository then still may have considerable quantities of toxic substances, such as lead, mercury, cadmium, arsenic, etc. Some of these substances have permanent toxicity, whereby the security claim is "permanently" justified.

3.8.4 Monitoring—Proof of the Function and Effectiveness of Geotechnical Environmental Structures for the Permanent Isolation of Radioactive Waste and Residues

In order to develop and implement a repository program for radioactive waste and residues, it is important that, in addition to engineering competence and a well-founded safety strategy, there is a growing understanding in society that social aspects such as acceptance, participation and trust in the regions and also special interest groups should be included equally. Monitoring can be used as an important tool for public communication that, on the one hand, shows the technical security strategy, the safety of the technology and its life time are all working as planned, and on the other hand promotes understanding of processes in operation and later in the decommissioning phase, and helps to build confidence in the planned repository development in the post-operational phase.

The monitoring of a repository concept reacts to the task to make sure that certain protection objectives with regard to the radiation exposure of humans and nature, binding safety regulations, see BMU [3] "Safety requirements for the final disposal of heat-generating radioactive waste", are respected. In the operation and decommissioning phase, this also includes the monitoring of personnel as radiation exposed persons, see also Sect. 3.7.2.

If deviations from these objectives are found, this may have different consequences depending on the stage of the procedure. The phase in which the repository is located is a decisive factor. If the repository is not yet operational, then decisions can be reversed that led to the potentially unsafe situation. This includes the task of the selected location or the termination of the selection process. In the phases of operation, decommissioning or closure of the repository, the reversibility of decisions becomes more and more complicated and thus more difficult. Technical and social monitoring should therefore be specifically integrated and used to prepare problem-solving decisions. The prerequisite, however, is that the monitoring provides the observing authorities with data that allows them to review safety. and evaluation procedures are integrated into governance processes, Brunnengräber [44], This process should ideally involve a small circle of delegates of government, formally competent officers, institutions, etc., as well as supervisory authorities and others who ensure or provide public control [45].

The focus here is on technical monitoring, see also IAEA [46]. Thus, the entire process of final disposal of radioactive waste and residues should be accompanied by a comprehensive monitoring program. The monitoring program should be part of the approval process. Such a monitoring program is divided into the three sub-areas:

- monitoring of the repository,
- monitoring of the environment,
- health surveillance of persons and groups of persons.

With regard to the service phases of a repository, the monitoring program is further specified:

- **Construction phase**—Monitoring compliance with the requirements of the planning approval decision, quality requirements for the structures, execution, radiation protection standards,
- **Operating phase**—the monitoring program should monitor all safety-relevant parameters of the repository during the whole operating period (heat generation, dissipation, stress conditions in repository, rates of convergence, sink holes above ground, machine speed, hydrogen concentration, efficiency of highly efficient filters, mine ventilation, etc. …), Included in this must be the monitoring of the external (and possibly the internal) radiation exposure of the personnel [47].

In addition to monitoring in the strict sense, parameters are monitored that can influence the development of the repository in the medium and long term: temperature of environmental air, extension of the facility (collapse of the tunnels), durability of the concrete, corrosion of the steel, etc. During the operating phase, the entire ECZ, including the barriers, should be extensively instrumented, programmed and provided with suitable power sources to support internal long-term monitoring.

If a **pilot project** is operated during the operating phase, this will be equipped with a comprehensive measuring program, the results of which will serve both to optimize the repository concept and the development of the repository in the medium and long term. The pilot project can be integrated into the repository.

If an **underground laboratory** is operated, it can be equipped with a similar measurement program as a pilot project. It can also fulfill the same functions as the pilot project, provided it is also operated during the operational phase. Excluded is the incorporation into the repository, which should not occur.

- **Decommissioning phase**—must be equipped with comprehensive decommissioning- (remediation-accompanying) monitoring so that all parameters listed in the planning approval decision are recorded and compliance with them monitored. The monitoring should ensure that the geotechnical environmental structure constructed, continues to separate the stored (deposited) radioactive waste materials from the material cycles of the biosphere for the required period of time with a probability bordering on certainty.
- **Post-closure phase**—in this phase, the function is to ensure, the compliance with the objective is monitored by both internal and external long-term monitoring. For external monitoring, an observatory will be set up to record, analyze and archive the relevant data over large networks of air, soil and water observations. This includes the monitoring of flora and fauna, agricultural production, etc. in the observation area and monitoring of physical-chemical and biological soil quality. The observatory should work according to the best current practice (best practice) and be regularly optimized during its working life. During the post-closure phase, health monitoring of the general population in the area, including a network for monitoring local dose rates and background radiation, will also need to be carried out. The monitoring of the general population should have started already in the operational phase [47].

On the issue of the duration of the long-term monitoring, there are very divergent views internationally. The monitoring concept described above is balanced and also practicable. With the current and indicated development opportunities, there should be hardly any limits with regard to instrumentation (including energy supply) and data management.

With regard to the duration of long-term monitoring, both in the area of the KrWG and DepV, concerning VLLW, as well as in the area of the BBergG, uranium mill tailings storage, there are clear regulations. The long-term monitoring is to be operated until the landfill is released from supervision of the environmental authority or the uranium residue storage released by the mining authority. This could be called a guard-principle.[13]

The monitoring concept described above has already been implemented, see Fig. 3.11. This is implemented in the safekeeping of the uranium mill tailings ponds of SDAG Wismut. This is structured as remediation monitoring, basic monitoring and long-term monitoring. Integrated into this is radiation protection, which is divided into operational radiation protection (occupationally radiation exposed persons) and radiation protection for the general population and is monitored by the authorities. The subranges listed above—Monitoring of the repository—Monitoring of the environment—Health surveillance of persons and groups of people—are fully illustrated, see also Chap. 8.

The following presents and discusses the current or possible solutions for the disposal of the categories of radioactive waste and residues shown in Chap. 2, Fig. 2.1. In particular, the necessary long-term stability and safety of the repository and the resulting requirements for geotechnical environmental structures will be addressed. For the assessment of the insulation potential at highly active, heat-generating wastes, the maximum allowable heat input that can be admitted and dissipated by a geotechnical environmental structure (repository) during intended operation without any

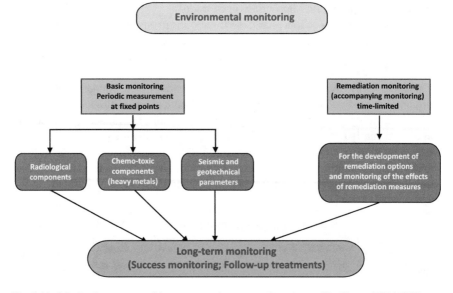

Fig. 3.11 Monitoring concept of the permanently storage of uranium mill tailings of SDAG Wismut. *Source* Lersow and Schmidt [48]

---

[13] A virtually unlimited permanent storage of the waste is known as "guards" or "guard-principle".

safety-related impact plays an important role to develop the most suitable geotechnical environmental structure. But the nature of geological formation is not the only deciding factor.

References

1. COUNCIL DIRECTIVE 2013/59/EURATOM of 5 December 2013: Laying down basic safety standards for protection against the dangers arising from exposure to ionising radiation, and repealing Directives 89/618/Euratom, 90/641/Euratom, 96/29/Euratom, 97/43/Euratom and 2003/122/Euratom.
2. Published by ICRP – International Commission on Radiation Protection Annals of the ICRP: ICRP Publication 103 – The 2007 Recommendations of the International Commission on Radiological Protection. Elsevier, Amsterdam 2007, ISBN 978-0-7020-3048-2.
3. The safety requirements governing the final disposal of heat-generating radioactive waste of 30.09.2010 by the Federal Ministry for the Environment, Nature Conservation and Nuclear Safety (BMU).
4. Law on Search and Selection of a Site for a Repository for Heat-Generating Radioactive Waste (Site Selection Act - StandAG); of 23 July 2013 and its development of 23 March 2017.
5. Abschlussbericht der Kommission Lagerung hochradioaktiver Abfallstoffe „Verantwortung für die Zukunft", Deutscher Bundestag (2016); English translation: Report of the German Commission on the Storage of High Level Radioactive Waste; July 2016; https://www.gruene-bundestag.de/fileadmin/media/gruenebundestag_de/themen_az/Gorleben_PUA/Report-German-Commission-Storage-High-Level-Radioactive-Waste.pdf.
6. Plan approval decision approval decision for the construction and operation of the Konrad mine in Salzgitter as a facility for the disposal of solid or solidified radioactive waste with negligible heat generation of 22 May 2002; Lower Saxony Ministry of the Environment (2002).
7. Decommissioning of the industrial tailing plants (IAA) of the SDAG Wismut - Federal Mining Act (BBergG), Regulation for ensuring on nuclear safety and radiation protection (VOAS), Order for ensuring of radiation protection in dumps and industrial tailings ponds and in use of therein deposited materials (HaldAO), Radiation Protection Ordinance (StrlSchV), Atomic Energy Act (AtG) etc.
8. SSK 92: Volume 23: Radiation protection principles for the storage, use or release of contaminated materials, buildings, areas or heaps from uranium mining; 1992/12/30, in German only.
9. The Safety Case and Safety Assessment for the Disposal of Radioactive Waste for protecting people and the environment; No. SSG-23 Specific Safety Guide; International Atomic Energy Agency (IAEA), Vienna, 2012.
10. IAEA Safety Standards – Geological Disposal Facilities for Radioactive Waste Specific Safety Guide No. SSG-14; published 2011.
11. Horizon 2020: European Joint Programs in Radioactive Waste Management - EC Policy change, and Joint Programming: the issue and aim; JOPRAD programs document workshop, London, 04 April 2017.
12. U.S. NRC: PART 20—STANDARDS FOR PROTECTION AGAINST RADIATION; https://www.nrc.gov/.
13. Antonia Wenisch, Wolfgang Neumann, Gabriele Mraz, Oda Becker: Disposal Strategy of Slovak Republic; Expert Opinion on Strategic Environmental Assessment; Federal Environment Agency of the Republic of Austria, Vienna 2008, in German.
14. U.S. Department of Energy (DOE); Savannah River Site, http://www.srs.gov/general/about/history1.htm.
15. Repository for Radioactive Waste (ERAM) - operational radioactive waste, Federal Office for Radiation Protection (BfS), 2009.

16. David von Hippel and Peter Hayes: Deep Borehole Disposal of Nuclear Spent Fuel and High Level Waste as a Focus of Regional East Asia Nuclear Fuel Cycle Cooperation; Nautilus Institute; 12/8/2010; http://nautilus.org/wp-content/uploads/2012/01/Deep-Borehole-Disposal-von-Hippel—Hayes-Final-Dec11-2010.pdf.
17. US Environmental Protection Agency EPA US Environmental Protection Waste Isolation Pilot Plant (WIPP), Withdrawal Act; Amended September 23, 1996.
18. Conceptual and Safety Issues of Final Disposal of Radioactive Waste, Host Rocks in Comparison; Federal Office for Radiation Protection, Salzgitter 04.11.2005.
19. M. Lersow: Safe and long-term stable storage of tailings ponds, in particular from uranium ore processing; geotechnik 33 (4/2010), p. 351–369 - September 2010.
20. Responsibility for the future a fair and transparent procedure for the selection of a national repository site, Commission for the Storage of Highly Radioactive Waste in accordance with § 3 of the Site Selection Act, StandAG, K-Drs. 268, May 2016, German only.
21. B. Baltes et. al.: Safety requirements for the final disposal of high-level radioactive waste in deep geological formations - draft of GRS; GRS - A – 3358; January 2007.
22. T. Metschies: Conceptual Site Modelling, Hydraulic and Water Balance Modelling as basis for the development of a remediation concept; Wismut GmbH; Chemnitz, 04.12.2012.
23. Conceptual Site Model for disposal of depleted Uranium at the clive facility; Neptune And Company, Inc.; NAC-0018_R1; Los Alamos, 05. June 2014.
24. IAEA-TECDOC-1403; The long term stabilization of uranium mill tailings; Final report of a co-ordinated research project 2000–2004; VIENNA, 2004.
25. U.S. NUCLEAR REGULATORY COMMISSION REGULATORY GUIDE OFFICE OF NUCLEAR REGULATORY RESEARCH REGULATORY GUIDE 3.64 (Task WM 503-4: CALCULATION OF RADON FLUX ATTENUATION BY EARTHEN URANIUM MILL TAILINGS COVERS; https://www.nrc.gov/waste/mill-tailings.html.
26. TECHNICAL REPORTS SERIES No. 419; EXTENT OF ENVIRONMENTAL CONTAMINATION BY NATURALLY OCCURRING RADIOACTIVE MATERIAL (NORM) AND TECHNOLOGICAL OPTIONS FOR MITIGATION; IAEA, VIENNA, 2003.
27. US ENVIRONMENTAL PROTECTION AGENCY, Code of Federal Regulations Title 40: Protection of the Environment (1996ff.).
28. SSK 98: Recommendation of the German Commission on Radiological Protection: "Release of materials, buildings and soil surfaces with low radioactivity from handling subject to notification or approval"; Adopted at the 151st meeting of the Commission on Radiological Protection on 12 February 1998; Reports of the Commission on Radiological Protection, Volume 11, in German only.
29. Final Report of the KFK: Responsibility and Safety - A new disposal agreement - Final Report of the German Commission to Review the Financing of the Phase-out of Nuclear Energy (KFK); Berlin, 2016/04/27, in German only.
30. COUNCIL DIRECTIVE 96/29/EURATOM of 13 May 1996: Laying down basic safety standards for the protection of the health of workers and the general public against the dangers arising from ionizing radiation.
31. U.S. Nuclear Regulatory Commission (NRC); 10 CFR Part 40, Appendix A: https://www.nrc.gov/waste/mill-tailings.html.
32. SECY-99-100 attachments: of Civilian Radioactive Waste Management geologic repository program - performance of a geologic repository for high-level radioactive waste (HLW); https://www.nrc.gov/reading-rm/doc-collections/commission/secys/1999/secy1999-100/attachments.pdf.
33. Radioactivity, X-rays and health; Bavarian State Ministry for the Environment, Public Health and Consumer Protection; Munich; October 2006, in German only.
34. Forecast of the generation of conditioned waste with negligible heat generation up to the year 2060 (cumulated) (as of September 2014); http://www.endlager-konrad.de/Konrad/DE/themen/abfaelle/entstehung/entstehung_node.html.
35. Frank Peiffer (GRS), et. al.: Waste specification and quantity structure as basis for lifetime extensions of existing nuclear power plants (September 2010), report on work package 3,

preliminary safety analysis for the Gorleben site, GRS - 274 ISBN 978-3-939355-50-2; Juli 2011.

36. Gesetz zur Neuordnung der Verantwortung in der kerntechnischen Entsorgung"; vom 27. Januar 2017 (BGBl. I S. 114, 1222).

37. Long-term safety for the final repository for spent nuclear fuel at Forsmark_ Main report of the SR-Site project, Volume I, Svensk Kärnbränslehantering AB, March 2011.

38. Bernhard, G.; Geipel, G.; Brendler, V.; Nitsche, H.: Speciation of uranium in seepage waters from a mine tailing pile studied by time-resolved laser-induced fluorescence spectroscopy (TRLFS); Radiochimica Acta 74(1996), 87–91.

39. U. Schmocker: The influence of transversal diffusion/dispersion on the migration of radionuclides in porous media - Investigation of analytically solvable problems for geological layer Structures; NAGRA Technical Report 8O-06, July 1980.

40. U.S. Government: Code of Federal Regulations No. 10 CFR 61.55: Waste classification, U.S. Government, https://www.nrc.gov/reading-rm/doc-collections/cfr/part061/part061-0055.html.

41. Buhmann, D., Mönig, J., Wolf, J.: Investigation for the determination and evaluation of release scenarios, GRS-233, Gesellschaft für Anlagen- und Reaktorsicherheit (GRS) mbH: Cologne 2008; pp. 115.

42. Freiersleben, H.: "Final disposal of radioactive waste in Germany", TU Dresden, Dec. 2011, available on the Internet at: https://iktp.tu-Dresden.de/IKTP/Seminare/IS2011/Endlagerung-IKTP-Sem-01-12-2011.pdf.

43. Safety requirements for the long-term interim storage of low- and intermediate-level radioactive waste; Recommendation of the RSK; Version of 2002/12/05, in German only.

44. Brunnengräber, Achim [Hrsg.]: Problem-trap repository: Social challenges in dealing with nuclear waste; Nomos; Berlin 2016.

45. REPORT FROM THE COMMISSION TO THE COUNCIL AND THE EUROPEAN PARLIAMENT - on progress of implementation of Council Directive 2011/70/EURATOM and an inventory of radioactive waste and spent fuel present in the Community's territory and the future prospects; Brussels, 15.5.2017.

46. IAEA-TECDOC-1208; Monitoring of geological repositories for high level radioactive waste; IAEA, VIENNA, 1991; ISSN 1011–4289.

47. IAEA-TECDOC-630; Guidelines for the operation and closure of deep geological repositories for the disposal of high level and alpha bearing wastes; IAEA, VIENNA, 1991; ISSN 1011-4289.

48. M. Lersow (2006), P. Schmidt; The Wismut Remediation Project, Proceedings of First International Seminar on Mine Closure, Sept. 2006, Perth, Australia, p. 181–190.

Chapter 4
Disposal of Waste with Very Low Radioactivity (VLLW)

The International Atomic Energy Agency (IAEA) has recognized Very Low Level Waste (VLLW) as a category that provides both practical and economic benefits, see [1]. The criteria for national and international low active and very low radioactive (LLW) disposal classifications are described in the EPRI Final Report [2]. This report provides the technical basis for a waste category of very low level waste that applies to any existing low level waste classification system. The generic VLLW category is defined in terms of individual radionuclide specific activity limits, see Chap. 2, Fig. 2.1. From the individual radionuclide-specific activity limits, an effective dose limit of no greater than 50 μSv/a (5 mrem/a) to the maximally exposed individual (MEI) is given in the final EPRI report. The U.S. Resource Conservation and Recovery Act (RCRA) Subtitle C; [3] has been developed in the USA for the disposal of this waste in accordance with the EPRI report.

In Germany, the classification according to both activity limitation (radioactive waste and residues not requiring monitoring) and the effective dose limitation for personnel, as well as the classification according to the origin of this waste and residues with the associated disposal routes, are clearly regulated by law.

In the USA, the VLLW is declared to be a hazardous waste, but not in such a way that a clear classification is possible either via the waste code or activity limit values.

The U.S. Environmental Protection Agency (EPA) has now developed a comprehensive program to ensure that these "hazardous wastes" can be safely managed from the time they are generated until their final disposal. The problem in the USA, however, is that radioactive waste is not classified on the basis of its radiological properties, but on its origin. This has led to some wastes being perceived as more hazardous than others—an assessment that has sometimes led to waste being treated inconsistently.

Subtitle C [3] empowers the EPA to delegate this task to the federal states instead of the federal government; namely to implement important provisions on the requirements for hazardous waste management. There are therefore no uniform regulations for the disposal of waste with very low radioactivity (VLLW) in the USA, such as the Landfill Ordinance (DepV) and the Closed Substance Cycle and Waste Management Act (KrWG) found in Germany.

© Springer Nature Switzerland AG 2020
M. Lersow and P. Waggitt, *Disposal of All Forms of Radioactive Waste and Residues*, https://doi.org/10.1007/978-3-030-32910-5_4

The following considerations are therefore based on the existing regulations and laws in Germany, from which geotechnical environmental structures for the long-term safe and stable storage of waste with very low radioactivity (VLLW) can be derived.

4.1 General Description of the Type of Waste

As described in Chaps. 2 and 3, the origin of radioactive waste and residues has similar sources worldwide, but the limits and storage conditions are not uniform. The limit values which in Germany for radioactive waste and residues with very low radioactivity (VLLW) can be found in Fig. 2.1. For residues, the specific activity must be ≤ 0.2 Bq/g (for underground storage ≤ 0.5 Bq/g) and for wastes with radioactivity (much less-than) $\ll 1.0E+11$ Bq/m^3 (when the De Minimis dose of less than $10 \, \mu Sv/a$),[1] see also [5]. Then they can be deposited in Germany in accordance with the KrWG in accordance with DepV.

In the USA as well, for example, large quantities of waste will be generated during the decommissioning of nuclear power plants. As already described, in the USA it has not yet been possible to establish a uniform regulation for handling waste with very low radioactivity (VLLW).

The current situation is that the limit values and storage conditions of LLW class A as prescribed in [6] are used as a basis. A summary of classes of Low Level Waste (LLW) can be found in Table 3.3 according to [6]. The system captures the slightly radioactive solids in category LLW in the lowest grade A and describes them as follows: Slightly radioactive solid materials—debris, rubble, and contaminated soils from nuclear facility decommissioning and site cleanup. They arise in very large volumes but produce very low or practically undetectable levels of radiation. They are classified at the very bottom of U.S. NRC Class A (the lowest of the classes). U.S. NRC has commissioned a study to improve the management of large volumes of VLLW. In this way the U.S. NRC is reacting to the large quantities of this waste, which result from the increasing decommissioning activities for NPPs. The study has the following mandate: "The U.S. Nuclear Regulatory Commission (NRC) is conducting a very low-level radioactive waste (VLLW) scoping study [7] to identify possible options to improve and strengthen the NRC's regulatory framework for the disposal of the anticipated large volumes of VLLW associated with the decommissioning of nuclear power plants and material sites, as well as waste that might be generated by alternative waste streams that may be created by operating reprocessing facilities or a radiological event." Similar events are imminent in Germany, see Fig. 4.2. Safety regulations for the permanent storage of VLLW have also been presented by the IAEA [8].

The largest LLW storage site for the lowest class A is located in Clive, Utah, USA. It is a repository only for this type of waste, licensed by the state of Utah;

---

[1]Corresponds to one population risk per year of 5.0E−07, see [4].

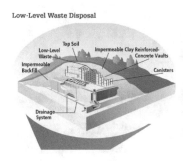

Fig. 4.1 Left: schematic of low level waste disposal design, right: disposal site. *Source* U.S. NRC

however, Clive accepts waste from all regions of the United States. The U.S. NRC has specified the following design for near-surface storage for permanent disposal of LLW class A as well, see Fig. 4.1.

It becomes clear that the design of the geotechnical environmental structure for the storage of LLW is linked to requirements that clearly overestimate the real hazards associated with the VLLW, see Fig. 4.1. The results from the Very Low-Level Radioactive Waste Scoping Study [7] should help to develop appropriate geotechnical environmental structures that will meet the requirements of safe storage of VLLW [1].

The following remarks refer in turn to the permanent storage of radioactive waste and residues in Germany. The permanent storage of VLLW envisaged in the individual countries can be mirrored in this system. What stands out, however, is that the American solution has an obvious advantage, namely that dust formation is almost completely avoided. In the case of large high-waste landfills (several tens of meters high) in open landscapes, wind erosion from the uncovered landfill body poses a health risk that should not be underestimated.

4.2 Deregulated Residual Materials, Residues Released from Supervision and Residues not Requiring Monitoring

4.2.1 Origin

As described in Chap. 2, the classification of a substance as radioactive is linked to the legal definition of radioactivity. There are two basic concepts:

– Without release—institutional monitoring of waste is necessary
– With release—The concept of release was developed by the IAEA [9], and internationally established. This may also be considered as deregulated waste. A summary under German law is given in FS [10].

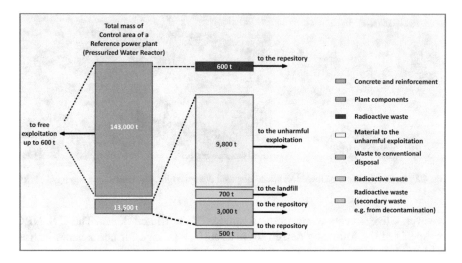

Fig. 4.2 Quantity balance of a reference-NPP (pressurised water reactor) of the controlled area (estimate). *Source* according to VGB PowerTech 2011

Radioactive residues from nuclear energy, medicine, research or industry are usually first examined to see whether they still need to be monitored as radioactive substances or whether they have only very low levels of radioactivity and which half-lives and what types of radiation they have and, should it be required, if they can be released from institutional monitoring. Such released residues are, within the meaning of German nuclear legislation, materials with a radioactivity in the range of natural background radioactivity (AtG). In most cases, these released materials are building rubble, scrap metal and other solid and liquid substances from the decommissioning of nuclear power plants and research facilities, industrial plants or medicine, see also Fig. 4.2.

4.2.2   Legal Foundations

Section 2 (2) of the Atomic Energy Act stipulates that the activity or specific activity of a substance within the meaning of subsection 1 sentence 1 may be disregarded if the material is in accordance with an ordinance issued on the basis of this Act, that is to say:

1. the substance falls below the defined exemption limits,
2. if the substance is produced within the scope of an activity subject to authorisation under this Act or under a statutory order issued on the basis of this Act, and if it falls below the specified release values and the substance has been released,

3. as far as it is a substance of natural origin, which is not used for its radioactivity, as nuclear fuel or for the production of nuclear fuel, it is not subject to monitoring under this Act or an ordinance issued under this Act.

In Germany the release of radioactive substances (§ 29 StrlSchV) or the dismissal of radioactive substances (§ 98 StrlSchV) can be respectively unrestricted or restricted for a certain disposal route (for example, the reuse of buildings or disposal at landfills).

4.2.3 Radiological Basics

The conditions for the release are based on recommendations of the International Atomic Energy Agency IAEA [9], and EU directives and are set nationally in the Atomic Energy Act and the Radiation Protection Ordinance, FS [10]. The radiological criterion of insignificance[2] per release option is set out in § 29 (1) StrlSchV with the range of 10 μSv of effective dose per year for individuals of the population (including employees at the landfill or disposal facility); in accordance with the provisions of Council Directive 2013/59/EURATOM [11]. For residues, an effective dose of 1 mSv[3] per year for individuals of the population is set as the radiological criterion according to § 98 StrlSchV.

Following Germany's exit from the peaceful use of nuclear energy for energy generation, this regulation is of considerable importance for the decommissioning of nuclear power plants. Every year, large quantities of waste not requiring monitoring will be produced, see Fig. 4.2, which must be disposed of in surface landfills. It is already foreseeable that not enough landfill volume will be available for this purpose. It must therefore be determined where, and to what extent, new landfill space is needed for this waste, also in order to help avoid transport of waste in large quantities over long distances. These steps have not yet been taken in Germany.

4.3 Intentional Release of Radioactive Waste and Disposal of Residues from Industrial and Mining Processes

Radioactive waste and residues requiring monitoring are subject to different regulations.

(a) **Release of residues from prescribed radiation-protection monitoring**

A prerequisite for the release of residues from legally prescribed radiation-protection monitoring is that an effective dose of 1 millisievert per year (mSv/a) for individuals of the population is not exceeded during the intended recovery or disposal.

---

[2]De Minimis criterion.

[3]Corresponds to one population risk per year of 5.0E−05, see [4].

Due to the different origin of the residues, the following legally different waste streams may arise for disposal in conventional landfills:

- Residues from activities for which a safe disposal or disposal route exists and which can be disposed of or recycled without further checks when complying with the monitoring limits. If the specific activity for each radionuclide of one of the nuclide chains is below 0.2 Bq/g, the respective nuclide chain is disregarded. The general monitoring limit for recovery and disposal (on surface landfill) is 1 Bq/g. The monitoring limits are dependent on the disposed—or recycled route and are thus between 0.2 and 5 Bq/g.[4]
- Residues requiring monitoring within the meaning of § 97 StrlSchV which is pursuant to § 98 StrlSchV, can be released from monitoring on application in individual cases and then deposited or utilised.
- Residues requiring monitoring which are not released according to § 98 StrlSchV, but for which a disposal has been ordered (according to § 99 StrlSchV).

The StrlSchV includes more detail in Part 3 Chap. 3 Regulations for the protection of the population from naturally occurring radioactive substances (§§ 97-102). The provisions of this chapter of the Radiation Protection Ordinance apply to residues listed in Annex 12 Part A StrlSchV (list of residues to be taken into consideration) if for at least one radionuclide of the nuclide chains U-238 or Th-232 the specific activity exceeds 0.2 Bq/g. The residues require monitoring if the monitoring limits according to Annex 12 StrlSchV Part B are exceeded. In addition, according to § 102 StrlSchV, the authority may also order measures for other materials with natural radioactivity, if this is necessary for radiological reasons.

Residues not requiring monitoring can be recycled in certain ways as specified in the Radiation Protection Ordinance or disposed of in landfills that are subject to the Landfill Ordinance (DepV) and the Closed Substance Cycle and Waste Management Act (KrWG). Such residues may be e.g. Red mud, copper slag, also phosphogypsum, sludges from processing, dusts and slags from the processing of raw phosphate (phosphorite) as well as scales, sludges and deposits from oil and natural gas extraction.

The recycling or disposal of residues requiring supervision is only permitted after release from compulsory monitoring. This can be done by the producer of the residues at the request of a competent radiation protection authority. A prerequisite for release is that the intended use or disposal will not result in an effective dose in excess of 1 mSv/a for individual members of the population. The proof of compliance must be supported with a radiological report. The principles for the determination of the radiation exposure for such a proof are contained in Annex 12 part D StrlSchV. Under certain conditions, a simplified proof according to Annex 12 Part C StrlSchV is possible.

The amount and specific activity of residues that may be disposed of per year and into landfill are limited in accordance with Annex 12, Part C StrlSchV. If more than

---

[4]Limit value in underground disposal.

2000 tonnes are recycled or landfilled every year, then according to § 100 StrlSchV a balance obligation exists.

The disposal takes place at a landfill permitted by waste law. In most cases, these are landfills classes II or III, see § 6 DepV.

(b) **Waste with very low radioactivity (VLLW)**

§ 29 StrlSchV regulates in detail the release procedure and the standards for the assessment of which substances are to be classified as harmless under which radiological boundary conditions are to be classified as harmless if various further uses are taken into account. In DIN 25457, DIN ISO 11929 are the definition of the boundary conditions for both the measuring technique and the measurement. In addition, there are recommendations (SSK), guidelines (BMU, ESK, supervisory authorities of individual federal states) to specifying the application. There are two main categories of residual materials and waste with very low radioactivity (VLLW):

- **Waste with a purposive-directed release**: **This is waste that can be stored mainly in landfills**.
- **Recyclables and conventional residues**: For the most part, these materials can be returned to the conventional recycling cycle. The majority of the dismantling residues from the decommissioned NPPs will be part of it.

In the case of controlled release of radioactive waste for disposal on landfills, which are subject to the Landfill Ordinance (DepV) and the Closed Substance Cycle Waste Management Act (KrWG)—depending on the size of the landfill—per year and per landfill, only activity quantities corresponding to a maximum quantity permitted under Annex 3 StrlSchV for the disposal of 1000 tonnes of released waste are stored. Landfilling can take place on landfills of classes DK I–DK IV permitted under waste legislation for construction debris, also solutions also include construction waste dumps (DK I landfills). Often, however, landfills of higher landfill classes (DK II or DK III) are preferred, see § 6 DepV in connection with AVV [12] and disposal contracts concluded on this basis.

4.3.1 Examples of Residues not Needing Monitoring in Landfilling Acc. DepV and AVV

In Tables 4.1 and 4.2 [13], the results are shown for radiological investigations of red mud and power plant ashes as a basis for guidance according to the guideline. Both tables provide examples of residues not requiring monitoring that can be disposed of in DK II and III landfills. The different specific activities in red mud result from the large variations in activity within the various bauxite deposits. In AVV 2001, the waste codes are for

Table 4.1 By [13]; Table 3-16: Specific activities of red mud (measurements: IAF GmbH/Dresden); Nuclides of the uranium and thorium series excepted K-40, see Chap. 2

Residue	Sample/origin	Specific activity [Bq/g]					
		U-238	Ra-226	Pb-210	Ra-228	Th-228	K-40
Red mud	Site 1	0.210	0.220	0.150	0.430	0.450	0.072
Red mud	Site 1	0.210	0.230	0.150	0.510	0.500	0.023
Red mud	Site 1	0.210	0.250	0.150	0.450	0.450	0.051
Red mud	Site 2	0.300	0.315	0.250	0.285	0.286	<0.020
Red mud	Site 2	0.270	0.330	0.265	0.290	0.285	<0.030
Red mud	Site 2	0.320	0.370	0.320	0.275	0.285	<0.040
$T_{1/2}$ in a		4.468×10^9	1,602	22.3	5.7	1.9131	1.277×10^9

Table 4.2 From [13]; Table 4.10: Investigation results of tailings piles with combustion residues from power plants in Lower Saxony pursuant to [14]

Origin/location	Residue of combustion	Type and origin of coal	Ra-226 in Bq/kg	Pb-210 in Bq/kg
HKW Herrenhausen	Fly ash		298	15
KW Emden	Granulate		363	44
KW Emden	Fly ash		127	1.980
KW Wilhelmshaven	Fly ash	Hard coal (USA)	232	26
KW Wilhelmshaven	Fly ash	Hard coal (Australia)	264	44

– Red mud—group 01 03 with waste code 01 03 09 (red mud from production of alumina—with the exception of red mud, which registers under 01 03 07[5]— residues from ore processing (tailings); with reference to no. Pkt. App. 5 DepV and

– contain various waste from power plants and waste co-incineration plants (except category 19[6]) of group 10 01.[7]

KrWG Part 6 sets out the obligations, acceptance conditions and monitoring criteria for both the operators of the landfills and the authorities.

---

[5]AVV § 3 Danger of waste (1) The types of waste in the waste list whose waste codes are marked with an asterisk (*) are classified as dangerous in the sense of § 48 of the Closed Substance Cycle Waste Management Act (KrWG).

[6]Category 19: Waste from waste treatment plants, public wastewater treatment plants and the treatment of water for human and industrial use.

[7]Kinds of waste marked with an asterisk (*) in the list of waste are hazardous waste within the meaning of § 48 of the Recycling Management Act, and are not listed therein.

From Table 4.1 it can be deduced that with a duration of the operating phase of a landfill in the specified normal operation of 30 years and an annual acceptance of 500 t red mud at the two sites in the landfill body, an activity from only this waste category of 25 gigabequerel (not uniform) would accumulate and the landfill would have to be shut down. The $T_{1/2}$ of U-238 is 4468 billion years.

4.4 Landfilling of Waste with Very Low Radioactivity

As stated in Sect. 4.1, residues requiring special monitoring, residues or other monitored materials that are not requiring monitoring (which have been released from the supervision) and released residues from authorized handling of radioactive substances, are not radioactive substances in the legal sense. However, they are still physically radioactive to a certain extent. To distinguish them from radioactive wastes, which are radioactive substances in the legal sense, we use the term "waste and products with radioactivity" for such substances, to clarify that they are not non-radioactive waste in the legal sense.

The objective for safe, long-term stable storage of waste with radioactivity is defined so that they are deposited and encapsulated so that transfer of radio-toxic contaminants to the air, water and/or soil pathways into the biosphere is largely avoided or kept within the respective legally formulated limits.

The requirements for the implementation of this objective are based exclusively on concepts of conventional landfill construction. It is implicitly assumed that these radioactive substances, based on the total amount of all waste, play only a minor role in the landfill concept. Furthermore, a selective deposit can be prescribed by the authorities in the permit notification, see Barriers 2 and 3.

The further explanations will show that, although the materials to be deposited and the legal bases differ substantially, there are great similarities with regard to the construction of the geotechnical environmental structures of

- Landfills acc. DepV and
- Residue storage, in particular from uranium ore processing (uranium tailings ponds) acc. Federal Mining Act (BBergG), see Chap. 5.

Proof-and closure-(security) concepts, based on applicable laws, are contrasted and the differences and requirements clearly highlighted in following sections. On the basis of the radiotoxicities of the respective inventories, the reader should obtain a well-founded overview of the degree of unequal treatment in the safety requirements of the respective geotechnical environmental structures during specified normal operation. For this reason, the multi-barrier concept gem. DepV is explained extensively here.

Multibarriers concept
A landfill is used for the long-term storage of waste, see § 6 DepV in connection with AVV [12]. Since the Landfill Ordinance came into force, a generally valid

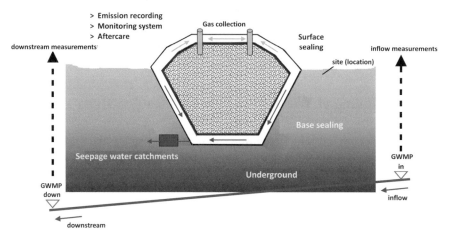

Fig. 4.3 Multibarriers concept of landfill classes II and III according to the German landfill ordinance (DepV)

definition of the various service phases of a landfill has been developed (§ 2 DepV). Accordingly, a distinction must be made between three phases:

• the disposal, • the decommissioning, and, • the post closure phase.

To limit the impact on the environment, landfills are designed in accordance with the multibarrier concept. This states that several safety concepts (in this case barriers) must exist independently of each other in order to prevent damage to the environment even if one barrier fails, see Fig. 4.3.

1. **Barrier**: **Site selection** (The site must be suitable with appropriate geology and hydrogeology, such as the presence of impermeable layers).
2. **Barrier**: **Waste pretreatment** (separation of particularly hazardous fractions and their chemical-physical pretreatment, for example combustion, in special plants).
3. **Barrier**: **Condition of the landfill body**. In the landfill body, chemical, biological and physical processes take place. Therefore, the landfill body must be constructed so that it is mechanically stable and gas emissions do not leak to the outside. The landfill gas may be collected and processed, or an active degassing system will be provided. The ingress of water should be prevented, so that a minimal amount of leachate is formed.
4. **Barrier**: **Landfill base sealing and leachate treatment** (to prevent contaminated leachate from entering groundwater or other waters).
5. **Barrier**: **Surface sealing** (to minimize the ingress of rainwater).
6. **Barrier**: **Maintenance and Repair** (The landfill must be monitored, even if it is completely filled in. All systems must be constructed in such a way that they can be repaired, e.g. the pipes for seepage water sampling. Facilities should exist

to enable sampling of water for analysis at various levels in order to assess the long-term behavior of the landfill.

Although the DepV stipulates a classification of the various landfills in the context of the GCU 2001, the planning approval decision is approved on the basis of an environmental impact assessment of the site with a site-specific solution, see barrier 1.

Description of individual barriers
Barrier 1—Site selection: The location should be determined according to set objectives and generally applicable criteria. After the site has been identified or a landfill site has been selected, a geotechnical environmental structure will be designed and constructed according to a site concept with site-specific refurbishment targets so that safe, long-term stable storage of residues and wastes contaminated with radioactivity can be achieved and guaranteed for the long-term. Waste arising from authorized handling that has been released from any monitoring requirement and, in particular discharged residues that are classified as hazardous waste, must often be disposed of in DK III type landfills. These will require a legally binding planning approval decision for the above-ground landfill on the basis of the DepV, the KrWG, the Federal Pollution Control Act (BImSchG) etc. to build, operate, monitor and document, shut down and carry out post closure operations.

Barrier 4—Landfill base sealing and leachate treatment: The main problem to be considered with regard to environmental assessments and long-term consequences for surface landfills of landfill class DK III is the sealing against groundwater. The DepV prescribes how the protection of groundwater should be structured, see Table 4.3 and Fig. 4.4.

For landfill sites DK III, locations should be found where a suitable geological barrier exists as a natural barrier against contamination transfer to groundwater, see Barrier 1. The legislation also permits the installation of a technical base seal, see Table 4.3 and Fig. 4.4. DepV Annex 1 lists the requirements for the site, the geological barrier, base and surface sealing systems for class 0, I, II and III landfills. When choosing the location, the following in particular must be considered:

1. Geological and hydrogeological conditions of the area including a permanently guaranteed distance of at least 1 m from the upper edge of the geological (geotechnical) barrier to the highest expected free groundwater level,

Table 4.3 Regular assembly of the geological barrier and the base sealing system DK III according to DepV (1), (2), (3) see DepV Annex 1

No	System-component	DK III
1	Geological barrier (1) (2)	$k_f \leq 1 \times 10^{-9}$ m/s; $d \geq 5.0$ m
2	Mineral sealing layer—at least 2-ply (2)	$d \geq 0.50$ m; $k_f \leq 5 \times 10^{-10}$ m/s
3	Synthetic geomembrane $d \geq 2.5$ mm	Required
4	Protection layer	Required
5	Mineral drainage layer (3)	$d \geq 0.5$ m; $k_f \leq 1 \times 10^{-3}$ m/s

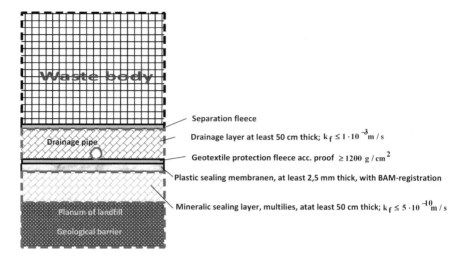

Fig. 4.4 Base sealing, landfill class DK III—composite liner systems acc. DepV; according to Table 4.3; BAM-Federal Institute for Materials Research and Testing (BAM)

2. Due to its low permeability, its thickness and homogeneity as well as its pollutant retention capacity, the subsoil of the landfill and the wider environment should be able to significantly hinder the spread of pollutants from the landfill (effect as a geological barrier), so that contamination of the groundwater or other adverse change in its nature is not permitted to happen.

From a radiological point of view, the legislation did not stipulate special requirements for above-ground landfills, in particular the DK III. The landfill should be formed as a tub, see also the concept of a settlement depression. In high landfills, this is the subsidence cavity caused by the sinking of the landfill body at the base. Proof that the required distance to the upper aquifer is met must be provided in the plan and application submitted for approval. Furthermore, reference is made here to compulsory monitoring of groundwater by the landfill operator and possible proof of exposure. The operator must manitain a permanent monitoring system for both leachate and groundwater and make the system and the data accessible to the authorities, see Fig. 4.3.

5. Barrier—surface sealing: In order to prevent washing out of pollutants by rainwater (impregnation of the landfill body), the waste top surface is covered and provided with a surface seal after each landfill section has been filled up, see Table 4.4 and Fig. 4.5, this process should be continued until the whole landfill body has been covered.

One speaks of a "dry" storage of the landfill body. The structure is shown in Fig. 4.4.

In order to achieve and ensure dry storage of the landfill body, site-appropriate seepage and surface water collection and treatment facilities are provided or prescribed. If composite systems of capillary barrier, geomembranes and geotextiles

Table 4.4 Regular assembly of the surface sealing system DK III according to DepV (1), (2), (3), (4) see DepV Annex 1

No.	System-component	DK III
1	Levelling layer (1)	d ≥ 0.5 m
2	Gas drainage layer (1)	where appropriate required
3	Mineral sealing (2) (3)	d ≥ 0.50 m; $k_f \leq 5 \times 10^{-10}$ m/s
4	Synthetic geomembrane	d ≥ 2.5 mm
5	Protection layer	required
6	Drainage layer (4)	d ≥ 0.3 m; $k_f \leq 1 \times 10^{-3}$ m/s
7	Recultivation layer, d ≥ 1 m	Required
8	Vegetation	Required

Structure of a surface sealing system landfill class DK III acc. DepV

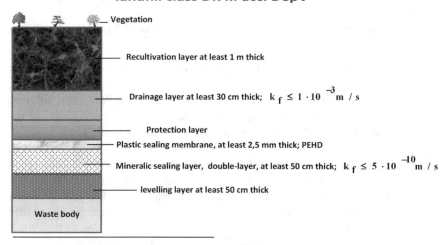

Fig. 4.5 Surface sealing for landfill DK III acc. to DepV; pursuant to Table 4.4

are approved for the surface sealing of these landfills, water catchments for both functional layers (waterproofing membrane and capillary layer) must be installed and kept in working order. Thus, should there be any occurrence of leakage water within the capillary layer—that is below the geomembrane, it can be managed. The design and construction of the entire system must be based on the conditions that are required for the functioning of the capillary barrier. The structure of a surface sealing system shown in Table 4.4 and Fig. 4.5 can be described as a multi-functional cover system and is in direct comparison with the uranium tailings ponds, see Chap. 5.

Methods of providing proof of success for covers such as: stability of the slope system, bearing capacity, behavior of the landfill subsoil, settlement behavior of the landfill contact area under action of the landfill body, stress-related barrier design, hydrological conditions, distance of the base seal to the highest free groundwater level, mass transport modeling for possible contamination spreads, settlement, consolidation of the landfill body and their influence on the applied surface seal, etc. will not be discussed in detail here.

In the case of landfills, in particular landfill class DK III, which were constructed before the entry into force of the DepV and the KrWG, in most cases a task-based base waterproofing has been dispensed with. This cannot be installed subsequent to the opening of operations at the landfill. In some particularly serious cases, it was only possible to excavate the landfill and relocate the material to other approved sites.

Decommissioning and post closure maintenance, see Barrier 6, are today one of the main tasks of landfill operators. In doing so, legal regulations must be observed. These are:

- The decommissioning announcement§ 40 (1) KrWG
- Application for final disposal of the landfill or landfill section § 10 (2) DepV
- Determination of the final decommissioning of the landfill or a landfill section § 40 (3) KrWG
- Determination of completion of post closure maintenance § 11 (2) DepV and § 40 (5) KrWG on application.

In particular, the determination of the completion of post closure maintenance on application is important in the case of the final disposal of "waste with radioactivity". The period from the final decommissioning of a landfill to the time when the competent authority determines the completion of the post maintenance phase for the landfills, in accordance with § 36 (5) of the Closed Substance Cycle Waste Management Act, cannot be responsibly predicted in particular for landfills with "waste with radioactivity". The competent authority decides on the date of the end of the required maintenance of the respective landfill only on the basis of facts. Under current law, in Germany a post maintenance requirement exists indefinitely. It is mandated that the locations where waste and waste from times long past are kept must be clearly announced. The existence of an accurate and reliable land registry system makes this possible. This has particular significance for landfills, in particular Landfill Class DK III, which were established prior to the entry into force of the DepV and the KrWG. To protect the groundwater in the absence of base waterproofing, well galleries were often designed and installed; these lower the groundwater level artificially and thus keep the landfill dry at the base. The operation of these well (pump) galleries has to be guaranteed permanently. Although this entails a considerable financial outlay for the operators, it is important that groundwater penetration into the landfill body is prevented and additionally the groundwater is protected against massive contamination.

4.5 Outlook

In Germany a long-term safety assessment in the case of a final disposal of radioactive waste is not planned for a landfill built and operated in accordance with DepV and KrWG (except for long-term storage and landfill class DK IV in rock salt). The security concept of landfills assumes that the safety of the operational phase (both for the staff and the population in the area) can be ensured by technical and organizational measures. For the post-operational phase, monitoring (in particular groundwater monitoring) must prove that there are no discharges into the groundwater that could lead to significant groundwater contamination. If the operator can prove this objective through a sufficiently long series of measurements, the competent authority may exempt the (old) landfill from (groundwater) monitoring at the request of the operator. The landfill site remains registered as an old deposit in the land register for an unlimited period of time.

The following important security aspects are legally defined:

- Location requirements
- Requirements for the technical barriers.

The following important security aspects have not yet been specifically defined:

- Criteria for the derivation of acceptance values (eluate values and other criteria according to DepV)
- Triggering thresholds for groundwater contamination related to environmental monitoring (see below).

With regard to the deposition of waste with radioactivity, it should first be noted that for the worker on the landfill, only the duration of the handling of the waste is relevant. The worker only comes into contact with the radioactive waste and residues for the short period of delivery and storage. At landfills according to DepV, personnel and persons of the surrounding population are treated equally. The limit value (maximum permissible dose) for the annual radiation exposure of a person of the normal population from activities according to § 2 of the Radiation Protection Ordinance is 1 mSv/a.

As a long-term relevant route of exposure, the water pathway through which radionuclides can be transported to a domestic or commercial well, assumes radionuclide longevity, see Table 4.1.

However, the long-term safety of depositing waste with radioactivity at a landfill has so far been classified very differently. For the release of waste, the dose is determined using conservatively designed radio-ecological models and neglecting all geotechnical barriers, which may arise after decommissioning the landfill through using groundwater in the effluent. The clearance values according to Annex 3 Table 1 StrlSchV ensure that the *De Minimis* dose of less than 10 μSv/a is maintained. The only assumption on which this evidence is based is the permanent existence of the deposit of waste on site and its location relative to an (fictitious) aquifer.

In the case of residues, a similar modeling approach applies both to the control limits (Annex 12 Part B StrlSchV) and the thresholds for simplified detection to protect the population, and for the landfill of residues released from the monitoring (Annex 12 Part C StrlSchV) see SSK [5] and AKNAT[15]. However, as the applicability of these schemes is in many cases not sufficient, in practice evidence must be provided on a case-by-case basis. Here, assumptions must be made and model parameters chosen for which no methodological framework conditions have yet been defined. This results in two questions:

1. What doses are to be expected for the storage of waste with radioactivity originating from residues requiring special monitoring with specific activities of more than 50 Bq/g on above ground landfills of class DK III over periods of more than 1000 years. Under what conditions may the target values for a long-term safety of 1 mSv/a effective dose possibly not be met?
2. What are the consequences of an unwitting settlement of such a landfill site in the future?

After the exit of the Federal Republic of Germany from the peaceful use of nuclear energy to generate energy and the associated dismantling of decommissioned nuclear power plants, the problem of long-term safety of disused underground landfills for waste with radioactivity must not be underestimated. This is shown by the strong increase in civil protests against landfill sites, which are to be used for such waste disposal. Against this background, answers to the questions set out above are needed. In particular how to determine the end point for post-closure maintenance upon request in connection with the maintenance of a land title.

It seems advisable to deal scientifically with the long-term behavior of landfill sites according to KrWG and DepV in order to improve confidence-building with the community. In particular, the following should be considered:

– it is better to avoid "waste tourism" within Germany and suitable disposal sites should be sought near the place of origin and
– disposal is avoided as far as possible and valuable waste materials and conventional residual materials are returned to the conventional recycling cycle to the greatest extent practicable.

Initial results have been presented [16]. This should clarify how landfills behave in the long term and whether, for example, substances with radioactivity could reach the environment in the long term via seepage or groundwater. The optimization of landfill sealing systems reduces the risk of unacceptable contamination spills, see Springer [17]. The importance of human intervention at a landfill that has been closed for decades or centuries was also examined. In order to calculate the radiation dose of affected persons for different scenarios and periods, long-term safety-analytical models and computer programs for the disposal of radioactive waste were used.

References

Classification of radioactive waste, general safety guide, ISSN 1020–525X; No. GSG-1,Vienna: International Atomic Energy Agency, 2009.

Generic Technical Basis for Implementing a Very Low Level Waste Category for Disposal of Low Activity Radioactive Wastes, Final Report, December 2013, Electric Power Research Institute, Inc.

Resource Conservation and Recovery Act (RCRA); U.S. Environmental Protection Agency; https://www.epa.gov/rcra/resource-conservation-and-recovery-act-rcra-overview.

Rolf Michel, Bernd Lorenz, Hansruedi Völkle: Strahlenschutz heute – Erfolge, Probleme, Empfehlungen für die Zukunft; Fachverband für Strahlenschutz e.V.; https://www.fs-ev.org/fileadmin/user_upload/09_Themen/Philosophen/Zukunft_des_Strahlenschutzes_Gesamt_20180921_fuer_FS_Direktorium.pdf.

SSK: Ermittlung der Strahlenexposition; Empfehlung der Strahlenschutzkommission; Verabschiedet in der 263. Sitzung der SSK am 12. September 2013, Veröffentlicht im BAnz AT 23.05.2014 B4.

U.S. Government: Code of Federal Regulations No. 10 CFR 61.55: Waste classification. U.S. Government,https://www.nrc.gov/reading-rm/doc-collections/cfr/part061/part061-0055.html.

Very Low-Level Radioactive Waste Scoping Study; A Notice by the Nuclear Regulatory Commission on 02/14/2018; https://www.federalregister.gov/documents/2018/02/14/2018-03083/very-low-level-radioactive-waste-scoping-study.

The Safety Case and Safety Assessment for the Disposal of Radioactive Waste for protecting people and the environment; No. SSG-23 Specific Safety Guide; International Atomic Energy Agency (IAEA), Vienna, 2012.

IAEA; Principles for the exemption of radiation sources and practices from regulatory control, Safety Series 89, Vienna 1988/ISBN 92-0-123888-6.

Fachverband für Strahlenschutz (FS) e.V.; Leitfaden für die praktische Umsetzung des § 29 StrlSchV (Freigabeleitfaden); Arbeitskreis Entsorgung Ausgabe 3 Stand: 8. Dezember 2005; FS-05-138-AKE.

COUNCIL DIRECTIVE 2013/59/EURATOM of 5 December 2013 -laying down basic safety standards for protection against the dangers arising from exposure to ionising radiation, and repealing Directives 89/618/Euratom, 90/641/Euratom, 96/29/Euratom, 97/43/Euratom and 2003/122/Euratom; Official Journal of the European Union; L 13/1.

Verordnung über das Europäische Abfallverzeichnis (Abfallverzeichnis-Verordnung - AVV), 10.12.2001, zuletzt geändert 4.3.2016.

Abschlussbericht zum Vorhaben StSch 4416 „Methodische Weiterentwicklung des Leitfadens zur radiologischen Untersuchung und Bewertung bergbaulicher Altlasten und Erweiterung des Anwendungsbereichs"; Teil B: Erweiterung des Anwendungsbereichs auf NORM-Rückstände; Bericht I: Vorkommen und Entstehung von radiologisch relevanten Bodenkontaminationen aus bergbaulichen und industriellen Prozessen (StSch 4416 Leitfaden NORM-Rückstände, Teil B, Komm.-Nr. 5.18.001.3.1), Auftraggeber: Bundesminister für Umwelt, Naturschutz und Reaktorsicherheit vertreten durch das Bundesamt für Strahlenschutz (BfS); 222 Seiten, 2 Anlagen, 06.10.2006.

J. Schmitz, H. Klein: Untersuchung bergmännischer und industrieller Rückstandshalden in Niedersachsen auf eine mögliche Freisetzung radioaktiver Elemente. KfK Karlsruhe September 1985.

AKNAT Positionen des Arbeitskreises „Natürliche Radioaktivität" (AKNAT) des Fachverbandes Strahlenschutz zur Richtlinie 2013/59/EURATOM vom 05.12.2013.

A. Artmann et. al.: Anwendung und Weiterentwicklung von Modellen für Endlagersicherheitsanalysen auf die Freigabe radioaktiver Stoffe zur Deponierung – Abschlussbericht; Gesellschaft für Anlagen- und Reaktorsicherheit (GRS) gGmbH; Köln, Aug. 2014.

Optimierung von Deponieabdichtungssystemen (Hrsg.: H. August, U. Holzlöhner, T. Meggyes), Springer Verlag, Berlin, 1998, ISBN 978-3-642-72062.

Chapter 5
Radioactive Residues of Uranium Ore Mining Requiring Special Monitoring

The following explanations relate to processing residues and their storage in various locations (tailings ponds) as a result of SDAG Wismut's uranium mining operations in Saxony and Thuringia. These are possibly the largest uranium mill tailings ponds in the world, so that they can be described as reference objects. Comparisons are made with remediation programs and legal requirements in the USA, so that the reader can make comparisons with corresponding site specific projects and thus one may also acquire ideas for options for site specific solutions.

5.1 Origin of Radioactive Residues—International, Historical Development

Definition—Acronyms

Radioactive residues are materials that arise from industrial and mining processes and are described in the Radiation Protection Ordinance (StrlSchV) mentioned in Annex XII Part A and those that meet the requirements specified therein. Their recovery or disposal is regulated in Part 3 of the (StrlSchV). These materials are substances that contain naturally occurring radionuclides or are contaminated with such substances. Mining or industrial processes often lead to residues containing Naturally Occurring Radioactive materials ("NORM") [1]. Directive 2013/59/EURATOM recognises mining for uranium and thorium and the extraction and processing of the resulting ores as a particular part of the nuclear energy cycle and thus these activities are given a special status, see Chap. 2. The residues must be subdivided into residues requiring monitoring and those that do not need monitoring. In the following section, only the residues requiring monitoring are dealt with. For information on "non-monitoring" residues, see Chap. 4. Further explanations such as classification, occurrence, custody methods, etc. can be found in Chaps. 2 and 3.

Uranium (ore) degradation and all related activities of the fuel cycle, see Fig. 2.5, produce large amounts of radioactive residues, as NORM—Naturally-Occurring

© Springer Nature Switzerland AG 2020
M. Lersow and P. Waggitt, *Disposal of All Forms of Radioactive Waste and Residues*, https://doi.org/10.1007/978-3-030-32910-5_5

Radioactive Materials or Technologically Enhanced NORM (TENORM),[1] see [3]. Other industries that are known to have NORM issues include:

- The coal industry (mining and combustion)
- The oil and gas industry (production and refining)
- Metal mining and smelting
- Heavy mineral sands (rare earth minerals, titanium and zirconium)
- Fertilizer (phosphate) industry
- Construction and Building industry
- Recycling operations for metals and process residues

The legacies of uranium mining may be summarised as heaps of tailings; uranium mill tailings ponds; underground mines, open cast mines, stockpiles of low level mineralised material and waste rock and huge volumes of contaminated waters.

The focus here is concentrated on uranium mill tailings facilities.

Historical development

Uranium minerals were known before the discovery of the element uranium in 1797. For a long time, however, miners did not know what to do with the pitchblende that appeared from time to time in processing plants. Targeted mining of uranium ores took place only after the beginning of the nineteenth century. The oxide was used to colour glass, usually yellow or black. However, the addition as a glass colourant is prohibited in Germany today.

Uranium minerals have been extracted in many parts of the world using different methods. With the discovery of nuclear fission in 1938 by Otto Hahn and his assistant Fritz Straßmann, the priorities of uranium mining changed worldwide. The extraction of uranium ores is seen as the beginning of the nuclear fuel cycle, see Fig. 2.5. It is significant that nuclear fission first led to the development of the atomic bomb and it was only some yeas later that nuclear fission was used to generate energy. The first nuclear power plant (NPP) constructed solely for the production of electricity from nuclear energy, was built on 27 June 1954, and is located in Kaluga, about 120 km southwest of Moscow. Today, more than 400 NPPs are in operation worldwide, see Chap. 7, and the application of radioactivity in medicine and research is now universal. These activities result in large amounts of radioactive residues. Without mining of uranium, the fuel cycle and the other peaceful nuclear related activities cannot be maintained.

Causes of the rapid development in the exploration and in the extraction of uranium ores

Germany had access to the Saxon-Bohemian uranium ore deposits, in particular to the world-known deposit of Joachimsthal (Jáchymov)—Johanngeorgenstadt. There was considerable fear in the international coalition against Nazi Germany that Germany would get access to a nuclear fission bomb during World War II. And so large-scale

---

[1] Specified limits by mining committed in North American mining industry see [2]; in the following becomes summarized TENORM with NORM to NORM.

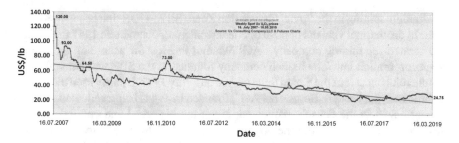

Fig. 5.1 Uranium price development since 1987, with historical price declines, according to UxC LLC and Future Charts

uranium mining began in the United States in 1942 as part of the Manhattan Project. A focal point was in the border area of the states Utah/Colorado/New Mexico and Arizona, on the Colorado Plateau. There, more than 500 uranium mines were established from which about 27 million tonnes[2] of uranium ores were mined. In total, 272,300[3] tonnes of U_3O_8 (yellow cake) were produced in the USA up to 1980, see also [4]. The work was halted in the Navajo Nation Reservation area in the 1980s, probably less because of the low yield, but more because of the huge drop in the price of uranium on the world market, see Fig. 5.1.

The mining area located in the Navajo Nation Tribal Land still shows many scars of uranium mining today, with significant health consequences for the residents. There are still no-go areas. The Navajo's sense of responsibility in World War II for the comprehensive mining of uranium ores in their tribal area has cost them dearly. The Navajos have the expectations that they will be fully compensated and that the legacy of uranium mining will be permanently eliminated by the Federal Government. Although the Navajo received compensation of US $554 million in 2014 in a settlement with the Obama administration, it does not meet all the compensation requirements; in particular the remediation of the legacy of uranium mining. The EPA has a partnership with the Navajo Nation, and since 1994 the Superfund program has provided technical assistance and resources to assess potentially contaminated sites and develop a solution. Since 2008, two five-year plans have been set up to gradually rectify the legacies. The US Department of Energy has also been cleaning up legacy sites through programmes including UMTRA (Uranium Mill Tainings Remedial Action) and DRUM (Defence Related Uranium Mines).

After the dropping of the atomic bombs on Hiroshima and Nagasaki and the end of World War II, nuclear escalation also began in the Soviet Union. Since no uranium deposits had been developed in the Soviet Union at that time, it concentrated on its areas of influence in which uranium deposits existed: Soviet Occupation Zone (SBZ)-GDR (East Germany); Czechoslovakia, Hungary, Central Asia etc. A significant uranium mining programme began throughout these locations.

---

[2] All data here in metric tons, 1 tonne approx. 1.103 short ton.

[3] Of which approx. 190,000 metric tons in the Collorado Plateau; https://nepis.epa.gov/Exe/ZyPDF.cgi/910088OG.PDF?Dockey=910088OG.PDF.

Uranium mining in the Soviet Occupation Zone (SBZ)-GDR first began in Saxony in the hydrothermal vein ore deposit of Niederschlema/Alberoda. In 1947 the Soviets had created the mining company SAG[4] Wismut in Moscow according to German stock corporation law. The branch company founded in the SBZ was entered in the commercial register of the Saxon town of Aue on 2 July 1947. It partly used the mining facilities of "Sachsenerz Bergwerks AG", founded in 1944. Unprecedented uranium mining began in the densely populated, important historical and cultural areas of Saxony and Thuringia. After the popular uprising of 17 June 1953 in the GDR, the GDR's reparation payments were discontinued with effect from 1 January 1954. On November 28, 1953, SAG Wismut was dissolved and on December 21, 1953, the Soviet-German stock corporation (SDAG) Wismut was founded. It took over all assets of SAG Wismut, but without becoming its legal successor. The overexploitation continued in this densely populated cultural landscape.[5] The miners and processing staff of SDAG Wismut were lured with very high wages and other benefits for the health-exhausting work.

Comparative starting situation for the sustainable processing of the legacies of uranium ore mining and all associated activities of the fuel cycle in the USA/Germany
Following the reunification of Germany, and after the Soviet Union had renounced its share in SDAG Wismut, SDAG Wismut was transformed into Wismut GmbH on 20 December 1991 as a federal enterprise which was commissioned to carry out the remediation of the uranium ore mining legacies in Saxony and Thuringia. The full extent of the environmental impacts is summarised in the following chapters.

In the market economy sense, however, SDAG Wismut was already bankrupt in 1990, see the historic fall in the price of uranium. It had been subsidized by the GDR with more than 10 billion euros[6] over the years. SDAG Wismut thus also contributed significantly to the decline of the GDR. Strictly speaking, the situation was the same as in the United States, but the situation there had already occurred some years earlier.

In order to formulate a comparable initial situation in the United States and in Germany regarding the handling of the legacies of uranium mining and uranium processing, here is a summary:

- The decline of uranium mining in the USA was completed around 1980, that in Germany around 1990;
- The size of the legacies was roughly the same, but they were distributed over a much larger, very sparsely populated area in the USA when compared to Germany;
- In both countries, mining operations have also disappeared with the decline. In Germany, the remediation of the uranium ore mining legacies in Saxony and Thuringia

---

[4]Soviet stock corporation (SAG).

[5]Now with a Soviet and a German general director (CEO).

[6]In the end, Germany will have spent around 30 billion euros on production and renovation.

Table 5.1 Production and costs of remediation of Titles I and II- uranium mills and related facilities [7]

Characteristics/ UMTRA	Number of sites included	Metric tons of ore processed	Metric tons of U_3O_8 produced	Total closure costs of all sites in US$	Average closure cost per site in UD$
Title I[a]	26	29,100,000	50,624	1695,000,000	56,900,000
Title II[b]	28	220,000,000	284,088	584,800,000	20,900,000
Total	54	249,100,000	334,712[c]	2279,800,000	42,200,000

[a]Title I Mills were abandoned, un-licensed mills operated during the Atomic Energy Act (AEC) existence
[b]Title II Mills were mills licensed by NRC or Agreement States in or after 1978
[c]bis 1992

is being carried out on the basis of the "Wismut Law"[7] of 16 May 1991 with an initial budget of DM 13 billion. In the USA, the Uranium Mill Tailings Radiation Control Act (UMTRCA) was passed by the US Congress in 1978, which entrusted the Department of Energy (DOE) with the responsibility for the stabilisation, disposal and control of uranium tailings and other contaminated material, in 10 EPA-regions with approximately 5200 associated properties. Within the UMTRCA, two discrete implementation programmes were agreed: Uranium Mill Tailings Remedial Action (UMTRA) Projects Title I and Title II, see [5–7]. The most important data are summarized in Table 5.1.

All abandoned uranium mine sites (AUM's)—Superfund sites—are listed on the National Priorities List (NPL) of U.S. Environment Protection Agency (EPA), see [8].

- In Germany, the financial resources for the entire project are provided by the federal government, the final operating plans are approved by the responsible mining authorities of the federal states, with the participation of the responsible water authorities, radiation protection authorities etc. A set of regulations for the uranium mill tailings ponds does not exist. The final operating plans for the individual remediation projects are understood to be site-specific solutions. For individual solution elements, the term standard cross-section, standard cover, is used, from which, however, there are deviations in individual cases.

In the USA, uranium mill tailings are recorded as radioactive residues by the US Nuclear Regulatory Commission. There is a uniform set of rules: "**Most of the regulations that the U.S. Nuclear Regulatory Commission (NRC) has established for this type of byproduct material are found in 10 CFR Part 40, Appendix A, "Criteria Relating to the Operation of Uranium Mills and the Disposition of Tailings or Wastes Produced by the Extraction or Concentration of Source Material from Ores Processed Primarily for Their Source Material Content.",** see [9].

---

[7]"Wismut-Law"—Law on the Agreement of 16 May 1991 between the Government of the Federal Republic of Germany and the Government of the Union of Soviet Socialist Republics on the termination of the activities of the Soviet-German stock corporation Wismut.

Table 5.2 Historical uranium production [2]

Country	Canada	United States	Kazakhstan	Germany[a]	Australia	South Africa	Russia	Nigeria
U_3O_8 produced up to 2014 (metric tons)	483,957	373,075	244,707	219,686	194,646	159,510	158,844	132,017

[a]Production includes 213,380 tU produced in the former German Democratic Republic from 1946 through the end of 1989

However, consideration must also be given to the legislation of the states;

• In case of Title I of UMTRCA it is a matter of abandoned mine sites. Congress set up a program to jointly finance rehabilitation measures for decommissioned uranium and thorium tailings storage facilities (public money). Especially where residues mainly resulted from the production of uranium for the American defence program. Under Title II of the UMTRCA, the State only provided supplementary financing as the mining companies were called on to finance the works themselves.

• The United States has the most experience in dealing with radioactivity, including NORM and TENORM. Through UMTRCA and in particular the UMTRA Title I program, both significant scientific and procedural knowledge of the long-term safe, long-term stable closure of uranium mill tailings storage facilities has been gained and made available for subsequent use. These data were used in Germany as an essential basis for their own closure concepts, see following chapters;

Significant uranium production took place in the former GDR as shown in Table 5.2.

Although there has been no significant uranium production since 1990, Germany is (and will remain for a very long time) still in 4th place in the historical uranium production table.

Around the year 2000 a further change in uranium mining and uranium processing began worldwide. On the one hand, this change resulted from the fact that uranium ore deposits being mined are now increasingly at both low grades and very high grades. In addition, the ISL (In Situ-Leaching) method, is increasingly being used, in addition to the traditional methods of underground and open-cast mining.

More than 80% of current uranium mining takes place in sandstone-bound deposits. In underground and open pit uranium mines, the ratio of overburden to ore has increased as less accessible and inferior ores are exploited. This increases not only the mining costs but also the closure costs, which are part of the operating permit in Germany, see Fig. 5.2 and [10].

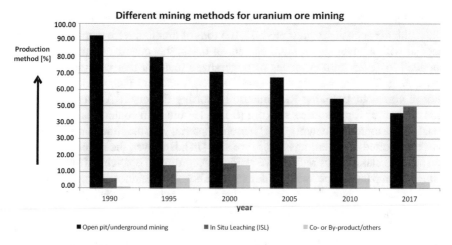

Fig. 5.2   World mining production—by different mining methods, see also [11, 12]

5.2   Origin of Radioactive Residues and Residue Storage Facilities, Basic Requirements for Their Decommissioning in Germany

A compilation of the origin and classification of radioactive residues is given in Chap. 2. By far the largest quantity of radioactive residues requiring monitoring in Germany originates from the enormous uranium ore extraction and processing of SDAG Wismut in Thuringia and Saxony between 1946 and 1991; Source: [13] Status and Results of the WISMUT Environmental Remediation Project; Vienna 2006:

- 231,400 tons of produced uranium ("Yellow Cake") by 1991;
- 37 km^2 of WISMUT-mining region in the mining areas of Saxony and Thuringia;
- 311 million m^3 tailings dumps—0.2 to 2 Bq/g (Ra-226), inventory: 20,000 t uranium;
- 30.6 million m^3/a mine drainage—25 t/a uranium discharged into receiving waters;
- 5.7 km^2 Industrial settlement plants[8] (tailings ponds), 178 million m^3 processing residues, inventory: 1800 TBq (Ra-226), 16,000 t uranium;
- 1.6 km^2 opencast mining area; 84 million m^3 hollow mould;
- 5 underground mines; 1.53 million m^3 of mine structures, tunnels and drifts with a length of 1470 km.

In case of radioactive residues in need of monitoring, from the former uranium ore mining in the new federal states, but to a lesser extent also in the older states, these are predominantly dumps of former uranium ore mining as well as tailings ponds of former uranium ore processing (Uranium Mill Tailings). To illustrate the full extent

---

[8]Storages for the uranium ore processing residues (Uranium mill tailings ponds) were designated by the SDAG Wismut Industrial settlement plants (IAA).

of uranium mining in Saxony and Thuringia, the following data on the Schneeberg-Oberschlema deposit are presented; 73,105 tons of uranium were extracted from the deposit with an average uranium content of 0.4%; this quantity of uranium corresponded to 18.25 million tonnes of mineralized material that was transported for processing. The ore was first pre-sorted underground and then carried to the surface in crates. In 1965, the Radiometric Presorting Factory (RA) went into operation at shaft 371, pre-sorting step ore (0.1 to 1% uranium content) and factory ore (0.01–0.1% uranium content). The "barren" material with uranium contents <0.01% was transported to the tailings dump, see also [14]. However, the material is not barren and it is classified as radioactive residues according to StrlSchV. This resulted in a huge "dump landscape" with 28 dumps of radioactive residues around the shafts of the Schneeberg-Oberschlema deposit. Their slope angle corresponded to the natural angle of repose at between approx. 34° and 36°. The two dumps at shaft 371[9] alone comprised a volume of 13 million m^3. The dump 371/I has a volume of 9.3 million m^3 and the dump 371/II has a volume of 3.7 million m^3. Both tailings dumps are currently in the final phase of remediation, see Figs. 5.3 and 5.4.

In this region, most dumps were flattened in the remediation program of the federal Wismut GmbH and covered with a standard cover of 1 m of inert material, so that the emanation of radiation and of the gaseous, radioactive decay product radon is largely

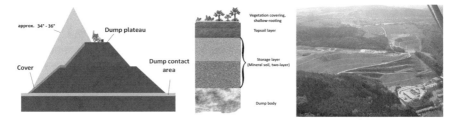

Fig. 5.3 left: Heap profiling and cover; **middle**: Standard cross-section of the dump of the Schlema/Aue region; according to Wismut GmbH; **right**: dump remediatio in situ, shaft 371, flattening, profiling and installing of the cover, Aug. 2009. *Source* Lersow, M

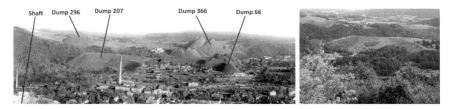

Fig. 5.4 left: dump landscape of the Schneeberg-Oberschlema deposit, 1967; *Source* Wismut GmbH; **right**: landscape after remediation. *Source* Lersow, M., August 2017

---

[9]The SDAG Wismut designated all components of the mining and processing plant as objects with a number, e.g. Shaft 371, this was the main production shaft of the deposit. This was part of the obfuscation of the activity of the SDAG Wismut in the GDR.

prevented in the biosphere and revegetation is encouraged. Radiation exposure and infiltration are also minimized. In addition to the installation of a path network, the heaps have been planted. The recultivated and remediated heap landscape fits in well to the landscape of "Erzgebirge" (ore mountains). Seepage water (leachate) and surface waters are collected over the long term and sent to a water treatment plant. In the water treatment plant, the waters are cleaned, thus permitting their discharge into the Zwickauer Mulde river in compliance with the permitted levels. The resulting radioactive and toxic residues of the water treatment are stabilized, conditioned in bags and stored at an approved installation location, usually within a surrounding dump, and safely disposed of. However, contaminated waters can also reach the groundwater via the tailings footprint, which has to be permanently monitored. However, by far the largest amount of contaminated water appears as excess mine water from the flooded pits and shaft; these are also fed to a treatment plant and cleaned. At present the treatment programme has no end date.

In most mining and processing areas of the Soviet-German stock corporation (SDAG) Wismut the permissible radiation dose limit of 1 Millisievert/year(§46 Radiation Protection Ordinance) was exceeded. As a consequence, after the reunification of Germany, there was an urgent need to look at the rehabilitation of these areas from a radiological point of view, see SSK [15]. The redevelopment of the areas designated in 1992 has meanwhile progressed considerably but is not yet completed, see Fig. 5.4.

The residues from the uranium ore processing at the central processing plants of the SDAG Wismut in Crossen and Seelingstädt, see Fig. 5.5, are the most protracted task for the remediation of the legacies of uranium ore mining in Saxony and Thuringia.

For example, the problem of radon exhalation occurring from the large waste heaps is not discussed here. However, this has no influence on the general statements here but leads to a focus on the topic.

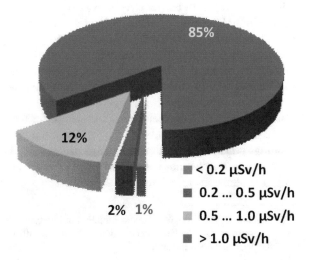

Fig. 5.5 Distribution of local dose rates in the uranium mining areas of Saxony and Thuringia; in state of 1992 according to Wismut GmbH

85%

12%

2% 1%

■ < 0.2 μSv/h

■ 0.2 … 0.5 μSv/h

■ 0.5 … 1.0 μSv/h

■ > 1.0 μSv/h

At the tailings ponds of the former uranium ore mining in Saxony and Thuringia, it should be noted that the locations where tailing ponds were built were initially selected so that the residues from the treatment process could be deposited there economically and with long term certainty. These sites had to be close to the processing plants, some of which were built close to the mining sites, so that in some locations a concentration of mining extraction, processing and storage of processing residues (tailings) from uranium ore processing can be found in just a few square kilometers.[10]

If one carries out an international comparison of the chosen locations, it is significant that the uranium ore extraction and treatment in Saxony and Thuringia took place in a very densely populated, historical European cultural landscape, partly in the immediate vicinity of drinking water production. This led to significant environmental impact in the affected regions.

The population density in the Schlema/Alberoda region (Saxony) is around 250 inhabitants/km^2. In comparable uranium production and processing regions, such as South Australia, Utah (USA) or the Athabasca Basin (Northern Saskatchewan and Alberta/Canada), the population density is sometimes less than 1 inhabitant/km^2. This must also be reflected in the closure technology for the residues requiring monitoring.

The deposition of residues from ore processing in surface storage facilities, in particular in tailings ponds, is practiced in many mining countries and many different climatic zones. Especially for gold, copper, zinc, lead and uranium ores, tailings ponds are created that assume considerable dimensions, see Fig. 5.6. It is the most commonly used deposition method for residues of ore processing (tailings). However, over the last two decades thickening of tailings and their underground disposal has also become established.

This is done primarily by dewatering and addition of binders. The processing residues (tailings) become a pumpable thickened paste and can be pumped into prepared underground cavities. To conclude this work, the storage chambers are reliably secured with barrier structures to protect the environment. The addition of proven immobilization reagents during the production of "paste tailings" Caldwell and Charlebois [16] should also be considered, see Fig. 5.7. In the case of paste tailings, which are deposited in containments on the surface, there can be no seepage of the pore waters and supernatant water, so that the tailings dams should be lined.

Generally, for the disposal of residues from ore processing, in the simplest case, a valley is sealed off with a dam. If no valley is available, an annular dam will usually be erected. This creates a hollow shape for the discharge and storage of the processing residues. In contrast with the construction of an earth dam, the slopes of a tailings pond may be built with tailings material that is not suitable for processing. Such slopes are also not compacted layer by layer and also contain no clay core, as is often done in earth dams. For the processing residues (uranium tailings) of the

---

[10]By this line-up the SAG Wismut was initially not interested. There took place an plundering of natural resources. Only in later years (SDAG Wismut) was the ore processing concentrated at two sites (Crossen and Seelingstädt).

Fig. 5.6 Discharge of
processing residues (tailings)
in an industrial settlement
plant (tailings ponds) of the
SDAG Wismut, 1963. *Source*
Quelle: ©Wismut GmbH

Fig. 5.7 Paste Tailings.
Source Mark T. Biesinger/
WesTech Engineering, Inc.,
[17]

Wismut SDAG mostly abandoned sand- or gravel quarries were used, which had been
designed for this purpose; these were prepared with a suitable slope structure (tailings
embankment) boundary, so that the planned hollow containment was formed, see Fig
5.10 and 5.14. Such structures may also be described as a "turkey nest" dam. Base
sealings to protect against seepage leakage, as required by regulatory agencies in
many countries today, were not implemented, see Vick [18].

Thus, two mined out uranium open pits Katzendorf (Trünzig) and Culmitzsch
were used by the SDAG Wismut as discharge basins for the uranium mill tailings
from the processing plant in Seelingstädt near Ronneburg (Fig. 5.8).

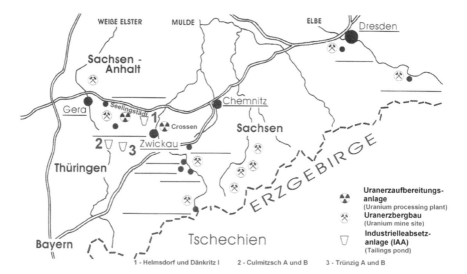

Fig. 5.8 Distribution of the uranium mill tailings ponds of the SDAG Wismut according to data of Wismut GmbH (There are considerably more, but smaller tailings ponds from the earliest uranium mining of SDAG Wismut as figure shows)

The uranium mining of Culmitzscher deposit took place between 1951 to 1967 in an open pit. This deposit belonged to the so-called "sandstone type".[11] In addition to oxidic uranium mineralization in this deposit sulphidic, unminable, mineralization of zinc, lead, copper, iron, arsenic, cobalt, nickel and antimony was also found. The components of this mineralization are now in the inventories of the tailings ponds at Trünzig and Culmitzsch and can be detected in the collected waters, TLUG [19].

Processing residues from the processing plant in Crossen, near Zwickau, were discharged in the former sand/gravel open pit in Dänkritz, now described as the tailings pond "Dänkritz I".[12] Another disposal location for the Crossen processing plant was created in the Oberrothenbach valley as a "valley impoundment". In terms of area it is the largest uranium mill tailings pond operated by SDAG Wismut, Helmsdorf, Jakubick and Hagen [20].

---

[11] Sandstones are also aquifers.

[12] Nearby, there is another smaller uranium mill storage Dänkritz II, which does not fall under the Wismut law.

5.3 Decommissioning/Long-Term Safekeeping of Radioactive Residues of Uranium Mining in Germany—Safekeeping of a Tailings Pond Based on a Site Specific Remediation Concept [Conceptual Site Model (CSM)]

When a tailings storage is full, it must be decided whether the deposited tailings will be reprocessed again (secondary deposit)[13] or whether the tailings pond will be safely closed out for the long term.[14]

In Germany, according to gem. § 53 BbergG, the operator has to submit a final operational plan, which has to be approved by the responsible mining authority. At the same time, the mining operator (operator) has an obligation of rehabilitation at mine sites and to fulfil the requirements of gem. § 66 BBerg. This obligation has a significant impact on the decommissioning/long-term safekeeping of the tailing ponds of the SDAG Wismut.

The uranium ore mining in Saxony and Thuringia ended abruptly with the German reunification. The activity had taken place in an old, densely populated European cultural area and the mining had been considered by many to be uneconomic and environmentally irresponsible. Although it had been the main shareholder of the Soviet-German Stock Corporation (SDAG) the Soviet Union was not willing to bear the necessary financial expense for remediation. The Federal Republic of Germany took over the 50% share of the Soviet Union and thereby assumed all obligations under the Federal Mining Act. In order to finance these obligations, the federal legislature initially planned for DM 13 billion, which in the meantime has increased to € 7.1 billion, see BMWi 2016, see also Sect. 5.1.

At the beginning of the 90s there was little experience in Germany regarding the safekeeping of uranium mill tailings ponds. When in 1991, on the basis of the Wismut Act, the renovation of the enormous legacy of uranium mining in Saxony and Thuringia began, this work could only be carried out on the basis of approved final operational plans according to BBergG. This requires a plan approval procedure including environmental impact assessments for the safekeeping of the uranium mill tailings ponds. Due to the concentration effect of the BBergG, at least the Water Act, the Radiation Protection Ordinance (StrlSchV) and the Atomic Energy Act (AtG) had to be used for the approval of the final operating plans. Indeed, the competent mining authorities as supervisory and approval authorities had experience with the operation of the mill tailings ponds of the SDAG Wismut, but not with their permanent decommissioning and the applicable legislation of the FRG. It was thus necessary to examine which laws of the GDR should be considered to apply. According to Article 3 of the Unification Treaty of 6 September 1990, the following laws of the

---

[13]This is unusual for uranium tailings, but is the rule for gold tailings.

[14]The SDAG Wismut has often no done remediation works for the decommissioned Tailings ponds from the period 1945–1963.

former GDR should continue to apply to the remediation of the legacies of the SDAG Wismut, GesfStrlSch [21]:

- Ordinance on Ensuring Nuclear Safety and Radiation Protection of 11 October 1984, together with the Implementing Regulation to the Ordinance on Ensuring Nuclear Safety and Radiation Protection (VOAS)
- Arrangement to ensure radiation protection of waste dumps and industrial tailings and the use of materials deposited thereon of 17 November 1980 (HaldAO).

Due to the lack of experience in the safekeeping of tailings ponds from uranium ore processing, international experience had to be used as a priority. There was a fortunate occurrence at this time. In the USA, due to the rapid decline in uranium prices, a large number of uranium mining and processing sites had been abandoned in the 1980s and 1990s, which also required closure, see Lersow and Märten [10]. Following the disappearance of the mining companies in many cases, the American government was obliged to plan for safe and long-term stable closure of these sites and commenced the UMTRA program (Uranium Mill Tailings Remediation Action Program). In addition to the provision of financial resources, the UMTRA program has generated significant, widely applicable, findings on the safe, long-term closure of uranium tailings ponds, which were incorporated into the remediation program of the tailings ponds of the SDAG Wismut, see Wels et al. [22], see also Sect. 5.1. Waggitt, P. gives in [11] an overview of the international status of developments in remediation of uranium ore mining and processing sites, worldwide.

Another partner involved in the process is the International Atomic Energy Agency (IAEA). It not only performs an oversight role, but also offers guidance and training in all matters relating to the handling of radio-toxic substances, including the processing residues of the uranium ores of the SDAG Wismut. The IAEA is not only endeavoring to collect and provide best practices from the various member states, but also finances research on controversial issues. With reports and, in particular, the "Joint Convention on the Safety of Spent Fuel Management and on the Safety of Radioactive Waste Management", it has created a legal instrument to directly address issues related to the management of radioactive waste and residues on a global scale [23].

For tailings dams, the ICOLD (International Commission on Large Dams) has formulated a set of guidelines that had to be taken into account and implemented in the permanent decommissioning of the uranium mill tailing ponds of the SDAG Wismut.

For the selection of closure technology for uranium tailings ponds it is crucial that, as a result, a geotechnical environmental structure is created, which will permanently eliminate the on-site hazards which result from the radioactive and toxic processing residues for humans and the environment. It is also important that the geotechnical environmental structure adapts harmoniously into the environment at the chosen location.

The site-specific boundary conditions and factors have a decisive influence on the choice of closure technology, on the geotechnical environmental structure to be erected and thus on the achievement of the remediation goal of safe, long-term safekeeping of uranium tailings ponds see Vick [24].

The safekeeping of a uranium mill tailings pond is often carried out worldwide on the basis of a site remediation concept [Conceptual Site Model (CSM)], see Chap. 3. In Germany, this corresponds to a final operational plan. The application for approval according to mining law, according to BBergG, § 54 is therefore a site-specific rehabilitation concept. An approved final operational plan for the decommissioning and long-term safekeeping of uranium mill tailing ponds of the SDAG Wismut can be according t. § 53 BBergG and amended. Thus the closure plans are practically updated.

The closure technology used is thus, strictly speaking, always unique. Therefore, the optimization criterion of any closure technology must always be based on the achievement of site-specific objectives. Long-term safety investigations and proofs must also follow site-specific objectives, but these must be based on objective criteria [23]. The resulting geotechnical environmental structures can be used as a reference on which modern requirements for long-term safe storage of radioactive substances can be mirrored and evaluated, see Lersow and Gellermann [25]. This also applies to a monitoring system to be installed, including long-term monitoring and environmental monitoring, which has been so far largely ignored during final disposal planning, see Chap. 9.

The following considerations are based on the general protection criteria for the long-term safe and stable disposal of abandoned uranium mill tailings ponds of the SDAG Wismut: "**To develop a closure concept for radio-toxic residues, with which permanently prevents radioactive contaminants overflowing into the Biosphere and if so, then only within the socially accepted borders**." The term "permanently" has here initially a time limit, which results from requirements of the regulatory mining authority. At which time a tailings pond is released from the supervision of the regulatory mining authority, the competent mining authority will decide. Based on current knowledge, it can be assumed that the uranium mill tailings ponds of the SDAG Wismut will remain permanently under supervision of the mining authorities. It is guaranteed by law that the mine plans and the operating chronicle according to the Documents–Mining Ordinance, § 67 BBergG, remain with the competent mining authority and thus also permanently accessible to the public.

Internationally, the problem of missing base sealing is not untypical. Where there are large seepage volumes of radioactive waters, there may be a possibility for the relocation of the uranium tailings or the large-scale lowering of the groundwater level in the vicinity of the uranium tailing spond.

During the relocation of the contents of a uranium tailings pond, the host structure is first prepared according to the rules of technology and, of course, provided with a base seal, possibly also with a watertight core in the dam body.

As a prominent example, the uranium mill tailings pond at Moab in the US state of Utah in the immediate vicinity of the Colorado can be considered. The uranium mill tailings pond was associated with the formerly largest uranium deposit of the USA. In 1984, the uranium mill tailings pond was closed and covered. Then significant seepages of leachate were discovered leaking to the Colorado River, which threatened to pollute the river in the long term. The river is regarded as a "lifeline" to the communities and environment downstream.

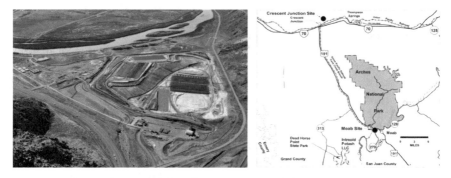

Fig. 5.9 left: uranium tailings pond Moab directly adjacent to the Colorado River; **right**: section of the transport plan. *Source* [26], public domain

The US Department of Energy declared the Moab talings pond as an abandoned mine site (mining without legal successors), and it was put under the UMTRA program. It was planned to relocate 10.8 million tons of radioactive residues, mostly by train, some 30 miles away to a disposal site at Crescent Junction, Utah, see Fig. 5.9. The transfer takes place mostly by train and should be completed in 2028. The cost of the relocation and remediation of the Moab uranium tailings, including the decontamination of the surrounding area, this relatively small amount of uranium tailings (compared to the tailings ponds of the SDAG WISMUT), is currently estimated at about US $750 million see [26, 27].

If the tailings ponds are far from civilization and the groundwater is already naturally highly contaminated, as in Western Australia—Kalgoorlie Goldfields— little aftercare may be operated next to the cover. Gold tailings are often re-processed once or twice, therefore these are frequently stored in uncovered locations.

The uranium mill tailings ponds in Canada which were covered with a clear water lamella (or with a mineral cover), see Fig. 5.14, must be observed in a similar way as in Germany. The regulatory authority is Natural Resources Canada. In addition to the radiological contamination of the waters, the uranium ore-containing residues contain the sulfide mineral pyrite, which can oxidize into sulfuric acid when exposed to oxygen. Acid drainage water from uranium residues is therefore a significant environmental problem in Canada, as uranium mill tailings ponds are often located directly alongside extensive surface waters. The acid mobilizes metals such as radium, copper, zinc, nickel and lead and this contamination must be prevented from escaping into surface waters in particular. In Canada, for example, the waters are captured, conveyed and treated. Examples include Elliot Lake (Algom's Stanleigh mine) in the Algoma District Ontario/Canada and McCabe Lake in Halifax/Nova Scotia/Canada, where there has been significant contamination from tailings ponds, see [28].

5.4   Properties and Radiotoxicity of Uranium Tailings, in Particular at SDAG Wismut

Uranium tailings are usually deposited into tailings ponds as a suspension (so-called sludge or slurry), consisting of the radioactive and non-radioactive constituents of the parent ore, chemicals, organic substances and process water (discharge), see Fig. 5.6. The way in which the tailings are deposited in the containment has an influence on the layers that develop there; layers will differ in terms of their grain fraction, consistency and mineral composition. The thickness of the tailings body and the radio-toxic inventory can take on considerable dimensions, see Table 5.10. Layout of standard and discharge hazards of uranium mill tailings are shown in Fig. 5.10.

- **Properties of the residues from the uranium ore processing, in particular the SDAG Wismut**

Depending on the processing methodology, tailings consist of solid constituents and process water from ore processing. The solid constituents are fine-grained sediments (typical grain fractions between 0.001 and 0.6 mm) of the ground ore and minerals. The goal of crushing is breaking the uranium ore to free the ore minerals from the gangue minerals.

The uranium ore processing residues of the SDAG Wismut in the uranium mill tailings ponds described here come from a wet-chemical treatment process (flotation). In flotation the different surface properties of the minerals have to be considered. To separate them, the ore must be crushed in several stages, finally in ball mills with the addition of water, to a specific fineness. By adding suitable reagents to the resulting suspension or "Trübe", the surface of certain minerals can either be

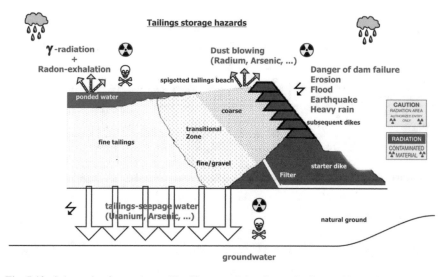

Fig. 5.10 Schematic of a uranium mill tailings pond showing main dispersal hazards. *Source* [29]

activated or passivated in this system. The activated ore minerals thereby become water-repellent-hydrophobic, while the others remain water-wettable -hydrophilic and become residues e.g. pumped as a suspension into a surface storage facility and stored, see Fig. 5.6.

Thus, it becomes clear that the physical and chemical properties of the tailings of a wet-chemical treatment process result from various individual properties—the properties of the remaining minerals, the properties of the added chemical reagents, the properties of reaction products, etc. The tailings from the wet-chemical treatment consist of solid components and process water. The solid constituents are fine-grained sediment (grain fractions see above) of ground ore and minerals resulting from leaching of the ores.

The residues from ore processing may be characterized by high chemotoxicity and/or radioactivity. The uranium mill storage facilities are usually designed to be active for several decades. It follows that the solid constituents of the slurry which are placed in the containment settle down and form a water-saturated layer of solids, the thickness of which increases over time. At the uranium mill tailings pond Culmitzsch A, shown in Fig. 5.6, water-saturated layer of solids is about 72 m thick.

- **Tailings ponds of uranium mining in Germany, radiotoxicity of inventories**

The legacy uranium mining sites in Saxony and Thuringia contain large quantities of radioactively contaminated materials, see Tables 5.1, 5.3 and 5.6. The long-term safe and stable storage of these legacies is carried out not only in compliance with geotechnical requirements, but also according to radiation protection criteria (StrlSchV, HaldAO, VOAS). This means under current requirements, the deposits would be classified as residues requiring monitoring. The SDAG Wismut's radiologically most significant deposits in terms of volume are the uranium mill tailings ponds; the details are shown below Table 5.3.

Table 5.3 Data on some of the Uranium mill tailings ponds of SDAG Wismut

Location/ Specification	Culmitzsch A	Culmitzsch B	Trünzig A	Trünzig B	Helmsdorf	Dänkritz I
Surface (ha)	159	76	67	48	205	19
Volume (Mio m^3)	61	24	11	6	45	5
Mass of solid (Mio t)	64	27	13	6	49	7
max. disposal thickness (m)	72	63	30	28	48	23

Source [25]

5.5 Site Remediation Concept—Conceptual Site Model (CSM)

5.5.1 Principles of Closure Planning

Initially, the closure planning of uranium mill tailings ponds is based on the conditions encountered at the site and the general, recognised, country-specific rules according to the state of the art: in Germany the Federal Mining Act (BBergG) and other applicable laws and ordinances, as well as the international requirements and guidance —Directive 96/29 EURATOM and the IAEA—Joint Convention Safety of Spent Fuel Management and on the Safety of Radioactive Waste Management as well as guidance such as IAEA Safety Report Series, etc.

- **Basic assumptions of closure planning**

Each closure plan is based on assumptions that are considered regardless of location.

- The design of the necessary closure measures (final operating plan) for the construction of the long-term stable, geotechnical environmental structure for the safekeeping of tailings in particular uranium mill tailings, is made on the basis of a risk-oriented assessment approach (attention paid n to structure-relevant properties of tailings and dam construction materials).
- The long-term stable, geotechnical environmental structure consists of a multi-barrier system consisting of three main elements—dam structure—covering of the tailings body—basic sealing. It is supplemented by a water capture and transfer system and a water treatment plant for the treatment of the collected contaminated ground, seepage and surface waters. This will allow treated waters to be discharged into the receiving waters in accordance with the relevant quality limits. The cover of the tailings body is designed as a multifunctional multiple barrier.
- The definitive requirements for the long-term stable, geotechnical environmental structure are determined in each case and are site-specific. In the context of the prescribed optimization [for example according to Atomic Energy Act (AtG)] and taking into account all relevant aspects, such closure measures are applied whose effectiveness are least likely to be impaired by erosion processes or natural events such as severe storms, earthquake etc.
- The long-term stability of the geotechnical environmental structure for the safekeeping of uranium mill tailings will be required for a minimum period of 1000 years. Special requirements arise from the relevant legal emission concentrations and the dose rate value, from the relevant legal regulations and standards. These are generally applied to all matters of handling and avoidance of the spread of radioactive and toxic substances, and from the point of view of optimization (site- and object-specific).
- The long-term stable, geotechnical environmental structure should follow the following objectives, irrespective of the site- and object-specific deviations:

• **Objectives of closure planning**

The site-independent objectives of a long-term closure of storage facilities for residues of from ore processing, in particular from uranium ore processing, can be summarized as follows:

– Long-term stable storage of radioactive and toxic residues

  geomechanical/geohydraulic long-term stability,
  geochemically long-term stability,
  landscape-compatible

– Reduction of:

  Emissions harmful to humans and the environment (emissions transfer to the biosphere) so that the concentration of pollutants remains within the legal acceptable range (air, water and soil pathways)
  radioactive radiation via air, water and soil pathways (radon exhalation, the local dose rate, etc.)
  Infiltration of surface water, protection against heavy precipitation

– Avoidance/reduction of:

  Material dispersal, harmful emissions to humans and the environment, in particular, toxic emissions via dust,
  Spreading of radioactivity via dust,
  Oxygen diffusion into the tailings body, in particular to limit the formation of acidity,

– Protection of groundwater and receiving waters, in particular by preventing the escape of toxic and/or radioactive contaminants via the soil and water pathways
– Necessary restrictions for post closure use (human intrusion impacts)

• **Site-specific types of closure uranium mill tailing ponds**

Tailings ponds, in particular uranium mill tailings ponds, pose significant hazards to public health and safety both during operation and after decommissioning, which must be mitigated and eliminated, see Fig. 5.10. In particular, the type of closure depends on the site conditions, which are dependent upon the climatic conditions and situation found at the site. The following types of closure of uranium mill tailings ponds are used worldwide, see Table 5.4.

– Dry in situ storage, covering of the residues inventory, mostly with mineral, inert material layers (containment, encapsulation);
– Wet in situ storage, covering of the stored material with a pure water lamella [30];
– relocation of the tailings body to a prepared new containment, see—UMTRA project: Atlas Mine in Moab (Utah, USA)—approx. 14.6 million t of uranium processing residues on approx. 53 ha [27] Moab (–), see Fig. 5.9;
– underground disposal, return of residues to underground mined out areas, see Fig. 5.1 [31];

Table 5.4 Comparison of closure strategies for tailing ponds

Closure strategies	pros and cons
Return of residues to underground mining areas	If this type of residue deposition is not to become a source of permanent groundwater contamination, the deposition of residues above the groundwater level must be carried out. Alternatively groundwater levels may be lowered. In such cases underground disposal may be implemented. Often the residues will be stabilized prior to disposal. With innovative mining and processing technology, this type of disposal is often used
Backfilling of residues in opencast mines	The filling of processing residues in former open pits, in particular uranium ore residues, is not uncommon. The potential to pollute groundwater depends on the disposal technology and the local hydrogeology. In particular, the redox potential should be limited and base sealing should be provided
Covering the residue storage facilities with a clear water lamella	Worldwide, large quantities of uranium ore residues are deposited in uranium mill tailings ponds and covered with a clear water lamella. This type of disposal is not suitable for the long-term storage of residues containing long-lived radioactive constituents (for example, for Thorium-230 the half-life is 80,000 years, for radium-226 the half-life is 1620 years). Often, a portion of the cover may become dry, allowing radon gas and radioactive dust to escape. If the residues are not removed and contained, these deposits are sources of permanent groundwater contamination. They represent a serious risk in earthquakes because of the danger of liquefaction of the sludge. The risk of uncontrolled overflow of dams is considerable
Dry closure method in situ	This type of disposal poses a risk of permanent groundwater pollution, especially in the absence of base sealing, and if the seepage and pore waters are not permanently captured and treated. Such locations requires permanent groundwater monitoring. There is a serious risk of containment failure in case of earthquakes

– Production of paste tailings, see Fig. 5.7 and storing in surface, near surface or underground cavities.

The primary objective of paste tailings production is to increase the stability of the geotechnical environmental structure and it opens up larger storage options, e.g. Underground storage, see Fig. 5.11. The loss of dam stability is very unlikely when using paste tailings, especially in seismically sensitive regions.

Dry in situ containment has prevailed in the US and Germany. Innovative mining companies today leave the majority of the mining waste underground. However, the backfilling of process residue containing structures on the surface resulting is still of major importance. Many residue stores are prepared for closure during the extraction and processing. In particular, the dams are profiled, secured and planted at this time. Therefore, today it is considered that the planning of the closure of the tailings storage facilities is an integral part of overall project planning. The closure structure itself has to guarantee the long-term stability, including the physical, chemical and biological stability [23, 32, 33]. Most containments have multi-layered, multifunctional covers which should provide the following functions:

– Radiation protection, limitation to background levels;
– Stabilization of the tailings body (static and dynamic);
– limitation of seepage and leakage of polluted tailings waters;

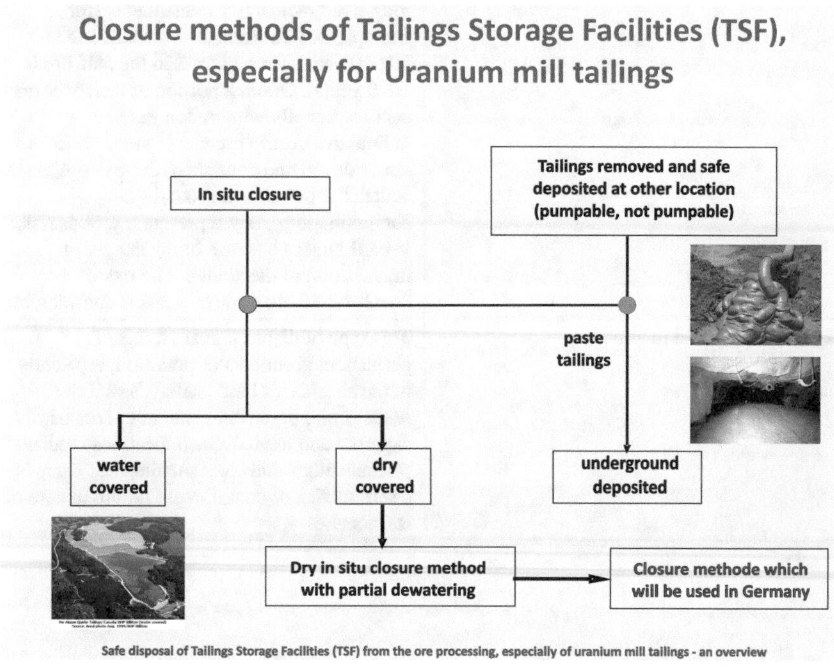

Closure methods of Tailings Storage Facilities (TSF), especially for Uranium mill tailings

Safe disposal of Tailings Storage Facilities (TSF) from the ore processing, especially of uranium mill tailings - an overview

Fig. 5.11 Closure methods of Uranium mill tailings ponds, location-dependent, also from the different climatic conditions

- limitation of thermal stress;
- promotion of vegetation;
- landscaping, adaptation of the geotechnical environmental structure to the natural environment;
- restriction of the entry to tailings body (protection of cover system).

The covering of the residue storage facilities with a clear water lamella and dry in situ storage are described in more detail in Sect. 5.5.3.3.

5.5.2   Long-Term Safe and Long-Term Stable Closure of Tailing Ponds

The residues from ore processing, which have been discharged into a prepared containment (tailings pond), settle down over time. They often form a depositional beach around the edges, consisting of coarse-grained material, and at the middle of the pond the fine-grained tailings settle (sometimes called slimes). The tailings body is water saturated. It forms a solid layer above the water-saturated zone, a liquid layer of process water and free surface water. At the centre of structures the open water is often referred to as the decant pond. The free ponded water is highly contaminated, but provides some radiation protection, since water reduces the escape of radioactive radiation. However, water evaporates, allowing radioactive contaminants to enter the atmosphere. This happens during the entire operating phase of the tailings pond and in the post-operational phase as long as the ponded water lamella is present.

In the absence of a base sealing, contaminated pore water can leak from the bottom of the containment across the whole dam footprint and into the surrounding area and from there into the groundwater.

In particular, the ponded water moistens the dam of the containment, resulting in development of a phreatic line. A phreatic line is the boundary between the dry and damp material of the dam structure where the tailings are stored. The phreatic line should ideally emerge in the dam foot area on the air side of the dam. This will allow any leaking waters or seepage to be collected using a leachate collection system. Alternatively, a leakage in the upper dam area leads to a high risk of stability including, potentially, dam failure. With heterogeneous dam construction, the permeability of the dam and its different areas plays a major role. Tailings dams should be designed so that they are denser on the water side than in the remaining part, so that the phreatic line in the denser area is pulled down and then can no longer escape on the air side. This significantly increases the stability of such dam constructions. Therefore, dams are sealed on the water side and dewatered on the air side by more permeable filter layers. For the dam constructions of the SDAG Wismut such protective measures were not available and had to be arranged in the final operational plan procedure, if this was possible, see Fig. 5.10. This problem does not arise with paste tailings.

- Primary requirements or the achievement of the following protection goals

(a) Elimination of potential hazards:

  – the radiological and chemical-toxic risks (air, water, soil)
  – the geotechnical/ conventional risks
  – the pollution of groundwater and surface water, and

(b) Minimization of material risks potentially associated with tailings ponds by:

  – the reduction of release and rate of spread (coverage, encapsulation),
  – the treatment of released pollutants (water treatment)

- **Secondary requirements**

(a) Adaption of the geotechnical environmental structure to the environment
(b) Sufficient availability of material suitable for the construction of the geotechnical environmental structure
(c) Socio-ecological acceptance of the geotechnical environmental structure

These requirements are valid worldwide. In any case, the final closure construction will be a site-specific solution that has to meet all the requirements for safe, long-term stable elimination of potential hazards, with emissions below the prescribed limits for radioactive and toxic pollutants and for socio-ecological acceptance. All of this has to be proven by an accompanying programme of monitoring for the service life of the closure containment. This is summarized in a site remediation concept for the respective tailings pond and must be approved by the relevant mining authority. The site redevelopment concept will also contain any restrictions that have to be complied with in case of possible reuse of the rehabilitated areas. It may be possible to prohibit reuse; "Human Intrusion Scenarios".[15] Quality control, monitoring, documentation (final documentation) are specifications that guarantee the reliability of the closure construction, the guarantee of the duration of protective functions over its service life, and show the need for any necessary follow-up services in good time. The timing for release of the site from supervision is part of the long-term safety plan for the Geotechnical Environmental Structure. However, the certification of the abandoned mining site, together with all relevant data remains in the registry of the competent mining authority.

- Evidence and objectives of the multibarrier concept for the long-term safe and stable closure of uranium mill tailing ponds.

The design of the site concept for the long-term safe, stable and site-specific closure of uranium tailings ponds is based on individual proofs (ENW), specification of target figures and objectives for ensuring compliance with socially responsible standards and for risk minimization.

---

[15]Particularly in sparsely populated regions of the USA, attempts were made to leave information on the hazard potential by means of appropriate information on monuments at the closure structure, and thus available even for periods for which administrative measures are no longer effective. This is intended in, particular, to prevent unintentional human intrusion into the tailings ("human intrusion scenario").

- Individual proofs for the interpretation of the multibarrier concept (ENW)

  - Individual proof of the long-term stability of the dam structure (ENW1)
  - Individual proof of the long-term stability of the multi-functional cover (ENW2)
  - Individual proofs of long-term stability of the base sealing (ENW3)
  - Individual proofs of protection of groundwater and run-off protection (ENW4)

- The multibarrier concept is designed to achieve, the reduction of

  - Environmental radiation
  - Radon exhalation,
  - Infiltration of surface water, protection against heavy rain (storm water), the leaching of radioactive and chemically toxic substances

- The multibarrier concept serves to avoid

  - the spreading of radioactivity via dust formation;
  - oxygen diffusion to limit acidity formation

- To ensure that the multibarrier concept permanently isolates the radio-toxic material apply:

  - restrictions for the reuse
  - base sealing for special protection of groundwater and receiving waters

In order to achieve the overarching objectives, the site remediation concept (CSM)] must be based on the site-specific factors (conditions). These are:

- Climatic conditions at the location
- Geological/Geomechanical/geochemical conditions and environmental risks at the deposal site
- Sufficient availability of correct specification, material (both grown and suitable tailings material)
- Implementation options for an appropriate closure strategy
- Socio-ecological acceptance and
- Cost.

5.5.3 Multibarrier Concept

5.5.3.1 General Structure

The geotechnical environmental structure for long-term safe and stable storage of tailings must generally consist of the following main elements, see Fig. 5.13:

- the dam construction
- the cover (encapsulation)

– base sealing with a geological barrier and/ or as a geotechnical barrier
– a water collecting and treatment system as well as the transfer of the treated water (in compliance with statutory discharge standards) to the receiving waters
– re-vegetation
– the monitoring system, long-term environmental monitoring with data management.

For the decommissioning/long-term safekeeping of the tailings ponds of the SDAG Wismut a comparison with landfill construction is obvious, see Chap. 4. Since the processing residues from the uranium ore preparation are radioactive residues requiring monitoring, the Federal Mining Act (BBergG), the Radiation Protection Act with Radiation Protection Ordinance (StrlSchV) and the Atomic Energy Act (AtG) replace the Landfill Ordinance (DepV) and the Recycling Management Act (KrWG). The Federal Mining Act develops a concentration effect during the decommissioning/long-term safe storage of radioactive residues, i.e. It is the authorities that participate, who carry the official responsibility for the objectives of protection (water, radiation protection, etc.).

The Site Remediation Concept (CSM) to be provided for the safekeeping of uranium tailing ponds is not a static concept and must have a number of established milestones at which the concept can be readjusted to match the closure progress. An approved final operational plan for the decommissioning/ long-term disposal of tailing ponds of the SDAG Wismut can be acc. § 53 BBergG be amended and be complemented. The final operational plans can be practically updated to reflect the state of progress of decommissioning. Thus, the final operational plans fulfill acc. BBergG also the concept of interfaces of the CSM.

The time at which a tailings pond is released from the supervision, is determined by the competent mining authority. Permanently means here, an indefinite period or a limited period, following an application for issuing of a release order by the competent mining authority.

Based on current knowledge, it can be assumed that the tailings ponds of the SDAG Wismut will remain under supervision of the mining authorities permanently. It is guaranteed by law that the mine plans and the operating chronicle according to the Documents- Mining Ordinance, § 67 BBergG, remain with the competent mining authority and thus permanently accessible to the public. This is also a significant difference to the DepV and the KrWG.

With the approval of the construction of a tailings pond the licensing authority, here the competent mining authority, must already have in mind that the facility must be shut down after the end of the operating phase. A distinction must therefore be made between the following phases:

• design, • construction, • operation (storage), • permanent closure and • post closure phase

In order to limit the environmental impact of tailings ponds from uranium ore processing, the geotechnical environmental structures are designed according to a multi-barrier concept. This means that several safety concepts (in this case barriers) must exist independently of each other in order to be able to prevent damage even if a barrier is limited in its function in time or if subsequent improvements are necessary to restore the barrier's function. This presupposes that all the barriers provided are also installed.

One barrier must already be installed before entering the operating phase, this is the base sealing. Options for consideration include:

– geological barrier,
– geotechnogenic barrier (e.g. clay and/or bentonite barrier),
– geosynthetic barrier or
– combination of sealing systems

These can be combined with each other, according to the site-specific conditions.

Basic sealing was not installed in the uranium tailings ponds of SDAG Wismut. A serious defect.

The operating phase also determines whether the tailings will be pretreated, e.g. with thickeners or by addition of barium chloride to precipitate radium. This results in so-called paste tailings (dewatered and thickened), which considerably increase the safety of the deposit, especially as a result of their low flow characteristics, see Caldwell and Charlebois [16].

5.5.3.2   General Description of the Individual Barriers

Barrier 1: Location selection with geological barrier (the site must be suitable due to its geology and hydrogeology, e.g. the presence of impermeable layers in relation to storage and isolation sites for large quantities of uranium tailings); and of a **dam construction (containment)** (Fig. 5.12).

The locations and selection criteria of the tailings ponds of the SDAG Wismut are described in Sect. 5.1.

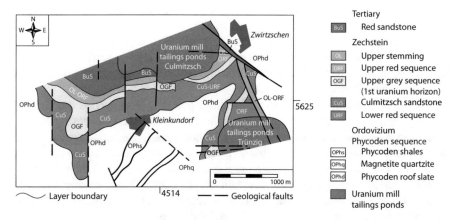

Fig. 5.12 Tailings Ponds Trünzig and Culmitzsch, Geological strata in the Ronneburger Horst [34]

However, they do not fully meet the barrier requirement, see also Fig. 5.19. The dam structures are profiled in accordance with the closure concept of Wismut GmbH; the containments have been stabilized, covered and planted with vegetation. The dam structures lack both surface sealing and base sealing; as a result seepage can take place along the foot of the tailings dam into the surrounding area.

Barrier 2: Tailings treatment before disposal

The uranium mill tailings of the SDAG Wismut were deposited as a slurry in the selected depressions, see Fig. 5.6. However, tailings could be also dewatered and thickened before deposition.

Barrier 3: Nature of the tailings body

The nature of the tailings deposited in the uranium tailings ponds at SDAG Wismut is determined on the one hand by the method of discharge of uranium mill tailings—see Fig. 5.6; and on the other hand by the removal of the open water lamella and by partial drainage using vertical drains (wick drains). A multifunctional cover is intended to prevent the mobilization of radionuclides in the tailings body.

Barrier 4: Installation of a base sealing in the prepared depression

The base sealing can be designed as a geological barrier and/or as a geotechnical barrier. The base sealing is intended to prevent polluted leachate from entering the groundwater via the footprint of the tailings body and thence to the receiving waters. Rules for the structure and cross-sectional specifications for base sealings do not exist in Germany. However, it is possible to base the suitability assessments of seals in accordance with the DepV and to adapt them to the site so that they can be approved, see Table 5.10. Depending on the requirements, this can be supplemented by (permanent) groundwater lowering, by leachate collection and installation and operation of a water treatment plant.

Barrier 5: Surface sealing (multifunctional cover)

On the one hand, surface sealing should prevent access of surface water and oxygen to the tailings body and, on the other hand, transfer of radioactive contaminants from the tailings body into the biosphere. Standards for the structure and cross-sectional specifications of surface sealing do not exist in Germany. However, it is also possible to develop a site-based solution based on the suitability assessments of seals in accordance with DepV, see Table 5.10.

The surface sealing of uranium mill tailings ponds is often carried out internationally using a multifunctional multi-layer system (multi-functional cover system). The surface sealing of the tailings ponds includes a water collection system and a water treatment system. The collected surface and leachate waters are fed to a water treatment plant. The radon exhalation rate and the local dose rate are permanently monitored in the vicinity of the uranium mill tailings ponds. This is in order to be able to detect possible radon exhalations and to be able to develop site specific solutions to eliminate or reduce them to levels required for compliance.

The surface seals of the uranium tailings ponds at SDAG Wismut are designed as a multifunctional mineral cover, including the tailings dam construction. The dam structure must have sufficient stability to totally encapsulate the tailings body. The phreatic line is also under specific observation as well as measures for the prevention of ingress of surface water and oxygen into the tailings body and the prevention of radon exhalations.

The reason is that, in addition to the radiological contamination of the waters, the uranium processing residues contain the sulphur-bearing mineral pyrite, which can oxidize into sulfuric acid when exposed to oxygen. The acid also mobilizes metals such as radium, copper, zinc, nickel and lead. In addition the mobilization of radionuclides from the tailings body must be prevented or at least significantly limited.

Barrier 6: Follow-up care, maintenance and repair

The geotechnical environmental structure for long-term safe and stable storage of residues from uranium ore processing, must be monitored extensively in all phases. All functional elements should be designed so they can be repaired, supplemented or replaced as required. The definition of action values should prevent large-volume contamination releases.

Thus, a monitoring network for ground and surface water with appropriate action values should be set up, as well as exhalation measurements and permanent monitoring of local dose rates. The resulting data are collected with appropriate routines, then collected, evaluated, made available to the supervisory authority and finally archived. Surveillance may lead to requirements that the operator has to implement.

5.5.3.3   Cover of the Uranium Mill Tailings Body

Two different methods of covering (encapsulating) uranium mill tailings have been developed for specific locations, see Figs. 5.13 and 5.14:

- Types of cover for uranium mill tailings ponds

  - Cover of a decommissioned uranium mill tailings ponds using a clear water lamella

Covering a decommissioned uranium mill tailings pond with a clear-water lamella has proved to be a particularly suitable method under the climatic and geographic, geomorphological conditions of Canada. It places special demands on the dam construction [30]. Also, escape of contamination via the clear water lamella is hardly possible. It represents a particularly good protection against radiation and radon exhalation.

- Dry in situ closure method

Dry in situ closure with or without partial dewatering, has proven to be a particularly suitable method under Central European conditions and is used in the closure of

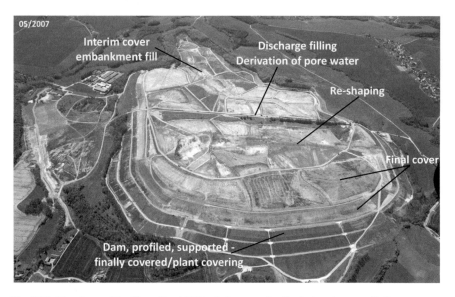

Fig. 5.13 Uranium mill tailings Pond "Trünzig", cover details during the remediation progress. *Source* [29]

Fig. 5.14 Covering of the uranium mill tailings with a clear water lamella, Quirke Tailings Facility at Elliot Lake, Ontario/Canada. *Source* [30]

SDAG Wismut- uranium mill tailings ponds. The main reasons for choosing this method are:

– Site-specific remediation solutions, which are sustainable and offer long-term stability[16]
– the option is most suitable under the conditions of Central Europe [temperate climate zone, high population density, etc.] It should be noted that comparable conditions exist, for example, in several parts of the U.S. and Canada)
– The long-term costs and the risks[17] associated with this closure solution are the lowest when compared to other closure technologies.

---

[16]The proof must still be furnished, however, since the periods of experience are still too short.
[17]According to current knowledge.

Some advantages that may be considered include:

- Low environmental impact of this closure solution (radioactive and toxic material remains on site)
- Low burden on staff and population
- Lowest residual risk over the service life of the geotechnical environmental structure

The tailings dam is already being errected with the construction of the residue storage facilities.

During the closure of the uranium mill tailings ponds, the dam structures are "repaired" for a long time with the following services: **profiling; stabilization (support dump); construction of water collecting facilities; Cover; Greening and planting. Special attention is paid to the phreatic line, which is related to the installed leachate collection system and the prevention of infiltration (oxygen diffusion).** This problem is not relevant with paste tailings.

Since dry in situ closure with partial dewatering of uranium mill tailings body has to fulfill several functions, **multi-functional cover** systems are developed specifically for each site, see Fig. 5.15.

The **multi-functional covers** for uranium tailings bodies have to meet the requirements described below, and compliance has to be substantiated by Individual proofs.

Individual proofs of the long-term stability of a multi-functional cover (ENW2)

- Long-term stability, geomechanical/ geochemical, against internal and external erosion
- Protection against infiltration (oxygen diffusion),
- Radiation protection (radon exhalation, γ-radiation),
- Emission protection (against radioactive and toxic material),
- Recultivation layer for revegetation and first planting etc.

The rules of the US Environmental Protection Agency (EPA) for a multi-functional cover as proposed in Fig. 5.15 focus on the advantages of both mineral material and geosynthetics. The long-term stability and safety of a multi-functional cover is determined in particular by the aging behavior of the related functional elements. A (partial) failure of one or more elements must not necessarily lead to the total failure of the cover system. The individual elements are also multi-functional. In particular, the related geosynthetic materials also act as a radon barrier. They are also particularly suitable as locations of measuring elements for the detection of contamination propagation, see [35].

The use of geosynthetics is also often justified by the fact that it provides greater protection against infiltration and thus against the mobilization of radionuclides from the tailings body. There is no final proof for this. Nevertheless, control of infiltration must be assigned a high importance for long-term safety. Since the uranium residues contain the sulfide mineral pyrite, which can oxidize into sulfuric acid when oxygen enters, a pH below 5 is likely to occur.

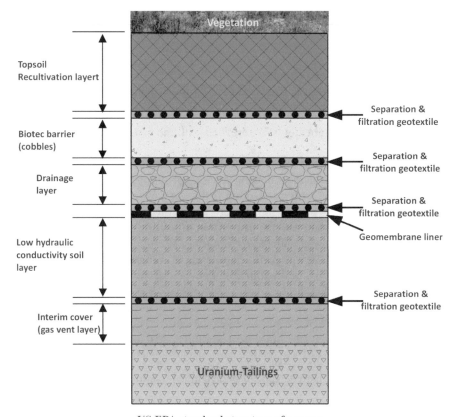

Topsoil
Recultivation layert

Biotec barrier
(cobbles)

Drainage
layer

Low hydraulic
conductivity soil
layer

Interim cover
(gas vent layer)

Separation &
filtration geotextile

Separation &
filtration geotextile

Separation &
filtration geotextile

Geomembrane liner

Separation &
filtration geotextile

Vegetation

Uranium-Tailings

US EPA standard structure of a cover
for tailings from uranium ore processing

Fig. 5.15 Multi-functional cover, rule structure according to the US Environmental Protection Agency (EPA). *Source* [23]

It is important not to create or counteract a mobility-promoting milieu in the talings body. Uranium is a lithophilic element and has a high affinity for oxygen. The most important minerals of tetravalent uranium are uraninite (idiomorphic UO_2 to U_3O_8), pitchblende (colomorphic UO_2 to U_3O_8), coffinite ($USiO_4$) and thorianite ($(Th, U) O_2$). The minerals of hexavalent uranium are essentially the uranium mica (e.g. from the group of phosphates: urococcite, torbernite, autunite).

Under oxidizing conditions, hexavalent uranium is the most mobile form of uranium, and tetravalent uranium is almost insoluble in water under "normal" pH/Eh conditions. Preventing infiltration involves controlling the influence of Eh value and pH on the stability of minerals, dissolved species and gases in complex reaction mixtures in such a way that the formation of hexavalent uranium (U6+) from tetravalent uranium (U4+) is largely prevented. The influence of the temperature is not taken into account here.

Another way of counteracting mobilization is to incorporate immobilisate in the encapsulation of the tailing body. It is known, for example, that the zero-valent

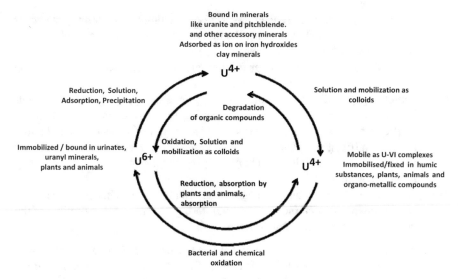

Fig. 5.16   Natural uranium cycle according to Boyle [37]

complexes must be taken into account for transport and sorption processes. They interact only slightly and are therefore hardly retarded [36]. Effective geochemical barriers for uranium transport in aqueous solution are Fe–Mn hydroxides, SiO_2–Al hydrolyzates, clay minerals and organic material (the product of soil formation). This leads to a reduction of U6+ to U4+ and associated loss of mobility, see Fig. 5.16. However, especially in organic material, chemical processes such as ion exchange, reduction or complex formation lead to a uranium enrichment.

Here, reference is only made to clay or bentonite barriers and the restraining effect of the vegetation of the cover. Regular sampling of the plants provides information about radon inputs to the root system of the plants and thus about the hazard of radon exhalation. Further explanations are can be found elsewhere [38].

The general correlations of reduction (mobilization by oxidation) from U6+ to U4+ are summarized in the natural uranium cycle according to Boyle [37].

- Process steps for the production of a long-term stable, safe cover for uranium mill tailings ponds in Germany

Dry in situ storage with partial dewatering of the tailings body was developed primarily in Canada and the USA in the 1990s and is carried out according to the following work steps, see Figs. 5.10 and 5.15.

(I)   pumping of the free and pore water, decontamination in a water treatment plant; dewatering and dehydration increase the shear strength in the area of fine-grain tailings; fine-grain tailings often demonstrate very low shear strengths (3–5) kN/m^2;

(II)   Creation of a working platform consisting of different layers of geosynthetics, insertion of vertical drains in a specified grid for near surface, partial drainage of the tailings body, subsequent layer-wise inserting of an intermediate cover (prevention of dust spread, consolidation of the covered tailings body);

(III)   contouring, profiling of the tailings dam construction (long-term stability); creation of a stable surface contour for discharge of surface water and protection against erosion, collection of surface water, creation of final cover;

(IV)   contouring, profiling of a stable surface, providing opportunities for collecting and removing surface waters; Covering (encapsulation) of the tailings, final coverage, prevention of radon exhalation (radon diffusion), γ-radiation, infiltration (oxygen diffusion), pollutant escape from the tailings body;

(V)   vegetation, first planting, use of regionally typical, flat-rooting plants, matching to the environment;

(VI)   gathering of surface and seepage water, treatment, discharge to receiving waters, environmental monitoring (monitoring system over the lifetime of the geotechnical environmental structure).

A multi-functional cover (see Fig. 5.15) has not been able to demonstrate success in the safekeeping of the tailings ponds at SDAG Wismut. The operator of the federally owned Wismut GmbH, was consistently focussed on mineral functional elements. The cover is thus a technogenic, loose rock body, in which weathering takes place. However, this should not be a disadvantage, especially since large thicknesses of mineral cover were chosen [39]. Geosynthetics were used in the creation of the working platform, which has enabled the use of heavy construction equipment. An intermediate covering is then placed, which consists mainly of tailings material at each site. This has two functions: as a load to consolidate the tailings body and to reduce the spread of contamination. This layer provides no more than a very slight barrier to radon emanation. The next layer consists of mixed-grained, mineral material, which is applied in connection with the dam profiling. The completion should be done with the application of a final cover of typical site-specific soils and a subsequent vegetation of the surface.

From the above work steps (see Fig. 5.17 and following [40]) the following standard design of cover for the uranium tailings of the SDAG Wismut can be derived; this may be varied on a site specific basis.

At the Helmsdorf site, a sub-aquatic pre-consolidation was carried out to stabilize the fine mud tailings deposited there, Illustration, Fig. 5.18, shows this situation although, in the author's view, it is lacking because it does not make a notable, substantial contribution to the long-term safe, storage of the uranium tailings ponds. The pre-consolidation carried out is also disproportionate to the length of the monitoring period, which is to be expected as a result of the escape of pore water over a very long time as the base of the pond is not sealed. Following the installation of vertical drains (usually from 5 m to 25 m deep), a partial drainage of the near-surface areas of the tailings body was effected. This, however, only partially drained the tailings body, see Fig. 5.17.

A description of the existing situation is given in Fig. 5.12. For long-term safety and stability of the multi-functional cover, a diagram (ENW2) is presented in Sect. 5.6.2.

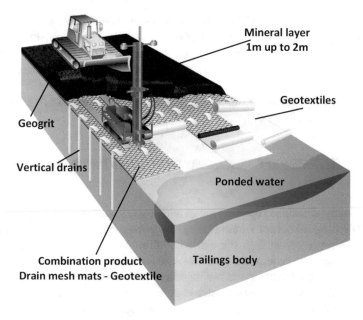

Fig. 5.17 Work steps for creating a multi-functional cover. *Source* [29]

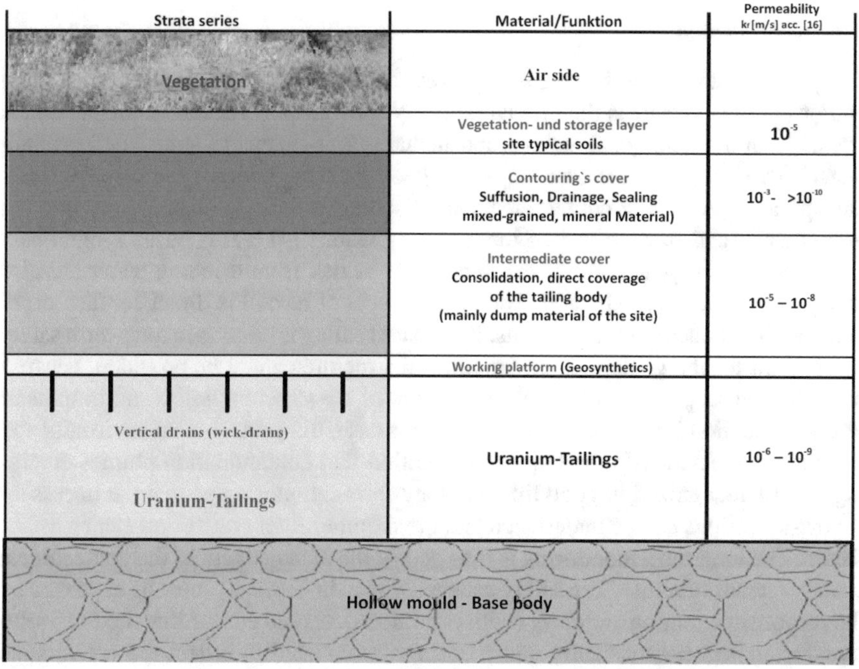

Strata series	Material/Funktion	Permeability k_f [m/s] acc. [16]
Vegetation	Air side	
	Vegetation- und storage layer site typical soils	10^{-5}
	Contouring`s cover Suffusion, Drainage, Sealing mixed-grained, mineral Material)	10^{-3} - $>10^{-10}$
	Intermediate cover Consolidation, direct coverage of the tailing body (mainly dump material of the site)	10^{-5} – 10^{-8}
	Working platform (Geosynthetics)	
Vertical drains (wick-drains) Uranium-Tailings	Uranium-Tailings	10^{-6} – 10^{-9}
Hollow mould - Base body		

Fig. 5.18 Standard cover design for a uranium mill tailings pond at SDAG Wismut according to [40]

Although the operator, Wismut GmbH, is a state-owned enterprise, permanent public participation in the construction of the geotechnical environmental structures for the long-term closure of the highly polluted uranium mill tailing ponds has not been provided for. Instead, reference is made to the state supervision by the competent mining office, which is supposed to represent public interests. This is in contrast to the perceived expectations of transparency in the disposal of radioactive waste, see in particular Chaps. 7 and 8.

5.5.3.4   Base Sealing of Containments for Uranium Mill Tailings Disposal

Internationally, abandoned open-pit mines with a clay barrier under the opencast mining floor (geological barrier—base sealing) are often used for the storage of residues from uranium ore processing. If a site without a geological barrier is chosen, the approval criteria will require the installation of a geotechnical barrier. Geosynthetic liner mats, clay and/or bentonite sealings in combination with other geosynthetic functional elements can be used as elements of a geotechnical barrier. Even asphalt concrete in various combinations has already been used as a geotechnical base sealing. The choice of final design is determined by local conditions. Ultimately, the design must be suitable for approval by the supervisory authority.

- **General design**

The purpose of the base sealing is to prevent contamination transfer into the subsoil and/or aquifers. Water is the crucial pathway for transport for contaminants and is available in the geological subsoil and includes soil-, seepage, pore-, and groundwaters. Particularly in the case of jointed geological formations, the seepage water can spread rapidly. The contaminants are discharged directly or indirectly into the groundwater and from there into the receiving waters [41]. This poses a significant hazard to the biosphere. Humans are not only at risk from drinking water but also from uptake via the food chain. The highest level of hazard is direct contact, especially with radioactive contaminants. If the base sealing is missing, a long-term safety assessment for the geological environmental structure cannot be provided, because uncontrolled seepage over the entire footprint of the structure will result in transfer of contamination into the environment. In this case, the geotechnical environmental structure is erected with the hope and intention that contamination plumes emerging will be recognized in good time via long-term monitoring so that the necessary follow-up actions can be undertaken and the required safety condition can be established. The long-term monitoring is thus part of the management of the geotechnical environmental structure. The safety concept, which in Germany must be approved by the responsible mining authority (with conditions), should ensure that the long-term monitoring will be constantly and permanently updated to reflect the actual state of the geotechnical environmental structure. The system must remain permanently under the supervision of the responsible mining authority (Fig. 5.19).

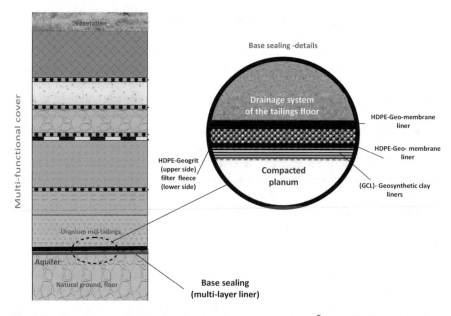

Fig. 5.19 Base Sealing, Geotechnical Barrier, Operating Plan for PIÑON RIDGE MILL [42]

The safety assessment (S) must thus be modified to the data situation and from which **S = S(t)** results.

- **Concept of permanent protection of groundwater on locations with uranium mill tailings ponds in Germany**

The six large uranium mill tailings ponds of SDAG Wismut are not only among the largest in the world, but they are also located in the very densely populated cultural landscapes of Europe and partly in a low mountain landscape. Therefore, there is a great responsibility to close them permanently and safely, and to reliably guarantee this for the public. The water pathway, especially over the contact area of the tailings body, is the critical path of the long-term safety of the Geotechnical Environmental Structure for the disposal of uranium ore processing residues at SDAG Wismut. This will be examined in more detail in the following section.

The uranium mill tailing ponds at SDAG Wismut are located either in mined out, abandoned opencast mines (Trünzig and Culmitzsch), laid in sand-/ gravel opencast (Dänkritz I) or placed as a "valley impoundment" (Helmsdorf). A base sealing has not been installed in the existing depressions or containments. Therefore, special attention must be paid to the possible spread of contamination into the subsoil (loose rock) as well as possible groundwater contamination in the closure process as well as considering safety in the long-term. The adaptation of the final operational plan in accordance with BBergG for the uranium mill tailing ponds at SDAG Wismut with regard to the arrangement of surface water (pore water) measuring points (GWMS), wells, drainage lines, etc. is an essential component guaranteeing the permanent safety, of the surface water, groundwater and soil.

The prevention of contamination transfer from the Wismut uranium mill tailings ponds into groundwater or surface water is practically impossible. This means that it must be constantly and permanently demonstrated that the contaminated seepage water (pore water) escaping from the floor of the tailings body is collected and fed into a water treatment plant so that it can then be discharged into the receiving water in compliance with the exemption limits. The extensive water catchment monitoring system consists largely of control wells and deep-drainage systems.

Here we will describe in greater detail the conditions in the contact- and transition zone—contact area of the talings body at the uranium mill tailings ponds of Trünzig and Culmitzsch; these are located on the Culmitzsch plain, see Fig. 5.12.

First of all, proof of the consistency of the related data sets is mandatory for a general assessment. The related data sets will be measured against relevant international literature.

The Lerchen- and the Pöltzschbach rivers flow through the Culmitzsch plain, where the Trünzig and Culmitzsch tailings ponds are located. The river "Weiße Elster", which can be considered the main receiving water here, and the river "Pleiße" are used to transport the mining waters from the Ronneburg mining area and thus also from the Culmitzsch plain, see Fig. 5.20.

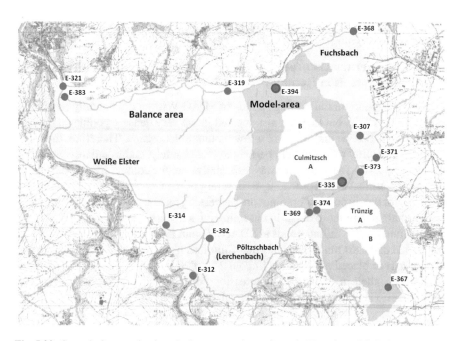

Fig. 5.20 Spread of contamination via the water pathway from the Trünzig and Culmitzsch tailings ponds—investigation area; E-xxx measuring stations of the special measuring network "Wismut"—example, not to scale [43], see also Fig. 6, p. 17 [44]

Tailings from uranium ore leaching with sulphuric acid (H_2SO_4) were flushed into the tailings ponds A of Trünzig and Culmitzsch, and tailings from uranium ore leaching with soda (Na_2CO_3) were flushed into the tailings ponds B of Trünzig and Culmitzsch. The leaching agents are found as essential components in the respective tailings waters, see Table 5.5.

Table 5.5 Tailings Ponds of SDAG Wismut—Pore water quality; according to [25]

Tailings components/ Locations	Culmitzsch A	Culmitzsch B	Trünzig A	Trünzig B	Helmsdorf	Dänkritz I
Surface of Tailings Pond (ha)	159	76	67	48	205	19
Volume total (Mio m^3)	61	24	11	6	45	5
Solid mass (Mio t)	64	27	13	6	49	7
Density (t/m^3)^a	2.20^b (1,94)	2.20 (2.64)	2.20 (2.00)	2.20 (2.00)	2.20	2.20
Pore volume (Mio m^3)^a	31.91 (28.40)	11.73 (13.40)	5.91 (6.50)	3.27 (3.20)	22.73 (n.d.)	1.82 (n.d.)
Drainable portion 50% of the total Pore water (Mio m^3)^a	15.96 (13.40)	5.87 (6.70)	2.96 (2.5)	1.64 (1.00)	11.37 (–)	0.91 (–)
Porosity (–)^a	0.52 (0.47)	0.49 (0.57)	0.51 (0.49)	0.55 (0.50)	0.51 (–)	0.36 (–)

Radioactive pollutants in pore water

U_{nat} (mg/l)	0.3–3.9	1.0–16.5	1.0–19.0	1.0–20.0	2.0–30.0	10.0–85.0
Ra-226 (mBq/l)	– 5000	– 2300	– 630	n.d.	– 2000	n.d.

Heavy metals contents in pore water (ppm)

Pb	Zn	Cu	Co	Ni	Mo	
60–800	250–800	250–300	15–40	25–500	20–70	
As	Bi	V	Cd	Cr	– ‰–	
68–168	5–30	200–800	10–30	30–580	– ‰–	

Catched leachate^c
– 9 g/l SO$_4$; 1.25 g/l Chlorides (Culmitzsch)
– 6 g/l SO$_4$; 1.6 g/l Chlorides (Trünzig)
– 5.6 g/l SO$_4$; 1.2 g/l Chlorides (Helmsdorf)

^{a}Specifications in brackets according to Schulze et al. [45]
^{b}[43] Average density for Uranium mill tailings 2.20 t/m^3
^{c}[46] Große Anfrage Deutscher Bundestag; Drucksache 12/3309; 24.09.92

For the following considerations, the consistency of the data sets underlying these considerations is of great importance. Table 5.8 is reviewed and compared with the data of [45].

The dewaterable proportion of pore water in Table 5.5 is limited to 50% according to [45, 47]. However, depending on the hydraulic gradient, it can be shown that a higher proportion may also be drainable. However, even if all pore water has leaked through the base (sole and adjacent areas of the dam footprint), the majority of the contaminants still remain in the solid tailings body and must be protected from water infiltration to avoid mobilization, see above.

The main aquifer in the Culmitzsch plain is the fissured pore aquifer of the Culmitzsch-sandstone and the fractured aquifer (Ordovician) situated beneath. The thickness of the aquifers can reach 20–30 m. They are separated from each other by low permeability clay and siltstone complexes. The waters are both confined and unconfined. In undisturbed areas the layering of groundwater has been proven. Different hydraulic potentials prove that the different major groundwater horizons are not hydraulically connected. In tectonic fault zones, however, the direct hydraulic connection was detected. For example, contamination can also reach the deeper-lying fractured aquifers (Ordovician), which makes water collection even more difficult.

The Culmitzsch-sandstone is cut steeply in the Zechstein bridge from Lerchenbach, see Figs. 5.11 and 5.19. Thus, the foot of the tailing body of the Talings ponds Trünzig and Culmitzsch may partially lie in the groundwater. The distance between the main aquifer and the tailings contact areas is—0 m (for the lower aquifers the distance is between 0 and +10 m [48], see Fig. 5.21). Especially in the case of floods, there are significant contamination escapes from the tailings ponds, which cannot then all be collected. The danger to the environment increases considerably for a short time in such circumstances.

Comprehensive investigations can be found in [47]. Winde expresses it this way in [47]: *"In doing so, the water-permeable Upper Red Series separates from the Culmitzsch-sandstone. Otherwise, it was completely mined as the highest ore horizon. The Culmitzsch- sandstone (CuS) uncovered in the Trünzig opencast mine forms the base floor of the tailings body, which the uranium mill tailings were filled in from 1960 onwards. There is thus a direct contact between the contaminated tailings waters and the main aquifer of the study area. The inclination of the CuS in direction of the Lerchenbach favors the outflow (left-sided) of tailings waters into the receiving waters* (Fig. 5.14)"

Contaminant escapes also come from the non-degradable mineralization of zinc, lead, copper, iron, arsenic, cobalt, nickel found in the Culmitzsch-deposit, see Table 5.5 and from the Culmitzscher sandstone itself. But here only the radioactive contaminants uranium and radium are considered.

In the Culmitzsch plain seepage and groundwater are collected by wells and drainage pipes and fed to the Water Treatment Plant (WBA) at Seelingstädt. WBA Seelingstädt currently has a treatment capacity for mining waters of up to 330 m^3/h, which corresponds to an annual effective treatment volume of approximately 2.3 million m^3; the service life will be over 20 years, see details published by Wismut GmbH.

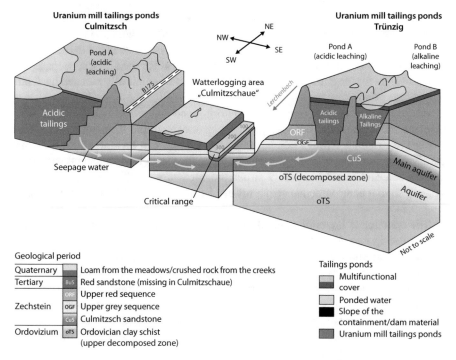

Fig. 5.21 Geological, hydrological situation in connection with leachate escapes from the tailings ponds at Culmitzsch and Trünzig into the Culmitzsch plain according to [48]

The crevasse formation of the geological base results in a variety of water pathways and thus flow paths for leachate (pore water) from the tailings ponds at Trünzig and Culmitzsch and consequent contamination spreads (contamination plumes). Reference [44] describes the situation as follows: *"The observations on the water balance show that annually between 2 and 3 million m³ of water inflow in the area of the Culmitzsch plain. In order to effectively limit the diffuse influx into the receiving waters and thus significantly reduce the pollution situation, especially in periods of low flow rates in the Culmitzsch,[18] groundwater of a similar size would have to be collected."*

The leachate leads to rising groundwater levels and to low to negative distances to the surface, so that again and again soil wetness areas can be observed in the Culmitzsch plain."

Since a long-term safety proof for the base sealing of the tailings ponds of the SDAG Wismut cannot be demonstrated, the method of using a functionality proof for each location of uranium tailings ponds has been selected. The principle of proof of function is based on a stable functioning system being in place. Here the water collection and treatment systems will be built, which fulfills both the self-monitored

---

[18]Here, the former name for the Pöltzschbach (Lerchenbach) was chosen. But the Culmitzsch or "Culmitzsch creek" is no longer a geographical term today.

and externally monitored environmental requirements. Authorization and monitoring are the responsibility of the responsible mining authority with the consent of the water authority. A time limit for the operation of the system does not exist. Initially, the design of the water collection system was planned for the surface, seepage and groundwaters for the entire site. The captured waters were initially pumped back into the tailings ponds.

At the same time, environmental monitoring and water treatment facilities were planned and built. After the system has been accepted, the long-term stable, safe closure of uranium tailings ponds, acc. Sect. 5.5.3.3, can be started. Start-up and certifying means the system guarantees compliance with environmental standards, and the relevant mining authority issues the necessary conditions to authorise the system, so that environmental standards can be met over the life time of the system (Fig. 5.21).

The operator of the tailings ponds is required to constantly adapt and update the system taking into account future conditions or repairs as well as considering wear and/or gain in knowledge and changes in technology. The permanent safety of the system, here limited by the lack of base sealing, is thus guaranteed and demonstrated by permanent monitoring and adaptation to changing conditions. Since an approved final operational plan can be supplemented and amended, so the tracking of the safekeeping of the tailings ponds is also legally secured and thus the operational period time is specified as "permanent", and is determined solely by the relevant mining authority, see also CSM (Fig. 5.22).

As a further consideration, it is important to distinguish here between leachate and pore water. Leachate is the term used to describe soil water that is in the soil pores (coarse pores) but under the influence of gravity and moving vertically downwards (seeping) soil water.

In the Culmitzsch plain leachate escapes from the tailings ponds (composed of ponded and pore water) and from the tailings dams (especially from into the tailings

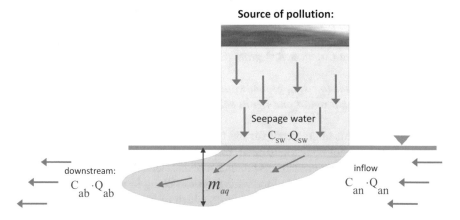

Fig. 5.22 Mass balance analysis during the transition from contaminated leachate GW (principle)

dams integrated low grade rock dumps). There are 3 low grade rock dump complexes in the study area. These are the area of the tailings ponds Culmitzsch, the Waldhalde (NE) and the Lokhalde (NW) and in the area of the Tailings ponds Trünzig, the Nordhalde (SW), see Fig. 5.20. Their base floors are also unsealed.

Since on the one hand the ponded water from the tailings ponds is largely pumped out and the surface sealings are applied, and on the other hand, the embankments are largely profiled and covered, the leachate consists largely of pore water from the tailings body and leachate on the tailings dam contact surfaces together. Since the pore water from the tailings bodies predominates in the Culmitzsch plain, often pore water and leachate will be seen as equal. However, this leads to a considerable underestimation of the size of the contamination load. In a water-saturated porous medium, the entire pore space is filled with liquid. The pore water amount corresponds to the pore volume.

Since the uranium mill talings ponds at Trünzig and Culmitzsch were created without effective sealing at the base and slope area of the tailings dam, the tailings bodies sit directly on the floor of the opencast pit. The leachate, fed here from the pore water of the tailings ponds, passes directly into the geological base below or into the surface water, see Figs. 5.20 and 5.21. Below is shown, the period of time the system of water collection and treatment should be operated when there is no base sealing. **The transportation of contaminants via the rivers Weiße Elster and Pleiße, the pollutants in sediments, dilution and contaminant adsorption of plants and animals etc. are not commented upon here. These items need to be the subject of a long-term safety assessment.**

First of all, the balance Eq. (5.1) applies for seepage to groundwater or surface water with respect to wells and balance measuring points.

$$Q_{Ab} = Q_{An} + Q_{SW} \tag{5.1}$$

Q_{SW}—amount of leachate; Q_{Ab}—a downstream GW; Q_{An}—incoming GW
with
Q—volume flow ($Q = k_f \cdot A \cdot I$) of incoming, of downstream or of seepage water;
k_f—coefficient of permeability;
A—incoming cross section ($A = b_a \cdot m_{aq}$)
I—hydraulic gradient
m_{aq}—contaminated Aquifer thickness;
b_a—contaminated downstream width;
c—concentration of a substance in incoming, downstream or of seepage water.

For the outgoing flow expected from the tailings ponds pollutant inputs, as a true volume exits from the tailings body as leachate (arising from the pore water) into the groundwater or surface water, taking into account concentrations and loads this transition can be described acc. Eq. (5.2), see Fig. 5.23 and [49].

$$c_{mix} = c_{SW} \cdot \frac{1}{1 + \frac{1m \cdot v_{GW}}{l_{mix} \cdot v_{SW}}} + c_{GW} \cdot \frac{1}{1 + \frac{l_{mix} \cdot v_{SW}}{1m \cdot v_{GW}}} \tag{5.2}$$

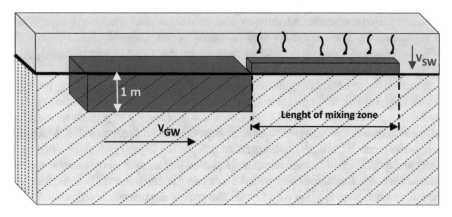

Fig. 5.23 Model of the transition of contaminated seepage water into the ground water (GW), acc. /23/

for $c_{mix} \leq c_{mix}\left(c_{SW_{Sätt.}}\right)$

c_{mix}—concentration in the admixture zone [g/l]

$c_{mix}\left(c_{SW_{Sätt.}}\right)$—concentration in the admixture zone at the saturated seepage water

c_{SW}—concentration in seepage water [g/l]

c_{GW}—concentration in groundwater [g/l]

v_{GW}—ground-water flow velocity [m/s]

v_{SW}—seepage water rate [m/s]

l_{mix}—extension of admixture zone in flow direction

1 m—top layer (1 m) of passing ground water (admixture layer)

Reference [44] records the outflow of ground water as (2–3) million m^3 per annum. In Table 5.6, acc. [44], the following trends can be shown.

Table 5.7 shows that pollutant inputs into groundwater or surface water and thus pollutant escapes from the tailing ponds in the Culmitzsch plain are not reduced until 2027. Note also the logarithmic course of time at measuring points E-335 and E-394, in Fig. 5.24. Some major transgressions of the environmental quality standard will

Table 5.6 Compliance with the Environmental Quality Standard (UQN) on an annual average (JD) and forecast until 2027 in the Culmitzsch plain—Excerpt, measuring points see Fig. 5.20

Parameter	UQN	Validity	Lerchenbach/Pöltzschbach measuring point E-382/result 2012	Lerchenbach/Pöltzschbach balance measuring points E-319 and E-369/prognosis 2027
Tl	0.2 μg/l	JD	0.2 μg/l	0.5 μg/l
U$_{unfiltriert}$	20 μg/l[a]	JD	131 μg/l	300 μg/l
SO$_4$	450 mg/l	JD	2460 mg/l	3000 mg/l

[a]Water-specific proposal of the states of Saxony and Thuringia, no UQN

Table 5.7 Determined exhaustion times from drainable pore water according to Table 5.5

Tailings pond/pore water escapes through the base	Culmitzsch A	Culmitzsch B	Trünzig A	Trünzig B
Pore water part[a] total in 10^6 m^3	31.91 (28.40)	11.73 (13.40)	5.91 (6.50)	3.27 (3.20)
Dewatering part[a] 50% of the total pore water in 10^6 m^3	15.96 (13.40)	5.87 (6.70)	2.96 (2.5)	1.64 (1.00)
Annual pore water contribution in m^3	~23.850	~11.400	~10.050	~7.200
Depletion period for pore water (years)	~670	~515	~300	~230

[a]Drainable portion and figures in brackets according to [45]

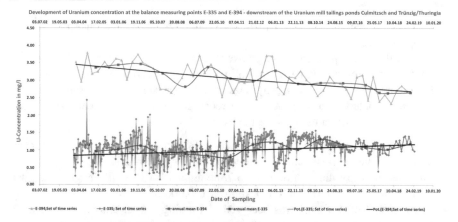

Fig. 5.24 Development of Uranium concentration at the balance measuring points E-335 and E-394—Culmitzsch plain downstream of the Uranium mill tailings ponds Culmitzsch and Trünzig; measured data from the Thuringian State Office for Environment, Minung and Nature protection (TLUBN), see also [44] and Fig. 5.20

probably have to be accepted, even though the overall quality of the environment at the site has improved significantly.

Figure 5.24 shows the development of uranium concentration at the balance measuring points E-335 and E-394 (water catchment Culmitzsch plain) for the mining-based waters downstream of the tailings ponds collected in the period 2004–2017. This can be scaled up to the entire water catchment in the Culmitzsch plain, where the level of uranium concentration will not decrease, at least until 2027.

For example [45] indicates an annual leachate volume (mainly due to porewater escapes) of 40 l/m^2 in the Culmitzsch plain. From recalculations and back data with Eq. (5.2) from [19, 44] and the information published by Wismut GmbH regarding the water treatment plants, an average value for the time period of pore water discharge across the tailings body contact area of approx. 15 l/m^2 can be determined. The value is very pessimistic. Using these data, the following depletion times of the chargeable pore water amount can be determined for the tailings ponds Culmitzsch and Trünzig.

In 2011, the Regional Water Authorities issued a guideline value of 10 µg/l as a threshold value for uranium. In the former uranium mining regions of Saxony and Thuringia, water quality standards for uranium were agreed with the water authorities, which were generally between 10 and 20 µg/l. Now the Federal Environment Agency (U-238) has submitted a proposal for an Environmental Quality Standard (UQN)—target value of 3 µg/L for Uranium (U-238) in surface waters [50].

All standards are considerably exceeded in the Culmitzsch plain and in some flowing waters, see Table 5.6, Fig. 5.24 and [44].

The depletion times for pore water escapes via the base of uranium mill tailings ponds in the Culmitzsch plain given in Table 5.7 are to be expected with the following basic assumptions:

– Dewatering share is 50% of the total, pore water; recent calculations show a share of up to 75%;
– Mean value of the porewater escape over the tailings body contact area of approx. 15 l/m^2/a remains constant over the entire period of observation; the slight concentration deviations. Reference [45] in the balance measuring points, see Table 5.6, can be attributed to lower groundwater recharge rates in recent years, and larger amounts of porewater escapes via the base.

It is obvious that the depletion times given in Table 5.7 are more likely to be longer than shorter. The times in Table 5.7 are therefore very optimistic

In order to verify the data in Table 5.7, an annual leachate discharge of at least 40 l/m^2, contributing at minimum of approx. 150,000 m^3 with an average uranium concentration[19] of 4.0 mg/l was accepted. Assuming a total annual volume of 2 million m^3 of mining-based waters as inflow to the receiving water a uranium concentration of 320 µg/l can be determined.

This value is compatible with the specification forecast from [44] in Table 5.5 (2027).

In addition to the geochemical concentration gradient, the steep hydraulic seepage water gradient in the tailings ponds Culmitzsch and Trünzig also contributes to the leachate supply and pollutant concentration remaining at the present level.

It is therefore very unlikely that, the water collection and treatment programme (water treatment plant—WBA Seelingstädt) for the Culmitzsch plain can be abandoned in this century. It is more realistic to expect that the water collection and treatment for the Culmitzschaue will need to remain in operation for an even longer period together with the environmental monitoring. In any case, the operating life of the current 20-year WBA Seelingstädt will be exceeded, so replacement investments should be planned for the future. The task of water collection and treatment considering the information from Table 5.7 may be described as a long term task,

---

[19]The weighted mean uranium concentration Unat in the pore water of the uranium mill tailings ponds of the SDAG Wismut in the Culmitzsch plain in Table 5.5 is 4.98 mg/l.

the operation of which will extend considerably beyond the original official time forecasts.

Even if all the dewaterable pore water volume had escaped from the uranium mill tailings ponds of the SDAG Wismut, the majority of the contaminant inventory still remains in the geotechnical environmental structure (see Table 5.8) and must be permanently prevented from loss to the biosphere. This makes it clear that even after the depletion period for the dewaterable pore water volume, the safety of the geotechnical environmental structure must be preserved. This applies in particular to the multi-functional cover of the tailings body, which is intended to prevent surface waters from infiltrating into the tailings body and mobilizing the tailings' radionuclides. Experience with the long-term behaviour of the cover systems described above arise not yet available. It is advisable to define a probability of failure for the cover system and to continuously update it as the data situation improves. This includes the monitoring of radon exhalations.

Determination of any additional measures required should be undertaken using data obtained from the long-term environmental monitoring. The weak points remain the base sealing, the bottom slope areas of the dam structure of the tailings ponds and in the case of the deposit of Culmitzsch, the Culmitzsch- sandstone (ore horizons OGF and UGF), see Fig. 5.21, which still contains significant amounts of uranium. The average uranium content in the crude ore was between 0.059 and 0.068%. Wilde [47] quotes an average of 660 ppm. In 1991, the SDAG Wismut identified that resources of 3350 tons of uranium remained in the Culmitzsch uranium deposit. Part of this resource is soluble. As a result, uranium continues to be discharged from the remaining ore horizons due to water ingress. For the same reason, in the medium term it is not expected that a significant decrease in uranium concentration in the collected water will be seen. For comparison, the uranium mill tailings pond Moab/Utah, USA also belonged to a sandstone-bound, tabular uranium deposit, named as the "Atlas Mill Uranium deposit".

To mobilize contaminant escapes from the partially dewatered tailings body, a transport medium is also required, usually water or gas. Both are available in principle; however, mobilization also requires a source of energy. A mobilization option, which has access to the appropriate energy, is the inflowing groundwater. This is true especially when the foot of the tailings body is standing in water. To investigate the spread of contamination, especially through the dam floor, mass transport models are generally used. Independent assessments can be made according to [51] using computer programs such as HELP and BOWAHLD 2D as well as the 3-D groundwater model SPRING. Models often fail to consider that the tailings body may cause significant deformation of the dam floor (tailings body contact area) and in the side slopes of the containment. As long as the tailings are water-saturated, this fact can be disregarded. When all of the pore water has leaked, the entire solid tailings body, including the cover, acts on the tailings dam floor. It can form a settlement depression, its maximum depth may be several meters depending on the existing thickness of tailings body. With the FEM program FEMPLER© [52] the settlement under the waste body of a heap landfill was calculated; the landfill was constructed up to 30 m

Table 5.8 Radionuclide inventories of SDAG Wismut Uranium mill tailings ponds—Extract [25]

Locations—standardized to the exemption limits/Radio nuclide inventories	FGi [Bq]	Culmitzsch A		Culmitzsch B	Trünzig A	Trünzig B	Helmsdorf	Dänkritz I
		Ai [Bq]	Ai/FGi	Ai/FGi	Ai/FGi	Ai/FGi	Ai/FGi	Ai/FGi
U-238+	1.00E+04	1.22E+14	1.22E+10	5.60E+09	3.80E+09	1.80E+09	1.27E+10	2.50E+09
U-234	1.00E+04	1.22E+14	1.22E+10	5.60E+09	3.80E+09	1.80E+09	1.27E+10	2.50E+09
Th-230	1.00E+04	7.90E+14	7.90E+10	2.40E+10	1.30E+10	5.00E+09	5.50E+10	4.00E+09
Ra-226++	1.00E+04	7.90E+14	7.90E+10	2.40E+10	1.30E+10	5.00E+09	5.50E+10	4.00E+09
U-235+	1.00E+04	5.56E+12	5.56E+08	2.55E+08	1.73E+08	8.21E+07	5.79E+08	1.14E+08
Pa-231	1.00E+03	3.60E+13	3.60E+10	1.09E+10	5.93E+09	2.28E+09	2.51E+10	1.82E+09
Ac-227+	1.00E+03	3.60E+13	3.60E+10	1.09E+10	5.93E+09	2.28E+09	2.51E+10	1.82E+09
Ra-223+	1.00E+05	3.60E+13	3.60E+08	1.09E+08	5.93E+07	2.28E+07	2.51E+08	1.82E+07
Summe		1.94E+15	2.55E+11	8.15E+10	4.57E+10	1.83E+10	1.86E+11	1.68E+10

The Th-232 series plays no role with regard to radioactive inventories

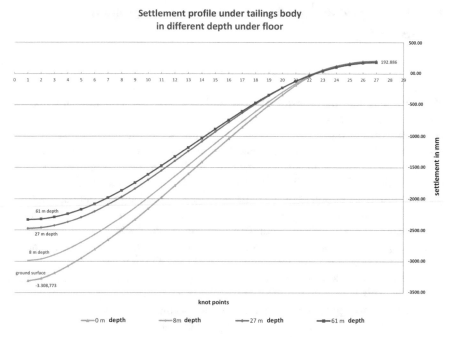

Fig. 5.25 Settlement results under the base of a heap landfill as basis for comparison [52] FEM-PLER © (2015) [Waste body: 2.8×10^6 m³; pile area ppr. 25 ha; Heap hight: 30 m; Base of landfill, length: 500 m; average density of waste 1.5 t/m³; underground: former brown coal opencast, Calculation model: plane strain state, transversal isotropic]

in height and covered, see Fig. 5.25. The largely dry uranium mill tailing ponds of Wismut GmbH behave similarly.

If the floor of talings ponds Trünzig and Culmitzsch were also lowered by several meters, the entire footprint of the respective tailing body would stand in water. Pollutant transport from the tailings body into the groundwater becomes very likely in this case.

In addition to mass transport models, tracer experiments with actively introduced tracers and environmental tracers [for example, isotopes and CFCs (fluorinated/chlorinated hydrocarbons)] are very suitable for determining the flow paths. Due to the deformation of the base and the sloped areas, weak zones are also to be expected through which pollutants can increasingly escape into the groundwater. The phenomenon of capillary action is neglected here, but this would reinforce the process.

The examples from the Culmitzsch plain used here are also transferable to other suspected contaminated sites in the inventory of Wismut GmbH. The merit of Wilde [47] is to show the transport mechanisms of uranium in mining-based disturbed landscapes of different climates. He has thus made a contribution to the site specific long-term storage and disposal of radioactive residues, in particular uranium tailings.

5.6 Definition of Long-Term Safety and Long-Term Stability

The Geotechnical Environmental Structure shall be designed to safely contain the residues from ore processing and ensure they are safely isolated and encapsulated in such a way that the transfer of toxic and/or radioactive contaminants into the biosphere on the air, water and soil pathways is prevented or can only occur within acceptable release rates. The construction must be so robust that it fulfills its functions even under changed environmental conditions. The proof is divided into two parts, which, however, are inseparable from each other the first requirement is to prove the mechanical stability of the geotechnical environmental structure and secondly to prove the tightness of the barriers so that no discharge of radioactive-toxic contaminants into the environment can take place. The structure should also ensure the prevention of pollutant transport (gaseous and/or liquid) through the barriers.

Although it is certain that a long-term safety certificate based on the above-mentioned principle for the uranium tailings of the SDAG Wismut cannot be presented due to the lack of base sealing, the basic features of the required proof are set out below. This is also due to the fact that the responsible mining authorities have approved the storage concepts for the uranium tailings at SDAG Wismut, subject to conditions. This can also be justified by the construction of efficient water collection and treatment systems as well as an associated, approved, water monitoring programme (special measuring network for groundwater and surface water). However, there are limitations to the efficacy of the works and (short-term) limit value exceedances are not entirely avoidable and remain as a long term risk to be managed.

- **Differences with the final disposal of radioactive waste with negligible heat generation:**

In the requirements formulated above, the geotechnical environmental structure—tailings pond—does not differ from the disposal of radioactive waste with negligible heat generation. Whereas dry storage in situ has been approved in Germany for the long-term safekeeping of tailings storage facilities from uranium ore processing, the German waste management concept for radioactive waste with negligible heat generation stipulates that a dry in situ storage is to be carried out in deep geological formations. However, repository structures with an incomplete barrier system were not be able to be approved for these radioactive wastes, see Chap. 7. These serious differences in the requirements for geotechnical environmental structures cannot be justified by the respective radioactive inventories. In any case, the hazard that results from dealing with residues from uranium ore preparation may not be adequately represented to the public. A look at the statistics of the responsible professional association RCI,[20] would make it clear what high radiological risks can be expected,[21] see also Table 5.8. Therein the radiotoxicity levels are presented, defined by the

---

[20]professional association of raw materials and chemical industry (RCI).

[21]Commercial employers' liability insurance associations (professional associations) are the statutory accident insurance providers for companies in the German private sector and their employees.

dimensionless quotient Ai/Fi of the radionuclides contained in the inventory of the tailings bodies of the individual tailings ponds as well as the total radiotoxicity.

Shaft Konrad in Salzgitter-Bleckenstedt is currently being upgraded for use as a repository for radioactive waste with negligible heat generation, see Chap. 6. The radioactive waste will be stored at a depth of approximately 1300 m. The design was made for a period of observation of 10^6 years. The consequence of this extremely long observation period is that today's generation cannot test the results of their investigations and that subsequent generations will have to "endure" them. Since all statements on this design can be justified only vaguely and thus with little certainty, it becomes clear that the requirement of this apparently high level of security is derived from a widespread uncertainty. These risks are not suspected in the approval procedure according to BBergG.

What applies equally to all geotechnical environmental structures for the long-term safe storage of radioactive waste and residues is the fact that they are constructed as a multi-barrier system consisting of geological and geotechnical barriers.

For tailings, the long-term safety case must cover the dams and their embankments as well as the cover. For the tailings dams and their embankments, these verifications can be made in Germany in accordance with DIN 19,700 part 10+ 15; DIN 4084; ICOLD Bulletins and other valid regulations (Radiation Protection, VOAS, HaldAO). The following items are required for the provision of a long-term safety assessment:

- Static long-term safety proof (BISHOP, JANBU, proof for 1.000 a)
- Dynamic long-term safety proof in case of operational earthquake (OBE,[22] proof for 500 a)
- Dynamic long-term safety proof for safe shutdown earthquake (MCE,[23] proof for 2500 a)
- Hydraulic safety, 1.000 a

  - Hydraulic failure
  - Safety against internal erosion (erosion channel)

- Long-term safety against external erosion, is recorded in the long-term stability certificates
- Avoidance of mass transport through the barriers (surface sealing, base sealing, dam construction) into the biosphere—safety concepts.

---

[22]Operating Base Earthquake (OBE): is used to demonstrate the serviceability and durability of the tailings dam. The tailings dam must withstand the OBE without restrictions on use (earthquake case 1).

[23]Maximum Considered Earthquake (MCE): design earthquake is the design case for which the safety of tailings dam must be proven. 10,000-year event (return period); see ICOLD Bulletin 82. In the newer version of DIN 19,700 (new), a return period of 2475 a is specified for the safe shutdown earthquake (SSE)—(design earthquake).

5.6.1   Mechanical Material Parameters, State Variables

The state variables S_i, with which the soil-mechanical state of both the dam structure and a multi-functional covering can be described sufficiently precisely, with the following functional relationship according to [53]:

$$S_i = S(x, y, z, t, \chi, \theta) \tag{5.3}$$

– x, y, z—Local coordinates; t—Zeit; χ—Status parameter; θ—Correlation parameters

The material behavior of a multi-functional cover under different scenarios can be explained by a functional link [53], of the different measured quantities in the form

$$F(\dot{\varepsilon}, \varepsilon, \sigma, T, S) = 0 \tag{5.4}$$

– F (…)—functional relationship between the measured quantities; ε—deformation;
– $\dot{\varepsilon}$—deformation rate; σ—stress; T—absolute temperature;
– S—status parameter

be described.

For the state variable S_i determined at the location x, y, z, it seems sensible to predefine the following dependency:

$$S_i = f(w, e_0, \dot{\sigma}_1, t) \tag{5.5}$$

– w—water content; e_0—start void ratio (stress free);
– $\dot{\sigma}_1$—effective largest main stress.

5.6.2   Long-Term Safety and Stability of Multi-functional Covers (ENW2)

The task of a multi-functional cover is the safe, long-term stable storage of the radioactive and toxic material with the following partial aspects:

– Reduction of radioactive radiation and radon exhalation in accordance with the exemption limits,
– Prevention of infiltration of surface waters and protection against heavy rain (storm water),
– Avoidance of oxygen diffusion to limit the formation of acidity,
– avoidance of the escape of contaminated waters,
– possibility of water collection and water removal,

– Landscaping and revegetation.

The proof and the successful control of the long-term stability of a multi-functional cover should be extended to:

– internal and external erosion (long-term stable: geomechanical/ geochemical)
– aging (formation of escape pathways, cracking, loss of function)
– Radiation Protection, Emission Control (Hydrological, Gas and Mineral Transport)
– consolidation and settlements (formation of pathways, cracking, loss of function)
– vegetation and contaminant adsorption.

Here, too, there is considerable need for research, in particular on the influence of aging of the material on the long-term stability and which parameters determine or dominate the aging of various cover materials (mineral substances, geosynthetics, etc.), see BMU [30].

5.6.2.1 Radiation Protection Function of the Multi-functional Cover

The radiation protection function of a cover results from the interruption or reduction of escape pathways (air path, water path, but also the carry-over of solids/pathway of minerals). If the cover loses its protective function, the function of the construction as a whole is disturbed. In doing so, the radiation protection function can be identified by threshold values to be observed or standard dose limit values. While in the USA (EPA), as already mentioned, a limit value for the radon emission rate ≤ 20 pCi m^{-2} s$^{-1} = 0.74$ Bq m^{-2}s^{-1} is prescribed, i. At Wismut GmbH a limit for external radiation exposure (ground radiation) of ≤ 0.15 µSv/h has been used with reference to earlier SSK recommendations (publications of SSK vol. 24), [54].

The proposal to demonstrate the long-term stability of a multi-functional cover for a period of 1000 years comes from the UMTRA programme. It is supported by a sufficient number of studies. In the regulations of the US EPA (EPA: 40 CFR 192) and the US Nuclear Regulatory Commission (NRC: 10 CFR40), this period is binding., However, there is an express note that in justified exceptional cases, fr the long-term safety proof the minimum period can be reduced to 200 years. The long-term safety proof of the multi-functional cover of the Wismut-Tailings Ponds requires that the operator will be responsible for a period of 1000 years, This has been approved by the mining authority and supported by the IAEA. It also requires evidence of retention of the long-lived α emitter Ra-226, the main supplier of radon, Rn-222 (Th-232 can be neglected here). The inventories of Ra-226 in Wismut tailings ponds are listed in Table 5.8. The long-term safety of tailings ponds from uranium ore processing always includes long-term environmental monitoring, as the geotechnical environmental structure is not constructed to be maintenance-free and precautions must be taken to prevent inadmissible escapes of contamination in the period under consideration. The period over which long-term monitoring is to be carried out is currently not defined as binding but will be decided upon following completion of the remediation works at Wismut GmbH.

The installation of a long-term monitoring system with integrated environmental monitoring should generally also be included in the repository construction, see Chap. 9.

For Trünzig, Pond A, the inventory of the isotope Ra-226 is given as 1.3×10^{14} Bq $= 130$ TBq. In addition, up to 630 mBq/l have been determined in pore water. (See also Tables 5.8 and 5.9). The dissolved activity is not considered in the total activity due to the small order of magnitude and also because of the seepage, Ra-226 is discharged from the inventory via the footprint of the tailings body, see Sect. 5.5.3.4. The leachate is collected and fed to a water treatment plant, cleaned and discharged to the receiving water in compliance with the authorized and approved exemption limits. Part of the seepage flows uncontrolled into the environment. The discharged load, in this case the radio-toxic contaminant, must be constantly re-evaluated using mass transport models and accompanying environmental monitoring [51]. Radium escapes by leachate drainage do not significantly reduce the Ra-226 inventory in the tailings body, because Ra-226 is constantly reproduced through the decay of U-238 in the tailings body (half-life of 4.468×10^9). This is countered by the immobilizing effect of secondary mineralization of the tailings (mineral crustal formation) [55]. In this dynamic process, a secular equilibrium is finally formed in which the amount of available Ra-226 remains constant and thus also the radon formation and thus the cause of radon exhalation. It is not expected that the activity of the main radon source in the inventory will be reduced significantly after 1000 years. However, this cannot be equated with radon emissions.

5.6.2.2 Consolidation and Settlement Behaviour of the Multi-functional Cover

The settlement following the consolidation of the sludge deposits in the tailing ponds can be considerable and must be permanently monitored. They may not stop even after placement of the multi-functional cover. It is particularly noteworthy that developing pathways (e.g. for radon), which can also be caused by the differential settlement behaviour of the cover layers, currently cannot be detected exactly. Continuous settlement monitoring is an indispensible part of the accompanying environmental monitoring. With increasing radon exhalation rates, however, a corresponding formation of pathways must be assumed. Aftercare is therefore difficult. There is a considerable need for research in this area. One solution could be to install suitable sensors in the cover in order to obtain better information about these processes, see [35].

When modelling soil and rock mechanical tasks with the help of FEM, two non-linearities have to be considered: the geometrical non-linearity (especially connected with the investigation of large deformations) and the physical non-linearity (connected with the modelling of material behaviour). However, the existing concepts for the long-term stability of multi-functional covers assume small deformations, i.e.

geometric linearity. Thus, the principle of linear superposition applies. The individually determined settlement components can thus be summarized:

$$s_g = s_e + s_l + s_{gw} \qquad (5.6)$$

s_g—total settlement
s_e—settlement of tailings body by own weight
s_l—settlement of tailings body by load
s_{gw}—settlement part of tailings body results by dewatering (groundwater resurgence)

Nonlinearity

1. Based on physical non-linearity in soil and rock mechanics tasks, it is essential to evaluate inhomogeneity and anisotropy with regard to the material properties of the tailings body. Correlation analysis, which uncovers linear correlations between different influencing variables, is well suited for this purpose. The statistical behaviour of the stated variables, Eq. (5.4), is represented with the help of the determined correlation coefficients of the sampled tailings body. For this purpose, the tailings body is covered with a sampling raster and depth-dependent stated variables at time t are measured at the nodes. According to Eq. (5.3), from the test results of the stated variable S_i at the different depth levels, we can determine the relief of distribution of the status level, see Fig. 5.26, see also [56]. However, anomalies cannot be recorded using statistical methods. These can be detected only with geophysical methods.

If stated variables are measured over a longer period of time or sufficiently closely along a longer profile, a trend in the data can be determined that is of great importance for long-term safety analyses.

A frequently used field test is the Cone Penetration Test (CPT), which is standardized in Germany according to DIN 4094. However, the procedure must be calibrated for the respective test field. The empirical relationship (not dimensionally accurate) is suitable as a calibration relationship for determining a constrained modulus:

$$E_S = a \cdot q_C + b \qquad (5.7)$$

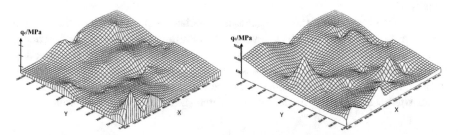

Fig. 5.26 left: 3-D visualization of the peak pressure q_C from regression analysis, depth 5 m, **right**: 3-D visualization of the peak pressure q_C from regression analysis, depth 30 m, see [29]

Table 5.9 Results of regression analysis of CPTs of a landfill site [29]

CPT-No.	2–13	24	24a	2–6	2–9	29a	31a	35
a/MN/m^2	−1.01129	2.17634	−7.87406	0.223683	3.43668	−0.959411	2.21298	−2.70202
b/MN/m^3	0.265345	0.161439	0.590386	0.244968	0.175812	0.213584	0.122971	0.230851
Correlation coefficients	0.955095	0.689819	0.849254	0.885242	0.949680	0.932856	0.765281	0.706560

E_s—constrained modulus in [MN/m^2]; q_C—cone pressure in [MPa].

Table 5.9 shows the results of regression analyses of cone penetration tests (CPTs) of a landfill site. The statistical analysis of the field tests is summarized in Fig. 5.26.

The procedure can also be used for the examination of tailings bodies.

Although the above compilation of physical phenomena appears initially to be very clear, it is difficult to adequately map them in a complex geotechnical model, due to the complex physical and chemical properties and consistency of the tailings. Long term monitoring over periods of time for which the forecasts (trends) of the long-term behaviour of the tailings bodies can be predicted are not yet known. Such monitoring goes beyond what can reliably be predicted. Thus, the results from one-dimensional consolidation models or the deformations determined with FEM models must be evaluated very critically with regard to their significance, especially if the phenomena described above are not taken into account. The necessity to monitor the settlement behaviour of a multi-functional cover in situ in the long run thus becomes clear. The models used in the settlement prognosis could in turn be validated from extensive measurement series used to determine the location-dependent deformations.

1D-Consolidation: Constitutive model considering nonlinear finite deformations in the Lagrange coordinates (a, t), according to SCHIFFMAN (1980), [57], Eq.(5.8)

$$\frac{\partial}{\partial a}\left[\frac{(1+e_0)k(e)}{\gamma_w(1+e)a_v(e)}\frac{\partial}{\partial a}\right] \pm f(e)\frac{\partial}{\partial a} = \frac{1}{1+e_0}\frac{\partial e}{\partial t} \tag{5.8}$$

with: $f(e) = -\left(\frac{\gamma_s}{\gamma_w}-1\right)\frac{d}{de}\left[\frac{k(e)}{1+e}\right].$

e_0—start void ratio; e—void ratio; k—permeability; γ_w—water density; γ_s—solid density.

The consolidation and the considerable associated deformations of the tailings body are accelerated by various measures taken during closure works, as described, or anticipated by loading(intermediate covering) and drainage[24] of the tailings body (vertical drains; this will not be discussed further here. Oedometer tests are usually used to determine k(e) in Eq. (5.8). It can be shown that for the respective tailings ponds site-specific results are determined which show considerable differences.

---

[24]The radioactive water from the drainage of the tailings body is collected, and decontaminated in a water treatment plant so that it can be discharged into the receiving water.

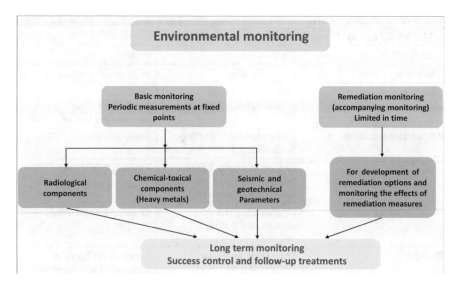

Fig. 5.27 Environmental monitoring

Possible effects on the multi-functional cover and on the settlement behaviour of the tailings bodies, and thus on the long-term behaviour of the geotechnical environmental structure, are shown here qualitatively, see also Fig. 5.27.

5.6.3 Long-Term Safety and Stability of Base Sealing (ENW3)

In Germany, base sealings must be installed according to the state of the art or selected as suitable geological barriers. For landfill construction, these are described in a uniform federal quality standard and in a suitability assessment based on this [58]; these requirements must be met when installing the suitability-tested building materials, sealing components and sealing systems so that they comply with the state of the art when installed. These performance criteria and evidence of suitability could also be used for uranium tailings ponds. These apply to all kinds sealings (Table 5.10).

Since base sealings have been not installed in the containments provided for the deposition of uranium tailings coming from the SDAG Wismut, appropriate countermeasures must also be taken to counteract the possible spread of contamination in the subsoil (loose rock body) and into the groundwater, and also in the storage process with regard to long-term safety. This does not correspond to proof of long-term safety and stability. When considering over what time period these countermeasures have to work, it must be assumed that this must be done permanently, see also Table 5.7.

Table 5.10 Performance and proofs of suitability assessment of seals made of mineral components [58] LAGA Ad hoc-AG "Deponietechnik" (LAGA)—German laws

Criteria/effects	Performance	Proofs[a]
Thickness	At least 0.50 m	In scope of construction work
Tightness	k-value-determination according to DIN 18 130	Coefficient of permeability k: $k \leq 5 \times 10^{-10}$ m/s at a pressure gradient of $i = 30$
Mechanical resistance	Permanently stable at slope inclination 1:3	Shear box test
	Deformable up to 200 m radius of curvature without increasing permeability	Flexural tensile test
	Hydraulically resistant (erosion and suffosion resistant)	Grain-size distributions
Stability	Long-term durability: (≥ 1000 years)	Based on the DepV in accordance with relevant GDA[b] recommendations for a site-specific solution
	Durability of the components influencing the stability	
	Resistant against aggressive Seepage water as a function of pH value	
Manufacturability	The construction must be possible and reproducible under construction site conditions with certainty	Test field
Other criteria (Note: Compliance is not always possible and necessary, additional measures and elements may be necessary, e.g. temporary frost protection)	System compatible	Test field Shearing tests
	Frostproof, until sufficient cover is achieved	Frost/thaw cycles, protective measures
	Environmentally compatible	Compliance with legal requirements; in case of using of substitute construction materials for landfills, it could be resorted on requirements, in accordance Part 3 DepV

[a]Which proofs are necessary, are explained in the following points
[b]GDA means: Recommendations for geotechnical engineering of landfills and contaminated sites, these are recommendations of the German Geotechnical Society (DGGT e.V.)

- **Countermeasures against contaminant escape from the tailings body/ mass transfer models**

Contaminants may migrate from the tailings body into the ground, and possibly into the aquifer, if the base can be penetrated. On the one hand, the design of the geotechnical structure makes it necessary to prevent the spread of contamination; on the other hand, forecasts have to be made as to which spreads of contamination may occur and which countermeasures can be taken. This may include construction of underground galleries and associated collection wells. In addition there should be a corresponding program of groundwater monitoring points and groundwater production wells which may collect groundwater and pump it to a water treatment plant or, if necessary, to be able to undertake large-scale dewatering and lowering of groundwater. Also mass transport models are often used as prognosis tools.

Mass transport models are used in 2D and 3D variants to simulate the distribution of dissolved (pollutant) substances on the basis of a hydraulic model. For this, the processes of convection, diffusion, dispersion, sorption, chemical reaction and degradation are taken into account, depending on the complexity of the problem and the data available. Mass transfer models can also be used to simulate historical pollutant discharges as well as possible remediation scenarios including the monitoring of Natural Pollution Retention (MNA). For the numerical simulation of mass transport processes in aquifers, a whole range of model systems are available. In conjunction with pre- and post processing modules, the mass transport, the ongoing hydrogeochemical processes, the development of the substance concentrations in the groundwater, etc. can all be visualized and compared with the determined measurement results further information may be found in Chaps. 6, 7 and 8.

Replica: Due to the lack of base sealing in the uranium tailings ponds at SDAG Wismut, long-term safety of the closed ponds cannot be demonstrated since the individual certificates (ENW3 and 4) can not be presented. It can be shown that over long periods of time (permanently), pore water can reach the groundwater through the base floor and in order to guarantee public safety must therefore be collected and treated over an extremely long period of time. This is currently being done to the extent necessary. However, it is very unlikely that it will be possible to shut down water catchment and treatment before the end of this century because the concentrations in the water catchment with radio-toxic elements will not allow this. A comprehensive monitoring system, in particular a special measuring network for groundwater and surface water, with a corresponding reporting regime, at Wismut GmbH is indispensable over this period and part of the assurance of the long-term safety of the uranium tailings stored.

As well as the possible escape of pore water through the floor of the tailings body there remains a need to store the radio-toxic inventory contained in the tailings mass. After the depletion of the pore water, a long-term safe surface seal should prevent this residual inventory from being mobilized. A risk of mobilization remains due to the inflow of groundwater. The foot of the tailings body must be prevented from becoming saturated. By forming a settlement mound of the now "dry" tailings body, the distance to the upper, free groundwater level may become so low that groundwater flows into the tailings body. This must be monitored and inflow prevented. In extreme cases, to avert damage it may become necessary to lower the ground water level by pumping. The long-term safety assessment for the multifunctional surface seal need not include this.

5.6.4 Monitoring

- **Environmental monitoring, part of the closure of tailings ponds of uranium ore processing**

For long-term monitoring, only a few suggestions are made in this work. The task of the surrounding monitoring, as part of the environmental monitoring, see Fig. 5.28, is to measure the effects of the individual objects themselves and the remediation measures carried out on the protected soil, air and water.

This applies not only to the period before and after the remediation, but also during the remediation works. A distinction is made between basic and remediation monitoring. In basic monitoring, the monitoring tasks necessary for successful control are summarized; these are carried out regularly at fixed measuring points and according to defined methods independent of the remediation works. In addition, the escapes and discharges of pollutants from the individual sites are measured. Basic

Fig. 5.28 Visualization of the settlements of the tailings body surface relative to the tailings thickness at the time of measurement [59]

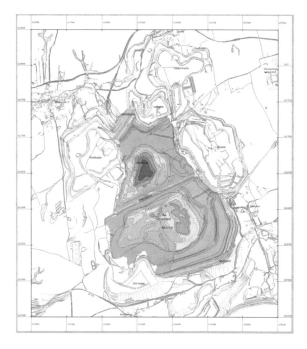

monitoring is intended as the basis for long-term monitoring, i.e. by continuing or further developing of this programme, the surrounding situation is monitored until the completion of remediation works in order to assess the remediation results. The post-closure works and the long term monitoring are not part of the basic monitoring.

Long-term monitoring is based on the basic monitoring. From the respective results in connection with the development prognoses, the necessary network of measuring points is set up. In due course, however, this network should be regularly reviewed and adapted to the development of science and technology.

The **remediation monitoring (accompanying monitoring)** accompanies the implementation of the remedial measures as a supplement to the basic monitoring. The defining characteristic of the renovation monitoring is its time limit. Remediation monitoring also includes radiological monitoring of workers. Since 1991, Wismut GmbH has provided environmental monitoring results to the public in the form of environmental reports.

- **Long-term monitoring—Part of the safety assessment of the geotechnical environmental structure and the run-off protection (ENW 4)**

The long-term monitoring of the stability analysis is on the one hand determined by the method described under Sect. 5.5.2 of individual proofs, and on the other hand by the underlying design variables for the tailings dam construction and the multi-functional cover according to the examples at (5.4), (5.5), (5.6) and (5.7). On the one hand, it is necessary to determine the time intervals at which sampling and test procedures must be carried out, and on the other hand to define the methods with which the determination of parameters should be carried out. Thus, the shear strength parameters are usually determined from direct shear tests and from triaxial tests (CU) in accordance with DIN 18,137. Using load-settlement tests (oedometer) according to DIN 18,135, the appropriate deformation parameters can be determined. The experiments are carried out taking into account the extraction depth of samples or the overburden stress present in situ for the expected stress ranges.

A recurring difficulty is to transfer the values determined in the laboratory to the respective engineering task being used. The tailings body consolidates further following the addition of the multi-functional cover. The tailings body and the cover form a coupled system, see Eq. (5.7). If the settlement differences between adjacent parts of the tailings body are very large (large settlement gradient), this can also be transferred to the cover. If there is such differential settlement the layers of the multi-functional cover can become offset from one another or water penetration may take place at the layer boundaries. As a result of such damage to the barrier system may be damaged, resulting possibly in the development of radon exhalation pathways. In contrast the use of geosynthetic separating layers can provide good protection, see Fig. 5.11. The radon exhalation rates and the local dose rates are measured and used to demonstrate the effectiveness, and thus the functionality, of the radon barrier within the multi-functional cover.

The location-dependent settlements measured on the surface of the cover are the sum of the settlement of the tailings body and the cover. Following the Boltzmann axiom of Eq. (5.6):

$$s_g = s_{tail} + s_{abd} \tag{5.9a}$$

s_g—on the surface measured total settlement;
s_{tail}—settlement of the tailings body;
s_{abd}—settlement of the multi-functional cover.

In order to minimize the effects of consolidation on the cover, extensive predictive calculations and accompanying monitoring for validation are necessary in order to derive specifications for the design of the cover (see Fig. 5.15). Load fillings and sub-areas of the intermediate cover take on the function of test fields. The following tasks can be identified, which serve the purpose of guaranteeing and proving the barrier effect of the multi-functional cover for the specified service life of the geotechnical environmental structure. The list is not exhaustive:

– Measures to homogenise the settlements of the surface of the tailings body,
– Determination of a reliable prognosis for the long-term behavior of the tailings body,
– Installation of a system with which the settlements of the cover can be reliably determined over long periods of time, with corresponding evaluation routines,
– Installation of a network of radiological measuring points, for the detection of inter alia radon exhalation rates, local dose rates, etc., with appropriate evaluation routines and correlations to the deformation behavior of the surface of the cover,
– Detection of potential radon pathways, e.g. by suitable tracer tests in conjunction with radiological measurements.

It appears advisable in areas of increased failure probability, e.g. during the tailings dam construction, to install an early warning system that could be coupled with elements of the above tasks. This will make it possible to identify weak zones and slip surfaces, etc., and thus to obtain information on measures necessary to avert hazards or to initiate more thorough investigation. Sites in the vicinity of residential areas, agricultural and animal production, transitions to receiving waters etc. are particularly worthy of protection. The installation of an early warning system is particularly recommended in zones with increased seismic activity and an increased number and duration of heavy rainfall events. Some suggestions already exist that can contribute to a solution. Glötzl, R. and Lersow, M., see one way to suitably join the mats-integrated Polymers Optical Fiber (POF) [60] and a stress measuring station [61]. The monitoring programme should also include radiological monitoring of the site. This includes monitoring the functionality of the multi-functional cover in minimizing the mobilization of radionuclides in the tailings body.

The cover of a tailings pond has various protective functions. It must ensure that in the long term escapes of pollutants into the ambient air ("air pathway"), the ground and surface water ("water pathways") remain limited to an acceptable level and encapsulated minerals are not exposed within the containment ("pathway of minerals"). If the cover loses only one of these protective functions, then the overall function of the containment is compromised. For the detection of chemical and biological substances, in particular of toxic and radioactive contaminants, development of a suitable procedure is shown [35] (see Fig. 5.29).

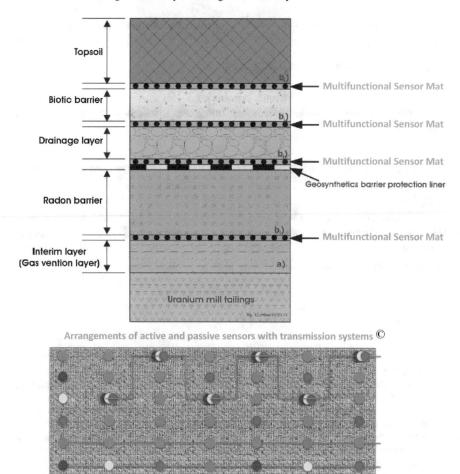

Fig. 5.29 Principle of a multifunctional geomat, © [35]

At Wismut GmbH, a local dose rate of ODL $\leq 0.15\ \mu$Sv/h for radiation exposure at the surface of the cover has been established for the assessment of the shielding effect from the encapsulated radioactive minerals.

5.7   Consideration of Extraordinary Events—Worst Cases

Unpredictable and extreme events may affect the geotechnical environmental structures for long-term safe storage of the uranium mill tailings ponds of the SDAG Wismut which could reduce the effectiveness of the final disposal construction, with the potential for release of significant quantities of radio-toxic pollutants into the biosphere.

- Earthquakes of unprecedented magnitude
- Aeroplane Crash
- Terrorist interventions (explosions, bombs, etc.)
- Interventions from outside (drill holes, holes, etc.), so-called "human intrusion scenarios"
- etc.

The third and fourth items on this list can be dangerous for the containment structures. Therefore, these should be deserving of particular attention. Matters relating to the "human intrusion scenarios" are covered by elements of the German mining code. The file (final documentation) of the abandoned mining areas (here Uranium-Tailings ponds) with all relevant data should remain permanently in the register of the responsible mining authority. This will hopefully exclude the possibility of human interference in the containment structure over the long term.

References

1. RICHTLINIE 2013/59/EURATOM DES RATES vom 5. Dezember 2013: zur Festlegung grundlegender Sicherheitsnormen für den Schutz vor den Gefahren einer Exposition gegenüber ionisierender Strahlung und zur Aufhebung der Richtlinien 89/618/Euratom, 90/641/Euratom, 6/29/Euratom, 97/43/Euratom und 2003/122/Euratom
2. Nuclear Energy Agency and International Atomic Energy Agency: Uranium 2016: Resources, Production and Demand; © OECD 2016; NEA No. 7301
3. NAS 1999a—National Academy of Sciences. Evaluation of Guidelines for Exposures to Technologically Enhanced Naturally Occurring Radioactive Materials. Washington, DC: National Academy Press, 1999.
4. U.S. Nuclear Regulatory Commission; https://www.nrc.gov/reading-rm/doc-collections/fact-sheets/fs-uranium-recovery.html
5. U.S. Energy Information Administration: Uranium Mill Sites under the UMTRA Project;
6. DOE/EIA-0592 Distribution Category UC-950: "Decommissioning of U.S. Uranium Production Facilities" February 1995
7. U.S. Environmental Protection Agency: Technical Report on Technologically Enhanced Naturally Occurring Radioactive Materials from Uranium Mining; Volume 1: Mining and Reclamation Background; Previously published on-line and printed as Vol. 1 of EPA 402-R-05-007, January 2006, Updated June 2007 and printed April 2008 as EPA 402-R-08-005
8. Environment Protection Agency: Abandoned Mine Site Characterization and Cleanup Handbook; EPA 910-B-00-001; August 2000
9. U.S. NRC: Uranium Mill Tailings; https://www.nrc.gov/waste/mill-tailings.html
10. M. Lersow, H. Märten; Energiequelle Uran-Ressourcen, Gewinnung und Reichweiten im Blickwinkel der technologischen Entwicklung, Glückauf 144 (2008)3, S. 116–122, Essen
11. M.V. Hansen: World uranium resources; IAEA BULLETIN, VOL. 23, No.2
12. Lersow, M.: Energy Source Uranium – Resources, Production and Adequacy; Glueckauf Mining Reporter 153 (3): p. 286–302, June 2017
13. Status and Results of the WISMUT Environmental Remediation Project; Vienna 2006:
14. Wismut Bergbausanierung – Landschaften gestalten und erhalten; Herausgeber Bundesministerium für Wirtschaft und Energie (BMWi); Öffentlichkeitsarbeit; 11019 Berlin; Juli 2015

15. SSK: Die Strahlenexposition durch den Bergbau in Sachsen und Thüringen und deren Bewertung; Beratung vom: 13./14. Dezember 1990 Veröffentlicht in: – Veröffentlichungen der Strahlenschutzkommission, Band 21

16. Jack Caldwell and Lawrence Charlebois, Robertson GeoConsultants, December 2012; http://technology.infomine.com/reviews/PasteTailings/welcome.asp?view=full

17. Tailings Paste Disposal – More than Water Recovery; http://www.westech-inc.com/blog-minerals/tailings-paste-disposal-more-than-water-recovery

18. Steven G. Vick: Planning, Design, and Analysis of Tailings Dams, John Wiley & Sons, New York, 1983, 369 p., ISBN 0-471-89829-5

19. TLUG 2001: Die Sanierung der ehemaligen Uranerzbergwerke und –aufbereitungsanlagen in Ostthüringen, Teil 1: Grundwasserüberwachung während der Sanierung, Teil I; Herausgeber TLUG, 2001

20. A. T. Jakubick, M. Hagen (2000); Wismut Experience in Remediation of Uranium Mill Tailings Ponds; 2000 Kluwer Academic Publisher, p. 93–105

21. Kommentar zur Novellierung der Strahlenschutzverordnung (Fassung vom 9. März 2001); Hrsg.: Gesellschaft für Strahlenschutz e.V., 21.04.01

22. Wels, C., Robertson, A. MacG., and Jakubick, A.T. (2000): "A Review of Dry Cover Placement on Extremely Weak, Compressible Tailings". Paper published in CIM Bulletin, Vol. 93, No. 1043, pp. 111–118, September 2000.

23. IAEA (2004); The long term stabilization of uranium mill tailings, Final report of a coordinated research project 2000–2004; IAEA-TECDOC-1403, August 2004

24. Steven G. Vick: Degrees of Belief: Subjective Probability and Engineering Judgment, Published January 1st 2002 by American Society of Civil Engineers, 455 p.; ISBN 0784405980 (ISBN13: 9780784405987)

25. Lersow, M.; Gellermann, R (2015).; Langzeitstabile, langzeitsichere Verwahrung von Rückständen und radioaktiven Abfällen – Sachstand und Beitrag zur Diskussion um Lagerung (Endlagerung); Ernst & Sohn Verlag für Architektur und technische Wissenschaften GmbH & Co. KG, Berlin · geotechnik 38 (2015), Heft 3, S. 173–192

26. Offizielle website Moab, Utah, UMTRA-project Moab; https://www.gjem.energy.gov/moab/documents/TransportationPlanRev11.pdf

27. Moab UMTRA Project; http://www.grandcountyutah.net/257/Moab-UMTRA-Project

28. Natural Resources Canada; http://www.nrcan.gc.ca/home

29. M. Lersow (2010); Sichere und langzeitstabile Verwahrung von Tailings Ponds, insbesondere aus der Uranerzaufbereitung, Geotechnik 33 (2010) Nr. 4, S. 351–369, VGE Verlag GmbH, Essen

30. Nand, Davé, Disposal of Reactive Mining Waste in Man-made and Natural Water Bodies Canadian Experience, Marine and Lake Disposal of Mine Tailings and Waste Rock, Egersund, Norway, Sept. 7–10, 2009

31. IAEA-TECDOC-1403; The long term stabilization of uranium mill tailings - Final report of a co-ordinated research project 2000–2004; August 2004

32. TAILSAFE (2004); Sustainable Improvement in Safety of Tailings Facilities; Report Implementation and Improvement of Closure and Restoration Plans for Disused Tailings Facilities, October 2004, http://www.tailsafe.com

33. Ritcey, G.M. (1989). Tailings management: Problems and solutions in the mining industry. Amsterdam; New York, Elsevier

34. Beckers, Nicole, Böden auf künstlichen und natürlichen Substraten der ostthüringischen Bergbaufolgelandschaft als Senken und Quellen bergbauinduzierter Stoffe, Regensburger Beiträge zur Bodenkunde, Landschaftsökologie und Quartärforschung, Band 5, ISBN 3-88246-275-2, 2005

35. Offenlegungsschrift, DE 10 2011 100 731 A1, DPMA 06.05.2012

36. Merkel B. J., Planer-Friedrich B.. Grundwasserchemie, Springer-Verlag, 2002

37. Boyle1982 Boyle R.W.. Geochemical prospecting for thorium and uranium deposits, Develop. Econ. Geol. 16, Elsevier-Scientific Publ. Co., Amsterdam- Oxford-New York, 498 S. 1982

38. „Methodische Weiterentwicklung des Leitfadens zur radiologischen Untersuchung und Bewertung bergbaulicher Altlasten und Erweiterung des Anwendungsbereiches"; Herausgeber: Bundesministerium für Umwelt, Naturschutz und Reaktorsicherheit; ISSN 1612–6386; Bonn, 2005

39. Lersow, M. (2010): Sichere und langzeitstabile Verwahrung von Tailings Ponds, insbesondere aus der Uranerzaufbereitung, Geotechnik 33 (2010), Nr. 4, S.351–369, VGE Verlag GmbH, Essen

40. Schmidt, P. (2007): Sanierung der Hinterlassenschaften der Urangewinnung in Sachsen und Thüringen, Technisches Seminar DESY Zeuthen, 13. Februar 2007

41. Metschies, Thomas, Conceptual Site Modelling, Hydraulic and Water Balance Modelling as basis for the development of a remediation concept; Regional Training Course on Remediation Infrastructure Development at a Test Site, Chemnitz, Germany, 3–7 December 2012

42. Operating plan Tailings Cells and Evaporation ponds PIÑON RIDGE MILL, Antrag der Energy Fuels Resources Corporation, zugelassen durch EPA, Aug. 2010

43. Tailings storage facility design report – Olympic dam expansion project – bhpbilliton, 2009; https://www.bhp.com/-/media/bhp/regulatory-information-media/copper/olympic-dam/0000/draft-eis-appendices/odxeisappendixf1tailingsstoragefacilitydesignreport.pdf

44. Stellungnahme zur zukünftigen Bewirtschaftung der von der Wismut GmbH beeinflussten Oberflächenwasserkörper in Thüringen in Umsetzung der EU-WRRL - Bewirtschaftungszeitraum 2015 bis 2021; Paul, M. et al., Wismut GmbH, Chemnitz, Mai 2014

45. Schulze, G. (1993): Bestandsaufnahme und Charakterisierung der stofflichen Auswirkungen des Uranbergbaus und der Uranerzaufbereitung (Standort Seelingstädt) am Beispiel des Wasserpfades. - In: MUSEUM FÜR NATURKUNDE GERA (Hrsg.): Beiträge zur Geologie, Flora und Fauna Ostthüringens. Naturwissenschaftliche Reihe 20

46. Große Anfrage Deutscher Bundestag; Drucksache 12/3309; 24.09.92

47. Winde, F.: Kontamination von Fließgewässern-Prozessdynamik, Mechanismen und Steuerfaktoren; Untersuchungen zum Transport von gelöstem Uran in bergbaulich gestörten Landschaften; Habilitationsschrift vorgelegt der Chemischen-Geowissenschaftlichen Fakultät der Friedrich-Schiller-Universität Jena, Erteilung der Lehrbefähigung 09.Juli 2003 an Dr.rer.nat. Frank Winde

48. Frank Winde, Urankontamination von Fließgewässern - Prozessdynamik, Mechanismen und Steuerfaktoren: Untersuchungen zum Transport von gelöstem Uran in bergbaulich gestörten Landschaften unterschiedlicher Klimate/Driesen, H. H. Dr.; Taunusstein; 657 Seiten, ISBN 978-3-86866-087-6, April 2009

49. BMU, 2015; Version 23.07.2015: 3. Arbeitsentwurf der Mantelverordnung Grundwasser/Ersatzbaustoffe/Bodenschutz, Bundesministerium für Umwelt, Naturschutz und Reaktorsicherheit

50. Wenzel, A.; Schlich, K.; Shemotyuk, L. und Nendza, M.: Revision der Umweltqualitätsnormen der Bundes-Oberflächengewässerverordnung nach Ende der Übergangsfrist für Richtlinie 2006/11/EG und Fortschreibung der europäischen Umweltqualitätsziele für prioritäre Stoffe; Umweltbundesamt, Dessau-Roßlau; ISSN 1862-4804; Texte 47/2015, Juni 2015

51. Andreas Artmann et. al., Anwendung und Weiterentwicklung von Modellen für Endlagersicherheitsanalysen auf die Freigabe radioaktiver Stoffe zur Deponierung; Gesellschaft für Reaktorsicherheit (GRS), Abschlussbericht, 2014

52. Finite Elemente Programm zur Lösung Geotechnischer Aufgabenstellungen, M. Lersow FEM-PLER ©, last update 2015

53. Backhaus, G. (1983), Deformationsgesetze, Akademie-Verlag, Berlin

54. Empfehlungen und Stellungnahmen der Strahlenschutzkommission 1990/1991; Band 24; ISBN 3-437-11519-7; Bonn, 1993

55. BGR (2007) Final Report „Mineral crustal (cement layers) formation on mine and mill tailings dumps", Federal Institute of Geosciences and Natural Resources (BGR), 30.12.2007

56. M. Lersow (2003); Exhaustive proof of the soil mechanical behaviour of loose rocks in stabilized extensive mining sites XIIIth European Conference on Soil Mechanics and Geotechnical Engineering, August 2003, Prague; CZ Republic

57. Schiffman, R. L. (1980): "Technical note: Finite and infinitesimal strain consolidation", Journal of Geotechnical Engineering 106(GT$_2$, 26), 203–207.
58. LAGA Ad-hoc-AG „Deponietechnik": Bundeseinheitlicher Qualitätsstandard 2-0 „Mineralische Basisabdichtungskomponenten - übergreifende Anforderungen" vom 04.12.2014
59. M. Lersow (2006), P. Schmidt; The Wismut Remediation Project, Proceedings of First International Seminar on Mine Closure, Sept. 2006, Perth, Australia, p. 181–190
60. Glötzl,R. (2009), Schneider-Glötzl, J., Liehr,S., Lenke,P., Wendt, M., Krebber, K., Gabino, L., Krywult, J.; First Results of a Field Test in Belchatow Brown Coal Mine to investigate Creeping Slopes; 2009, VIth International Brown Coal Mining Congress in Belchatow
61. Glötzl, R. (2002), Krywult, J.; Erfahrungen und neue Erkenntnisse zum Spannungs-verhalten einer Braunkohleschürfe; Bericht XXVI, Glötzl GmbH

Chapter 6
Disposal of Radioactive Waste of Low and Medium Radioactivity (Radioactive Waste with Negligible Heat Generation)

The following sections are consistent with the classification of radioactive waste previously set out in Chap. 2, Fig. 2.1 and also with the situation in Germany. The reader will notice that there are relatively large differences between individual countries in the long-term safe and long-term safekeeping of this class of radioactive waste. Comparing Germany to the USA, the differences are not only due to the different classification systems but also as a result of site-specific conditions. With low population density and often dry climate in parts of the USA, such conditions allow near-surface storage of this category of waste.

The US Radioactive Waste Classification System is set out in detail in various regulations of the Code of Federal Regulations (CFR). There are two main categories of radioactive waste in the USA. Low Level Waste (LLW) and High Level Waste (HLW), see Chap. 7. In addition, there are some special categories, such as waste containing transuranic elements (TRUW), and spent nuclear fuel (SNF); see Chap. 7 for information on these wastes while waste from naturally occurring radionuclides (NORM), is described in Chap. 5.

Only LLW will be discussed here. For LLW, near-surface disposal is prescribed in the USA. In Germany, radioactive waste of low and medium radioactivity with negligible heat generation is stored in deep geological strata, a clear difference. However, the requirement for long-term safety can be proven for both types of storage.

To be classified as LLW the waste must have a maximum activity of 3700 Bq/g of α-emitters with a half-life of more than 20 years.

Furthermore, for LLW, activity limits are specified for each individual β-emitter and for the sum of radionuclides referred to below. For simplicity, only the limits of the sums are given here. The figures in the regulations are in Curie per cubic meter of waste. In order to make them comparable to the European standards limits must be stated in Becquerel per gram, so these numbers have been converted. For this purpose, a density of the waste of 2 g/cm^3 was assumed, see also Chap. 3, Table 3.3

- For the sum of the short-lived nuclides Ni-63, Sr-90 and Cs-137 there is an upper limit of specific activity of 8.9 to $\times$ 10^7 Bq/g.

© Springer Nature Switzerland AG 2020

M. Lersow and P. Waggitt, *Disposal of All Forms of Radioactive Waste and Residues*, https://doi.org/10.1007/978-3-030-32910-5_6

– For the sum of the long-lived nuclides C-14, Ni-59, Nb-94, Tc-99 and I-129 is an
  upper limit of 9×10^5 Bq/g.

The examples for the remediation programs of other countries have been limited,
so that the reader has the opportunity to draw comparisons with appropriate site
specific projects and thus may also draw conclusions for site specific solutions or
there is the option to seek more information from international literature.

6.1 International Developments

According to the present classification, see Chapter 2, Low Active Waste (LAW)
is characterized by an activity $A* < 1E+11$ Bq/m^3 and a mean decay heat Q_{mean}
$\cong 200$ W/m^3; for medium radioactive waste (Medium Active Waste—MAW) the
corresponding numbers are activity $A* = (1E+10$ to $1E+15)$ Bq/m^3 and decay heat
$Q < 2$ kW/m^3. In addition it is noted that the heat generation during storage of the
inventory in underground cavities should be negligible, so that only a slight warming
of the surrounding rocks is to be expected.

In Germany, low-level and intermediate-level radioactive waste accounts for
around 90% of the total volume of radioactive waste. These wastes arise from the
operation and decommissioning of nuclear power plants, from research and industry
as well as from medicine, see Fig. 4.1, Chap. 4. Wastes may include, for example,
contaminated plant components, tools or laboratory equipment, protective clothing
from nuclear power plants, used filters, radiation sources from medical and other
engineering applications or radioactive chemicals and industrial process residues
requiring monitoring. For such waste, the German Federal Environment Ministry
(BMU) predicts a volume of just over 300,000 m^3 by the year 2080. A possible
future volume of waste to be considered for removal into another final disposal site
and conditioning of Asse II wastes is not included, see Sect. 6.3.4.

Internationally Low- and Intermediate-level radioactive waste, LAW and MAW,
or limiting heat generation (Negligible Heat-Generating Waste—NHGW), are generally deposited in near surface repositories, i. a. only few meters below ground.

In a global comparison, only a few countries, such as Germany, have decided to
dispose of low- and intermediate-level radioactive waste in deep geological areas.
Generally, no retrieval is intended for these radioactive wastes. Final repositories
for low and intermediate level radioactive waste have been in operation in many
countries for many years.

- **International overview**

Finland

In Olkiluoto/Finland a repository for low- and intermediate-level radioactive waste
has been constructed, that is comparable to German sites. The Olkiluoto repository
is assigned to the nuclear power plant of the same name and is located on the Olkiluoto
peninsula on the west coast of Finland, see Chap. 7. After initial site investigations in

Table 6.1 Nuclide inventory of repository Olkiluoto for low- and intermediate-level radioactive waste, acc. [1]

Total activity	[Bq]	4.2E+14		
α-activity	[Bq]	1.6E+10		
Bestimmende Nuklide up to $10^2/10^{4/}10^6$ years		Cs-137, Ni-63	Ni-59	Ni-59
Waste volume	[m^3]	8500 Thereof: 4960 for LAW and 3472 für MAW		
Specific total activity	[Bq/g]	24,700		
Specific α- activity	[Bq/g]	1		

1980, construction began in 1988. In May 1992, the first waste was put into storage. The operating license is valid until the end of 2051.

The repository consists of two separate cavities for separate disposal 60 to 100 m deep in the crystalline rock. These is a silo for dry LAW from the maintenance of the NPP and a silo for bituminized MAW. Each of these cavities is 34 m high and has a diameter of 24 m. The storage capacity totals 60,000 m^3. At the time of writing approximately 9500 m^3 have been used so far. The waste is stored in the form of concrete cubes, each containing 16 barrels. For a waste volume of 8500 m^3, the inventory has been listed in [1], see Table 6.1.

If the activity from Table 6.1 is calculated for 300,000 m^3, which is the storage volume of the future German repository Schacht Konrad, a total activity of 1.48 E+16 Bq would be achieved.

France

The Centre de l'Aube repository for—short-lived radionuclides LAW-SL and MAW-SL (SL stands for short-lived), is located in central France near the city of Soulaines-Dhuys within the departments of Aube and Haute-Marne, see Chap. 7.

Since 1992, LAW-SL and MAW-SL have been stored near the surface on a concrete platform. The waste barrels and concrete containers are stacked in six layers. The concrete platform is drained by small channels. When the final disposal is full, it is sealed to become impermeable to water and covered with soil. It is to be monitored for 300 years. Thereafter, the radioactivity of the inventory should have decayed to the level of natural radiation, see Fig. 6.1. Monitoring in the post-closure phase is a special feature and highly recommended for near-surface repositories.

The repository has a capacity of 1 million m^3 and is equipped with its own compacting plant. The plant was originally planned for operation up to 2040, with an annual delivery volume of 30,000 m^3. However, the quantities actually delivered are well below the anticipated 30,000 m^3 per year, so that an operating life of 60 years is now expected, see [2]. The operation at the Center de l'Aube repository is essentially automated and computer-monitored. For a waste volume of 1 million m^3, the following inventory may be found in [2] and is summarised in Table 6.2.

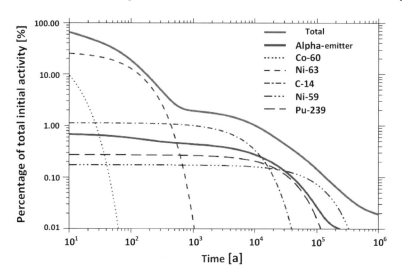

Fig. 6.1 Time course of the nuclide inventory of the Center de l'Aube repository. *Source* [2]

Table 6.2 Nuclide inventory at the Centre de l'Aube repository, according to [2]

Total approved activity	[Bq]	6.5E+17
α-Aktivity	[Bq]	7.5E+14
Determining nuclides: LAW-SL and MAW-SL		Co-60, Cs-137, Cs-135, Ni-63, Tc-99, Sm-151, Pd-107, mAg-108, I-129, Cl-36
Approved waste volume	[m³]	1.0E+06
Activity limit for α-emitters in the waste packages	[Bq/g]	3.7E+03

If the activity from Table 6.2 is calculated for 300,000 m³, which is the storage volume of the future German repository Schacht Konrad, a total activity of 1.95E+17 Bq would be achieved.

USA

In arid climates, near-surface storage of low- and intermediate-level radioactive waste is a common practice. As a typical example of disposal in trenches, can shown **the repository "Test Site" at the Clive location approximately 105 km north-west of Las Vegas, Nevada**, is shown in Fig. 6.2. The repository "Test Site" is designed to accommodate radioactive waste Class A/LLW according to 10 CFR 61.55 [3], see Table 6.3 and Chap. 3, Table 3.3. In this site, the containers of radioactive waste are stacked in several layers on a prepared, usually concreted, surface a few meters below ground level, see Fig. 6.2. They are often sealed with a cover (mostly concrete) and then covered with earth. However, special sealing systems may also be developed. This includes the use of geosynthetics. These are shown diagrammatically in Fig. 6.3.

Fig. 6.2 Disposal area of
repository "Test Site",
Nevada/USA

Table 6.3 Characteristics of Classes of LLW acc. Code of Federal Regulations, No. 10 CFR 61 §
55, [3], for storage by surface near final disposal sites

Radionuclide/Class LLW—activity limits per volume	Class A (Ci/m^3)	Class A (Bq/m^3)	Class B (Ci/m^3)	Class C (Ci/m^3)
Total of all nuclides with less than 5 year half life	700	2.59E+13	No limit	No limit
Tritium (H-3)	40	1.48E+12	No limit	No limit
Cobalt-60 (Co)	700	2.59E+13	No limit	No limit
Nickel-63 (Ni)	3.5	12.95E+10	70	700
Ni-63 in activated metal	35	12.95E+11	700	7000
Strontium-90 (Sr)	0.04	0.148E+10	150	7000
Cesium-137(Cs)	1	3.7E+10	44	4600
Carbon-14(C)	0.8	2.96E+10		8
C-14 in activated metal	8	2.96E+11		80
Nickel-59 (Ni) in activated metal	22	8.14E+11		220
Niobium-94 (Nb) in activated metal	0.02	0.074E+10		0.2
Technetium-99 (Tc)	0.3	1.11E+10		3
Iodine-129 (I)	0.008	0.0296E+10		0.08
Alpha emitting transuranic nuclides with a half life greater than 5 years	10 nCi/g	370 Bq/g		100 nCi/g
Plutonium-241 (Pu)	350 nCi/g	12,950 Bq/g		3500 nCi/g
Curium-242 (Cm)	2000 nCi/g	74,000 Bq/g		20,000 nCi/g

Low-Level Waste Disposal Site

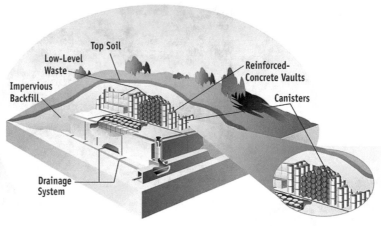

Fig. 6.3 Near-surface storage of LLW in the USA according to no. 10 CFR 61 [3]

The United States has the most experience in dealing with radioactive waste and residues. In Table 6.4, are represented the quantities of LLW, which have been stored at the four repository sites in the USA so far. Licensees may apply to an Agreement State for periodic renewal of a license. The quantity which is stored there is about three times greater than the total quantity of LAW and MAW with negligible heat generation planned to be stored in the Konrad shaft repository. The stored activities from the inventories differ insignificantly, see Table 6.7. The Clive/UT repository contains the largest volume but the lowest activity because only LLW class A is accepted.

Germany: Near-surface storage of radioactive residues

Table 6.4 LLW deposited at different sites in USA during the period 2005 to 2017, all data from Low-Level Waste Disposal Statistics, U.S. NRC, April 24, 2018

Disposal Facility	Volume (Cubic Feet)	Volume (Cubic meter)	Activity (Curies)	Activity (Curies)[a]
Andrews County[a, c], TX	114,100,000	3230.952	422,718.000	1.564E+16
Barnwell[a], SC	265,948.000	7530.809	2,774,177.000	1.026E+17
Clive[b], UT	31,007,890.000	878,045.664	197,430.000	7.305E+15
Richland[a], WA	400,878.000	11,351.601	207,856.000	7.691E+15
Total	31,788,816.000	900,159.026	3,602,181.000	1.333E+17

[a]Barnwell, Richland, and Andrews County accept class A, B, and C waste
[b]Clive accepts only class A waste
[c]Andrews County in operation since 2012

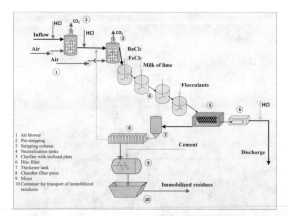

Fig. 6.4 Above: modified scheme of lime-Walhalla lime—water treatment process (WBA); [4]; bottom left: the storage field of the dump 371/1 with deposited big bags before sealing of the field; bottom right: Water treatment facility at Ronneburg. *Source* Figures bottom ©Wismut GmbH

A type of disposal site similar to the repository "Test site" Nevada/USA was developed by Wismut GmbH for treatment residues from their water treatment plants, see Fig. 6.4. This is an example of the near-surface disposal of residues requiring monitoring in Germany; the residues are being created in the remediation process for uranium mill tailings ponds. A classification as used in the USA does not exist. Substantial inventory activities are being disposed of in this way near-surface for the long term.

Wismut GmbH operates water treatment plants for the decontamination of mining waters at six locations downstream of their operating sites and thus ensures achievement of the protection goals for the receiving waters and aquifers used by mining, see Table 6.5. For this purpose, site-specific process management systems were developed and installed at the water treatment plants. The plant for the treatment of mine flood waters (AAF) at the Königstein mine has a pre-treatment in the form of a separate uranium separation by means of ion exchange/precipitation. The resulting uranium product is marketed. As an example: In 2014, 37.6 tonnes of uranium were recovered using this process [5]. If the AAF is operated without this separate uranium separation, the concentration of U_{nat} in the treatment residues will increase significantly, resulting in a significant increase in the specific activity A^* in the treatment

Fig. 6.5 Shaft 1 of Konrad mine in the conversation phase - used for the transportation of personnel and materials

residues. In general, it can be stated that the specific activities in the residues of the water treatment plants of Wismut GmbH (particularly Königstein) will diminish little in the future. This is due, on the one hand, to the consistently high levels of radioactivity in the mine derived waters and, on the other hand, to the improved methods of separation, particularly for uranium and Ra-226 in the water treatment plants. Figure 6.4 shows the scheme of a water treatment with lime precipitation, as used in the WBA Seelingstädt, see also [6, 7].

For the collection, transportation and final disposal of the residues from Table 6.5, these are initially filled into "Big Bags" of 1 m^3 volume and then transferred to a temporary storage site. From there, the big bags are transported to the repository site at the appropriate time. Special areas, mostly on dumps and on remediated tailings ponds sites, are planned and approved as emplacement areas for radioactively contaminated materials. The base surface of the emplacement areas will be concreted. The big bags will be stacked on the concreted areas. The emplacement areas are finally profiled, isolated and covered. Such sites then become final storage facilities for the treatment residues, with radioactive inventories which are comparable to those at the final disposal sites described here.

The storage location at waste rock dump Halde 371/1 in Schlema-Alberoda has been formed as a basin with storage sections, which are separated from each other. In Chap. 5, Fig. 5.1 right, the storage sections located in waste rock dump 371/1 are clearly visible. Since 2006, a pourable product from the residues of the water treatment plants has been made from the residues of the water treatment plants; when mixed with cement this product has very little water solubility. The material is then transported in Big Bags to the prepared storage site, compacted and initially covered with sand, see Fig. 6.4. The treatment residues from the water treatment plants contain

Table 6.5 Water treatment plants operated by Wismut GmbH, as of June 2017, according to Wismut GmbH—www.wismut.de; from [5] and according to annual reports of the Saxon State Office for the Environment, Agriculture and Geology

Water treatment plant	WBA Schlema-Alberoda	WBA Pöhla[d]	WBA Helmsdorf	WBA Ronneburg	WBA Seelingstädt	AAF Königstein[b]
Type and origin of water to be treated	Mine flood water from the Schlema-Alberoda mine and leachate from the slagheap 371/I	Mine flood water of Grube Pöhla mine	Surface-, seepage- and pore water of Tailings ponds Helmsdorf and Dänkritz I	Mine flood water of Mining areas of Ronneburg and collected surface water	Surface-, seepage- and pore water of Tailings ponds Culmitzsch and Trünzig	Mine flood water of Königstein mine, and collected surface and waste rock pile water
Water treatment technology	Modified lime-precipitation method	Chemical-physical precipitation	Modified lime-precipitation method	Lime-precipitation with partial sludge recirculation (HDS method[a])	Modified lime-precipitation method	Ion-exchange/lime-precipitation with partial sludge recirculation (HDS method[a])
Treatment capacity	1 150 m³/h	60 m³/h	200 m³/h	850 m³/h	330 m³/h	650 m³/h
Average water flow rate 2010–2014 [Mio. m³/a]	6.62	0.11	0.97	6.11	2.18	3.50
Amount of residues/year [m³/a][c]	1720	140	1 910	19 140	1 600	920
Separated main contaminants	U, Ra-226, As, Fe, Mn	Ra-226, As, Fe	U, Ra-226, As	U, Mn, As, Ni, Cd and some more heavy metals	U, Ra-226, heavy metals	U, Ra-226, Fe, Mn, Ni heavy metals
U_{nat} [mg/l]	1.53	7.02	1.43	0.0133	0.67	**14.50**

(continued)

Table 6.5 (continued)

Water treatment plant	WBA Schlema-Alberoda	WBA Pöhla[d]	WBA Helmsdorf	WBA Ronneburg	WBA Seelingstädt	AAF Königstein[b]
Ra-226 [mBq/l]	2 190	156	571	4 300	393	**4730**
Location of storage	Basin dump 371/1	mine Pöhla/Basin dump 371/1	Area of Tailings pond Helmsdorf	Area fill body opencast Lichtenberg	Area of Tailings ponds Culmitzsch	Dump Schüsselgrund - Leupoldishain

[a]HDS—High Density Sludge; [b]Will be rebuilt without uranium separation; [c]Reference period 2010–2014; [d]Until 2014 passive-biological plant, thereafter decommissioned and rebuilt as a technical WBA

the main contaminants listed in Table 6.5. The very low uranium concentration in waste from the water treatment plant Ronneburg is striking.

However, the specific activity values in the treatment residues from the AAF Königstein appear critical. Looking at the radioactivity loads of Table 6.5 and taking into account the daughter nuclides of U-238, U-234, Th-230, an activity concentration of $A^* > 3000$ Bq/g can be calculated. If the AAF Königstein is converted to have no uranium separation facility, the treatment residues for disposal will have activity concentrations of $A^* > 4000$ Bq/g. This should be viewed critically. Since the mine flood waters still have to be treated over a very long period of time, it can be assumed that the total amount of treatment residues to be shipped to the Schüsselgrund-Halde will amount to more than 300,000 m^3 and thus the storage volume of the repository Schacht Konrad repository for low- and intermediate-level radioactive waste will be insufficient.

Märten [7] has noted the following: "The special feature of the water treatment plant Königstein … consists above all in the prior separation of uranium from the flood water by means of ion exchange and the subsequent treatment to a usable product (instead of the transfer of uranium in a precipitated sludge and disposal into the waste rock dump at the site, which is very questionable for environmental reasons).

Storage of radioactive material in the open air is not permitted in any state in Western Europe. Even the outdoor storage of containers of radioactive material in the open air is problematic because of the harsher weather conditions and the influence of solar radiation which results in more aggressive corrosion of the waste container. In Central Europe, the permanent outdoor storage of containers of radioactive material is neither politically desirable nor legally permissible in any country.

Table 6.6 summarizes some examples of repositories for low and intermediate level radioactive waste in different countries. The diversity of the repository-concepts in each country is clear.

6.2 Description of the Current Situation for the Long-Term Storage of Low to Medium Level Radioactive Waste (LAW, MAW) in Germany

There are two sites in Germany where low and intermediate level (LAW and MAW) waste is deposited in geologically deep strata. One is the Morsleben Radioactive Waste Repository (ERAM) in Saxony-Anhalt and the other is the Asse II mine near Remlingen in Lower Saxony. These two sites exist from before the reunification of Germany.

In 2016 with the "Act on the Reorganisation of the Organisational Structure in the Field of Disposal of Radioactive Waste" [8], the Federal Government reorganised the responsibilities in relation to the disposal of radioactive waste. A Federal Company for repository mbH (BGE), based in Peine/Lower Saxony, was founded [8]. This

Table 6.6 Examples of storage of low and intermediate level radioactive waste worldwide/status: 2017

Country	Location	Type	Status
Belgium	Mol-Dessel, Province of Antwerp	Near surface	In operation
Denmark	Location not specified	Above ground temporary storage	A repository designed for a period of 100 years is being tested
Germany	Morsleben, Saxony-Anhalt	DGR—rock salt	Closed, approval procedure decommissioning
	Konrad, Lower Saxony	DGR—iron ore	Approved, construction (conversion)
	Asse II, Lower Saxony	DGR—rock salt	Closed, operation for keeping the mine open
Finland	Loviisa, Uusimaa	Cavern (granite)	In operation (since 1997)
	Olkiluoto; community Eurajoki, Satakunta	Cavern (granite)	In operation (since 1992)
France	La Manche near La Hague	Near surface	Closed
	L'Aube bei Soulaines-Dhuys	Near surface	In operation
	Centre de l'Aube	Near surface	In operation until 2080
Great Britain	Drigg not far from Sellafield	Near surface	In operation
	Dounreay/Scotland (County Caithness)	Near surface	Closed, should be cleared out
Japan	Rokkasho Muri, Präfektur Aomori	Near surface	In operation
Canada	Kircadine, Ontario	In lime stone	Approval stage
Sweden	SFR Forsmark not far from NPP Forsmark		In operation
		Near surface/cavern (granite)	
	Oskarshamn	Near surface	In operation
	Ringhals	Near surface	In operation
	Studsvik	Near surface	In operation
Switzerland	Decision open until 2050/2060	In mudstone	Location selection in progress
Spain	EL Cabril near Cordoba	Near surface	In operation, waste retrievable
USA	*Exemplary*		

(continued)

Table 6.6 (continued)

Country	Location	Type	Status
	Barnwell, South Carolina	Near surface	In operation
	Richland, Washington	Near surface	In operation
	Clive, Utah	Near surface	In operation
	Andrews County, Texas	Near surface	In operation
	Waste in 120 locations in 39 federal states		

[a]DGR—Deep Geological Repository

company assumes the tasks of Asse-GmbH, „Deutsche Gesellschaft zur Bau und Betrieb von Endlagern für Abfallstoffe mbH" (DBE) and the operator tasks of the Federal Office for Radiation Protection (BfS). Furthermore, the law provides that the state tasks of supervision and licensing in the field of nuclear technology, temporary storage, site selection and repository monitoring are to be bundled in one authority— the Federal Office for the Safety of Nuclear Waste Management (BfE). Thus, the BfE became the central licensing and supervisory authority independent of the repository operator. The states of Lower Saxony and Saxony-Anhalt remain the nuclear licensing authority and remain responsible for the supervision of the Asse II mine, Konrad repository and ERAM project. The responsibility of the state ends for the Shaft Konrad repository with its commissioning, and for the Morsleben repository with the conclusion of the ongoing planning approval procedure for decommissioning.

The BfS then concentrates on the state tasks of radiation protection and the monitoring networks for environmental radioactivity. This has to be taken into account in the following section which describes the processes with historical accuracy. However, it should be noted that this no longer corresponds to the current organisational structure for the disposal of radioactive waste.[1] Moreover, the new organisational structure for the disposal of radioactive waste still has to prove its efficiency. The alignment and correlation with international processes is an ongoing task.

---

[1]The bibliographical references are difficult, in each case the BfS is marked as author, and the literature produced by the BfS, even if the BfE has in the meantime taken over the tasks and thus inherited the library estate of the BfS.

6.3   Requirements and Their Implementation
for a Long-Term Safe and Long-Term Stable Storage

6.3.1   Limits of the Following Considerations

The long-term safety assessment of a repository for low-level radioactive waste also aims to permanently prevent the escape of radionuclides from the inventory into the biosphere. However, due to the situation described above, it is not appropriate to present here a long-term safety record for the final disposal of radioactive waste. Instead, reference is made to the available assessments (proofs) and this is evaluated. This is because, *inter alia*:

– The choice of location and thus the type of final disposal has been made; in each case storage in deep geological formations has been chosen,
– In the respective planning approval documents no information has been provided regarding accessibility to the storage areas for the duration of operations as these are unnecessary. Reversibility/retrievability/recoverability plays either no, or only a minor role in the disposal concepts of the ERAM and the Konrad mine.
– The determination of a probability of failure is not available for any repository site. Only deterministic methods were used.
– The classifying of the repository containers as a technical barrier in the multi-barrier concept of a repository was not adopted. The contribution of the waste matrix and the containers to the isolation of the radionuclides, and thus the limitation of the release of the radionuclides from the waste matrix and from the container into the effective containment zone, was not implemented in the repository concepts, see K. Kugel [9].

An integrated concept for low- and intermediate-level radioactive waste has not been presented separately. In the approval documents to be submitted, this manifests itself in a series of individual claims, some of which were further clarified within the administrative procedure or were substantiated with incidental provisions. One reason may be that Germany has not pursued the principle of one-repository-site. For this reason, reference is made to the concept of determining long-term safety for HAW presented under Sect. 8.4.3, which is largely applicable to low- and intermediate-level radioactive waste and possibly will be applied in the respective planning approval documents or in the safekeeping of stored waste at the Asse mine.

However, one advantage of the procedure outlined in 8.4.3 can be seen: the repository sites for low- and intermediate-level radioactive waste are located in different host rocks:—rock salt, a ductile host rock, the ERAM and in an iron ore horizon (coraloolite—sedimentary oolitic iron ore minette type)—see the Geological cross section of the Konrad mine, at Fig. 6.6. In this way experiences with different host rocks were gained.

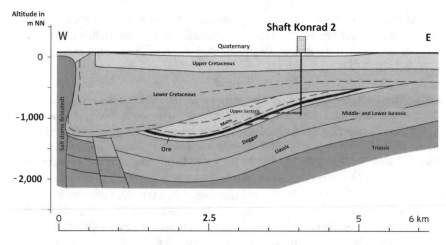

Fig. 6.6 Geological cross section in the area of the Konrad mine; Iron ore deposit, thickness of (4–18) m with a clay barrier appr. 400 m thick. *Source* BfS [10]

6.3.2 Shaft Konrad/Repository Shaft Konrad—for LAW and MAW with Negligible Heat Generation

In Germany, the disposal of radioactive waste comes down to a "two-repository-solution" in deep geological formations. These are a repository for low- and intermediate-level radioactive waste with negligible heat generation and a repository for heat-generating, high-level radioactive waste. However, it is difficult to predict future developments.

Shaft Konrad is a disused iron ore mine in Salzgitter/Bleckenstedt and is currently being converted into a repository for radioactive waste with negligible heat generation. It was selected as the best possible site for the disposal and long-term safe storage of radioactive waste with negligible heat generation, [11] by the Federal Ministry of the Interior (1983). In accordance with the planning approval decision of 22 May 2002, which finally confirmed the ruling of 26 March 2007, the mine will be converted into a repository for low- to medium-level active waste, so that a nuclear facility is created with a maximum storage capacity of 303,000 m^3 for "radioactive waste with negligible heat generation". The disposal will take place at depths between (800 and 1300) m.

In contrast to the ERAM and the Asse II mine, the Konrad mine will be expanded according to a repository concept, operation, decommissioning and final storage; these elements are listed in the planning approval documents, [12] published by the Lower Saxony Environment Ministry as Planning Approval Decision, 2002. The process will be optimised for normal operational and in the decommissioning phase. Recent findings from repository research will be incorporated into the optimisation procedure.

The radioactive waste disposal requirements for the Konrad repository specify the maximum activities of ten relevant radionuclides and two radionuclide groups that can be stored at the end of the operational phase of the Konrad repository, see Brennecke/BfS [13].

As already mentioned, Schacht Konrad is a disused iron ore mine in Salzgitter/Bleckenstedt, see Fig. 6.5. It was selected as the best possible location for the disposal and long-term storage of radioactive waste with negligible heat generation.[2] At the time of selecting the Konrad shaft as the location for a repository for low and intermediate level radioactive substances, the selection criterion "best possible location" did not exist. This term also has no relevance to the long-term safety assessment, because it only has to prove that a repository site will be secure in the long-term or not. However, in the case of the Konrad site, there was a selection process, although it did not consider the whole of Germany and may therefore be considered imperfect. Decisive in the selection of Schacht Konrad were the extremely favorable hydrological conditions in the Gifhorn trough, see Fig. 6.6. From a specialist point of view, the location Salzgitter/Bleckenstedt is therefore considered to be suitable for an installation for the reception and long term safekeeping of radioactive waste.

The Konrad mine will now be converted to the repository for the storage of a maximum of 303,000 m^3 of "radioactive waste with negligible heat generation" in accordance with the planning approval decision. Thus, a facility for receiving this radioactive waste will become available. The planning application was accompanied by a geoscientific long-term safety forecast for a period of observation of at least 100,000 years. This long-term safety prognosis integrated the forecast of the stresses of the shaft backfills, the drifts and shaft seals as well as the possible interactions between rock, backfill material and solutions and gases.

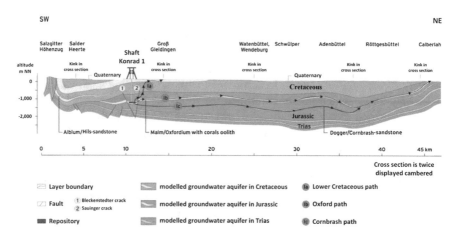

Fig. 6.7 Representation of the dispersal paths in the area model for Schacht Konrad. *Source* BfS [10]

---

[2]The increase in the temperature of the host rock caused by the decay heat must not exceed on average $\Delta T \leq 3$ K.

As part of the licensing procedure for the Konrad repository, it was found that the Konrad mine is suitable for conversion to a repository for radioactive waste with negligible heat generation, in particular because:

- The geo-mechanical conditions in the area of the chambers and the entire mine workings should not cause any rock damage nor permit gas pressure build-up,
- The properties of the geological formations serve as barriers against the spread of radionuclides (permeability, sorption behavior),
- A long-term seismic stability of the site can be predicted,
- Because the mine is exceptionally dry,
- etc.

The site characterization shows that above the storage fields the overburden rock layers of claystone form exceptionally good and effective natural barriers against the transfer of radioactivity into the biosphere (Fig. 6.6). Reference [10] describes the site situation as follows:

> The occurrence of iron ore does not occur anywhere at the earth's surface and is covered over a large area by a layer of up to 400 m thick of clayey rocks of the Lower Cretaceous. These are overlain by several hundred meters of limestone of the Upper Cretaceous. The claystone forms an effective natural (geological) barrier to the groundwater and thus prevents transfer of contaminants to the biosphere.

The hydrology of the site is quite complex; the well water, triassic sandstone and limestone formations (Keupersandstein, Upper Muschelkalk), which are found at a depth of about 2000 m, are responsible for the deep-water movement around the Konrad mine. These are located in the south of the mine, at an elevated topographic level in the Salzgitter mountain range, at the surface, see Fig. 6.7. For the long-term safety proof, fresh water conditions were assumed in the model calculations. Under this boundary condition, which does not map the conditions in situ, an increased hydraulic potential is identified, which leads to a small, northward movement of water within the Coraloolith. The long-term safety analysis calculations showed that even with this tendency to overestimate the permeability of rocks, the water pathways emanating from the mine workings can reach the biosphere at the earliest after about 300,000 a, and so long lived radionuclides from the inventory may be safely stored on this water pathway, see Table 6.5. This may lead to a potential radiation exposure of the population, which in the long term will remain well below the fluctuation range of the natural radiation exposure of ≤ 0.3 mSv/a. The system of natural and geotechnical barriers thus ensures that an unacceptable burden on the biosphere emanating from the Schacht Konrad repository is unlikely to occur, even in the long term. The long-term safety could thus be considered proven and resulted in the approval of the plan for the Konrad mine repository, see [12] Lower Saxony Ministry of the Environment (2002).

In order to achieve implementation of the approved plan for the construction of the Schacht Konrad repository, a formulation of the so-called 3K criterion was crucial. The 3K criterion means limiting the thermal impact of the host rock to 3° K at the sidewall of the mine workings. It therefore follows that only radioactive waste with negligible heat generation may be stored there and thus the radionuclide vector of the future inventory is largely prescribed.

The requirements for the radioactive waste destined for final disposal at the Konrad repository include the maximum storable activities of ten relevant radionuclides and two radionuclide groups at the end of the operating phase of the Konrad repository (Table 6.7), see Brennecke/BfS [13]. Since these radionuclides and radionuclide groups correspond to those set out in approval issued under water law, non-compliance would also lead to a violation of the approval issued under water law,[3] see [13] Accounting Standard (2010).

- Radionuclides H-3, C-14, I-129, Ra-226, Th-232, U-235, U-236, U-238, Pu-239 and Pu-241 and
- Radionuclide groups total-Alpha emitters and total-Beta-/Gamma emitters; see Table 6.7.

The balance sheet of the activity values for radionuclides and radionuclide groups, total α-emitters and total β/γ-emitters is stipulated in the approval decision for the Konrad repository. The deviations in the activity data in Table 6.7, between the sum values (total α-emitter and total β/γ-emitter) and the individual values are due to the fact that the radionuclides from the Th-232 decay series and the U-238 decay series in total α-emitters and total β/γ-emitters, see Chap. 2, Fig. 2.2. Any influence on the radiotoxicity index (right column, Table 6.7) is low.

The comparison with the nuclear installation Asse II shows that the activity indices are of the same order of magnitude, see Table 6.8.

According to the acceptance criteria, the conditioned containers, see [9], are to be stored in storage chambers, whereby several storage chambers will form a storage field. The storage fields will be located in the area of low grade iron ore that is partially mined out, see Fig. 6.6. For the excavation of the storage fields (theoretically up to nine units), six main levels are available at 800, 850, 1000, 1100, 1200 and 1300 m. The Konrad shaft consists of two pits, Konrad I (about 1232 m deep) and Konrad II (about 999 m deep).

Underground, the infrastructure, transport routes and a special ventilation system for the storage operation will continue to be created and the two shafts rehabilitated and converted. In parallel, the above-ground infrastructure will be advanced; this includes items such as the construction of roads and railway connections as well as building facilities for handling and checking the waste packages to be stored.

However, as the Konrad repository has not yet been completed and therefore is currently without radioactive waste, the interaction of stored radioactive waste—repository structures cannot be discussed here. From a specialist point of view, Salzgitter/Bleckenstedt is a sufficiently suitable location for a facility for the reception of radioactive waste and for their final long-term safekeeping. The completion date has been regularly extended. Currently completion is expected around 2030.

---

[3]The approval issued under water law is granted for a limited period of 40 years, [14].

Table 6.7 Maximum storable activities (Radiotoxicities) of relevant radionuclides and radionuclide groups at the end of the operational phase of the Konrad repository

Radionuclide/radionuclide group	Half-life [a]	Exemption limit FGi [Bq]	Activity Ai [Bq]	Radiotoxicity Ai/FGi [–]
H-3 (β^-)	12.3	1.00E+09	6.0E+17	6.00E+08
C-14 (β^-)	5700	1.00E+07	4.0E+14	4.00E+07
I-129 (β^-, γ)	1.60E+07	1.00E+05	7.0E+11	7.00E+05
Ra-226 (α)	1600	1.00E+04	4.0E+12	4.00E+08
Th-232 (α)	1.40E+10	1.00E+04	5.0E+11	5.00E+07
U-235 (α)	7.00E+08	1.00E+04	2.0E+11	2.00E+07
U-236 (α)	2.30E+07	1.00E+04	1.0E+12	1.00E+08
U-238 (α)	4.40E+09	1.00E+04	1.9E+12	1.90E+08
Pu-239 (α)	24,000	1.00E+04	2.0E+15	2.00E+11
Pu-241 (β^-)	14.4	1.00E+05	2.0E+17	2.00E+12
Total α-emitters			1.5E+17[a]	2.20E+12[b]
Total β/γ-emitters			5.0E+18	6.41E+08[c]
Total-Sum	–	–	5.15E+18	2.20E+12

Source Brennecke and Steyer (SE-IB-33/09-REV-1) [13]

[a]Including radionuclides from the Th-232 decay series and the U-238 decay series in total

[b]Sum of radiotoxicities of α-emitters without radionuclide groups

[c]Sum of the radiotoxicities of the β/γ-emitters without radionuclide groups

Table 6.8 ERAM: Radionuclide vectors of relevant radionuclides [15, 16]

Radionuclid	Half-life in a	Activity in Bq, stored until 01.07.1991	Activity in Bq reporting date: 30.06.2005	Activity in Bq reporting date: 31.12.2013
Am-241	432.6	1.5E+11	2.2E+11	2.3E+11
Am-243	7.365	6.4E+05	9.5E+07	9.5E+07
Pu-239	2.41×10^4	6.2E+10	6.8E+10	6.9E+10
Pu-240	6563	5.8E+10	6.6E+10	6.6E+10
Pu-242	3.73×10^5	1.1E+06	1.2E+08	9.9E+07
Ra-226	1600	2.3E+10	3.9E+11	2.3E+10
C-14	5730	3.0E+12	3.4E+12	3.2E+12
Ca-41	1.03×10^5	1.8E+07	7.3E+07	7.3E+07
Co-60	5.27	6.4E+13	2.5E+14	5.4E+12
Cs-134	2.07	9.1E+12	8.3E+10	9.4E+09
Cs-135	2.0×10^6	1.5E+08	3.7E+08	3.7E+08
Cs-137	30.17	6.4E+13	1.4E+14	6.3E+13
Ni-59	7.5×10^4	6.1E+10	1.8E+11	1.8E+11
Ni-63	100.1	9.8E+12	1.8E+13	1.4E+13
Pu-241	14.35	1.6E+12	1.4E+12	9.0E+11
Ra-228	5.75	1.6E+12	8.2E+11	3.6E+08
Sr-90	28.7	4.6E+12	5.9E+12	4.8E+12
Am-242 m	141	5.4E+06	1.2E+08	2.3E+08
Stored Volume				
Total:	36,754 m^3	1.58E+14	4.21E+14	9.19E+13

The retrofitting would have taken about 22 years and will have provided much information, some of which may be used for the construction of a repository for heat-generating HAW, see Chap. 7. The planning approval assumes an operating period of up to 80 years. During the post-operational phase (decommissioning phase), of course, an optimization is also to be provided.

The company commissioned to operate the Schacht Konrad repository will accept the radioactive waste for a fee. If the costs are calculated appropriately, the Konrad repository should be cost neutral.

6.3.3 Repository of Radioactive Waste—Morsleben (ERAM)

6.3.3.1 Description of the Location of the ERAM and the Inventory to Be Stored

The Morsleben (ERAM) repository for radioactive waste has been developed at the mine of the former Burbach-Kali AG, using the Bartensleben and Marie shafts. "In Upper Allertal there is a salt deposit of 40–50 km in length and an average of 2 km in width, which was developed at the end of the 19th century. Potash and rock salt were mined here for about 70 years. The resulting mine is 5.6 km long and 1.7 km wide. The Bartensleben mine was sunk to 522 m and the pit was laid out with four main mining levels. Mining took place by the open stoping-method without backfilling (Fig. 6.8).

This created cavities up to 120 m long with a width or height of up to 40 m. The chambers used until 1998 for final disposal are located in the peripheral area of the mine, see Fig. 6.9. The Marie shaft serves exclusively as the 2nd exit and the ventilation shaft." See illustration DBE TEC mbH.

A total of 36,754 m^3 of radioactive waste and 6621 sealed radiation sources were deposited on various levels and in various minefields. Due to the activity restriction, the overall activity of the ERAM is relatively low.

The ERAM—the final repository for radioactive waste—was available from for the disposal of low and medium radioactive waste 1971 in the former GDR. After the reunification of Germany, it served for the reception of this waste from the reunified Germany until September 1998, with some interruptions. Waste from the nuclear power plants as well as waste from the fields of research, industry and medicine were received for storage.

Following a re-evaluation in 2001, the BfS irrevocably renounced further storage in Morsleben, as it was no longer acceptable in terms of safety.

Due to the activity restriction, the overall activity of the ERAM is relatively low. The radionuclide vectors summarized in Table 6.8 contain only the radionuclides which make the biggest contributions to the total stored activity:

Alpha-emitters: Am-241, Pu-239, Pu-240, Ra-226

Fig. 6.8 Arial photograph ERA Morsleben. *Source* BfS 2009

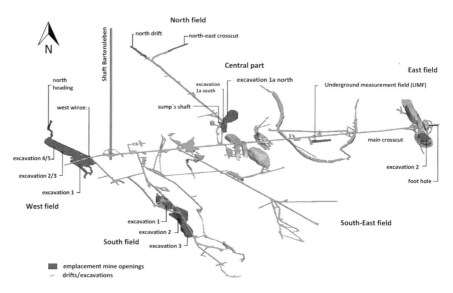

Fig. 6.9 Overview of storage areas in ERAM. *Source* Federal Office for Radiation Protection [17]

Beta/Gamma-emitters: Cs-137, Co-60, Ni-63, Sr-90, C-14

According to [15] the total stored activity was appr. 4.21×10^{14} Becquerel, the activity of alpha emitters was about 8.0×10^{11} Becquerel (reporting date: 30.06.2005), BfS [15].

In September 2005 the BfS[4] submitted to the Ministry of Environment, Agriculture and Energy (MLU) of the state of Saxony-Anhal a suite of design documents for the decommissioning of the ERAM; these included plans, such as the decommissioning plan for the ERAM, and documents for the environmental impact assessment. The procedure for plan approval has been in progress since then.

The long-term stable storage of radioactive waste is generally defined such that these materials are deposited, isolated, so that any escape of toxic radioactive contaminants to the air-, water- and soil pathways into the biosphere, are only within the socially and legally accepted limits. As can be seen from Fig. 6.10, the host rock, the salts between the salt base and the cap rock present a maximum thickness between 400 and 600 m, with the salt table located at a mean depth of approx. 140 m. The minimum thickness of cap-rock of the salt body is about 150 m [18] (Table 6.9).

Stabilization measures

As a former production facility, the mine structure is quite deep, especially in the central part of the mine where many cavities were created by the excavation of salt. Since these cavities had existed for a long time and some damage existed, it was necessary to provide support to the surrounding rocks, see Fig. 6.10. In total, 27

---

[4]Ownership has been transferred to the BGE.

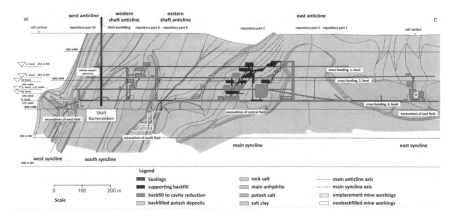

Fig. 6.10 Schematic section through the backfilled and closed pit cavities of ERA Morsleben. *Source* Federal Office for Radiation Protection [17]

Table 6.9 Compilation of the total activity Ai and the value of the activity indicator Ai/FGi (FG—exemption limit according to Annex 4 Table 1 Sp. 2 StrlSchV) for the radiologically most important radionuclides in the inventory of the Morsleben repository

Radionuclid	Half life	FGi	ERAM (30.06.2005)		ERAM (31.12.2013)	
			Ai	Ai/FGi	Ai	Ai/FGi
	[a]	[Bq]	[Bq]	–	[Bq]	–
C-14	5700	1.0E+07	3.4E+12	3.4E+05	3.20E+12	3.20E+05
Co-60	5.3	1.0E+05	2.5E+14	2.5E+09	5.40E+12	5.40E+07
Ni-63	100	1.0E+08	1.8E+13	1.8E+05	1.40E+13	1.40E+05
Sr-90	28.5	1.0E+04	5.9E+12	5.9E+08	4.80E+12	4.80E+08
Cs-137	30.2	1.0E+04	1.4E+14	1.4E+10	6.30E+13	6.30E+09
Ra-226	1600	1.0E+04	3.9E+11	3.9E+07	2.30E+10	2.30E+06
U-234	250,000	1.0E+04	1.1E+09	1.1E+05	1.10E+09	1.10E+05
U-238	4.40E+09	1.0E+04	4.3E+08	4.3E+04	4.30E+08	4.30E+04
Pu-238	87.7	1.0E+04	k.A.	0.00	7.80E+10	7.80E+08
Pu-239	24,000	1.0E+04	6.8E+10	6.8E+06	6.90E+10	6.90E+06
Pu-240	6600	1.0E+03	6.6E+10	6.6E+07	6.60E+10	6.60E+07
Pu-241	14.4	1.0E+05	k.A.	0.00	9.00E+11	9.00E+06
Am-241	432.6	1.0E+04	2.2E+11	2.2E+07	2.30E+11	2.30E+07
Cm-244	18.1	1.0E+04	6.6E+09	6.6E+05	4.80E+09	4.80E+05
Summe			4.18E+14	1.72E+10	9.19E+13	7.72E+09

Sources [15, 16, 19]

chambers of the central part were backfilled at a depth of 370–500 m, see Fig. 6.9. For this purpose, entrances to the workings had to be closed using dams, see [20].

Hydraulically setting salt concrete is used for backfilling. These stabilization measures prevent failure of the pillars and roofs, providing support for the layers between two superimposed excavating chambers. However, the actual closure measures can only commence once the plan has been approved.

Safety barriers Internationally, for final disposal of radioactive waste in deep geological strata, multi-barrier concepts are often implemented in existing repository projects, see Chap. 8. This is also provided for in the case of ERA Morsleben. The long-term safety concept must consider the totality of the various barriers. Thus, a distinction has to be made between the safety of the subsystems (individual components) and the safety of the geotechnical environmental structure (repository). These assessments must be coordinated. Within the safety barriers, the effective containment zone (ewG) is of particular importance. The effective containment zone is the part of the repository which, in conjunction with the technical closures (shaft seals, chamber sealing structures, dam structures, backfilling, …) ensures the embedding of the waste and for which, based on geotechnical knowledge, a statement about the long-term stability (integrity) of the repository is possible. The individual components of the system are:

– Technical barriers (waste product, waste packaging—packages). In the case of the ERAM, the repository containers are not designed as a technical barrier.
– Geotechnical barriers (backfill, sealing structures—filled and closed mine cavities of ERA Morsleben), see Fig. 6.10.
– The effective containment zone (ewG) ensures the encapsulation of the waste. Based on geotechnical knowledge, a statement about the long-term stability (integrity) of the repository is possible. In the case of the ERAM an ewG has not been designed. This is because the high excavation ratio already present means the ewG will only work with limited effectiveness in accordance with the criteria
– Geological barriers (host rocks, dirt bed/overburden—site conditions for which quantification of long-term properties is not possible). In the case of ERAM, the geological barrier consists of rock salt. The dirt bed/overburden is characterized in particular by the cap rock. The minimum cover of the salt body is about 150 m. In the case of the ERAM, the effectiveness of the isolation is also limited by the rock strata present in the Allertal structure and their barrier-effective characteristics and, in particular, by the former use of the area as a salt mine, see Fig. 6.10.
  The geological barrier at the site has different principles of effectiveness with regard to restraining and delaying effects (safety functions):

  – Retardation (delay)
  – Retention (retention—isolation)
  – Protection (protection—mechanical, chemical)
  – Dispersion (dilution).

– The overall safety of the geotechnical environmental structure (repository) must ensure the safe isolation of the stored radioactive waste, so that a escape of toxic,

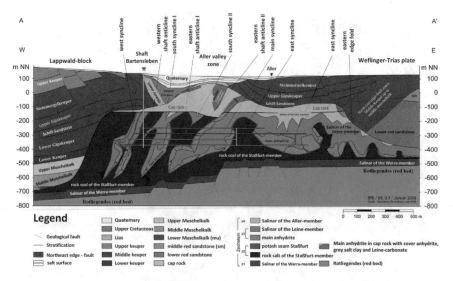

Fig. 6.11 Schematic geological SE-NW-section through the saline structure Allertal in the area of the Bartensleben mine. *Source* Federal Office for Radiation Protection [17]

radioactive contaminants to the air, water and earth pathways into the biosphere is avoided or is kept within the socially and legally accepted limits of pollution, see [21].

Figure 6.11 shows a profile section in the area of ERA Morsleben. This profile section shows that the overburden is not built up homogeneously but is traversed by some discontinuities. The Morsleben site can not be described as the best possible for a repository of low and medium radioactivity waste with negligible heat generation. The term barrier in a repository concept describes the geological conditions and/or technical or geotechnical measures to impede or prevent the release of contaminants of the waste into the biosphere, see [22].

The decommissioning of the Bartensleben and Marie, shaft involves insertion of seals which isolate the mine shafts from the biosphere, see Fig. 6.12. The shaft seals also function as geotechnical barriers, see also [23].

The long-term safety concept for the ERA Morsleben is based on the effects of the installed technical and geotechnical barriers and on the sum of the resisting and retarding effects of the geological barrier (effect of all geological layers of the geological barrier) under the site-specific boundary conditions (e.g. former potash and rock salt mines, at Bartensleben and Marie).

6.3.3.2 Costs

The total costs for the process are the responsibility of the federal government. The fees collected for storage commenced after reunification until 1998 can be used to

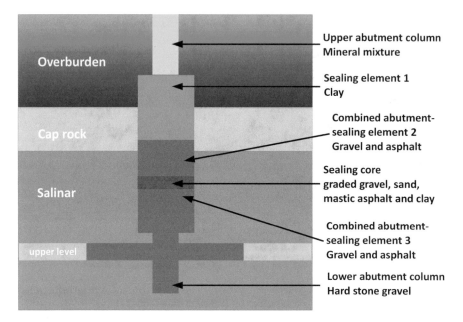

Fig. 6.12 Sealing plug of the Bartensleben and Marie shafts, used in decommissioning of the ERAM, according to data from [17]

offset some costs. The costs for the orderly closure of a nuclear installation on the basis of the pending planning approval may be up to 3.0 billion euro according to reputable estimates.

6.3.3.3   Illustrative Considerations on the Long-Term Safety Proof—The Post-closure Phase of the ERAM

A proof of long-term safety also appears as an analogy in Chap. 8. However, significant exemptions are to be granted because the storage was carried out in a former potash and rock salt mine; also approval was granted under GDR law. Most of the key security requirements for a repository system to be developed in Germany listed in Chap. 8 cannot be subsequently introduced, because:

– the ERAM is an old mine,
– the former mine shows extensive effects of mining activity,
– the cavities already had displayed some damage,
– the support function of the surrounding rocks had already been impaired.

Currently, ongoing stabilization measures are being implemented. The actual closure measures can only be started after the plan has been approved. The central safety requirement: "For an assumed observation period (detection period), it must be shown that only very small amounts of pollutants can be released from the repository." The prerequisite for this is that the integrity of the isolating rock formation is demonstrated and the risk of contaminant escapes from the repository can be assessed and presented in the plan.

Considerably less than 1 million years can be set for the ERAM as the detection or observation period since it can be shown that the activity of the stored inventory after 10,000 years will have dropped to approx. $A_{10,000} = 8E+11$ Bequerel and after 100,000 years to about $A_{100,000} = 5E+08$ Bequerel.

The following statements on long-term behavior using the example of ERA Morsleben can also be transferred in an exemplary manner to Chap. 8 because they are generally valid where the host geology is rock salt. The long-term safety assessment must show that the value of 0.3 mSv/a for the effective dose for an individual of the population (stated in §47 of the StrlSchV for the operation of nuclear facility), is not exceeded aa a result of the discharge of radioactive substances into the environment, including in the period after closure of the ERA Morsleben. The safety criteria define the final disposal of radioactive waste as a maintenance-free and indefinite disposal of this waste. "Therefore, the decommissioning of ERA Morsleben will be carried out in such a way that no control and monitoring program will be required after completion of all decommissioning measures", see [17]. Paragraph 55 (1) no. 5. Federal mining law (BBergG) requires that sufficient care be taken to protect the current land surface so that, in the long term, there can be no particular subsidence on the surface which may have unacceptable effects on the protected materials. A long-term safety assessment for a repository mine in saline rock, consists of considering:

(a) the stability of the partially backfilled and closed mine workings, the influence of the convergence of the saliferous rock and the influence, and in particular the convergence, on the deformations (subsidence and possibly uplift) of the terrain surface. The item to be looked for is that any possible subsidence on the earth's surface as a result of the convergence of the remaining cavities in the former mine is to be limited.

(b) evidence that in the post closure phase radionuclides enter the biosphere due to transport processes which can not be completely excluded, which lead to individual doses exceeding the values of § 47 StrlSchV (an inadmissible radiation exposure or contamination of persons, material goods or the environment is excluded or limited to a minimum). The threshold is ≤ 0.3 mSv/a as the effective dose of an individual from the population.

- **Exemplary proof of the robustness and integrity of the repository structure (ENW 2) and the individual barriers of the multi-barrier system (ENW 3)**

This proof is to be reproduced here as an example, the process presented can also be applied to other repository structures where the host rock is "salt". This not claimed to be a complete proof. According to (a) the prognosis of the stresses of the shaft backfillings, the drifts and shaft closures as well as the interactions between rock, backfill material and possibly upcoming solutions and gases is carried out using geomechanical and fluid mechanical model calculations. The determined stress states are compared and evaluated with the given safety criteria, the permissible stress and deformation states, the distances to the limit states and failure states. The outcome is presented from the operator's perspective in such a manner that the admissibility of the decommissioning procedure may be reasonably assured.

- Temporal changes of the stated variables were included in the considerations. This implies that a parameter variation in the predicted range is included in the calculations.
- According to the project proponent, "salt concrete" is primarily used as the material for the backfilling measures, consisting of: binding agent (cement), concrete additives (rock flour, hard coal fly ash or similar), aggregates (salt breeze, quartz sand or similar), mixing liquid (water, salt solutions) (a), see [17].
- The stresses that occur can lead to contaminant distribution pathways should be avoided by introducing appropriate countermeasures.
- According to the project proponent, a magnesia concrete is used as the building material for the closure dams (dam structures), which is produced from the following basic materials: binder (magnesium oxide), aggregates (quartz sand, anhydrite powder), concrete admixtures (microsilica), superplasticizer; Mixing liquid (magnesium chloride solution). The magnesia concrete also has an immobilizing effect, but also has a tendency to corrode, see [17].

The following geotechnical tests and assessments must be carried out and submitted to the licensing authority:

- Reliability and confidence limits of the stated variables and their temporal changes,
- The geomechanical model calculations determined temporally for non-constant stress conditions according to the given safety criteria, the permissible stress and strain conditions, the distances to the limit and failure states as well as the comparisons and evaluations,
- The fluid dynamics model calculations determined, temporally variable flow paths, whether and how fluids can penetrate to reach the stored waste containers and whether and how any dispersal of possibly leaked radionuclides from the repository structure would flow and what amount of radioactivity could thereby escape into the biosphere.

In order that the decommissioning procedure can be justified, it must be shown that the specified protection objectives can be achieved with a high probability of reaching the required safety standard. This is done using the sensitivity analysis and probabilistic analysis described in Chap. 8.5 and Fig. 6.13. The sensitivity analysis is used to analyze the reliability and confidence intervals of the stated variables as well as the boundary conditions; these are fed into a calculation procedure that will determine which results are to be expected for any specified input values. For this purpose, the sensitivity analysis determines the dependence of the result on any change of the input values. This indicates those input values to which the result reacts most sensitively. With the sensitivity and uncertainty analysis, the input values and boundary conditions are specified more precisely, the confidence in them is increased and thus the statements of the calculations are verified. Verbal assessments may also

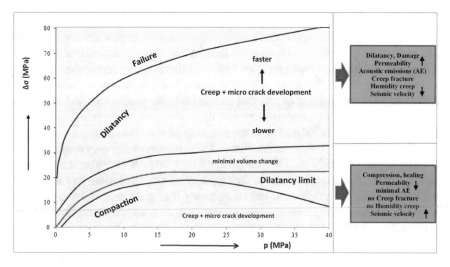

Fig. 6.13 Mechanical behavior of salt from laboratory tests. *Source* According to the BGR—Federal Institute for Geosciences and Natural Resources, [24]

be helpful. When designating the long-term safety for the ERAM repository mine, the ultimate aim is to identify the possible solution and assess the influence of the uncertainties, so that the licensing authority feels able to issue the planning approval. Reference is made to Chap. 8.

There is a need to prove the stability of the mine workings to be closed, the drifts and shaft seals of the ERAM and the influence of the convergence of the in situ salivary rock and residual cavities and the resulting impact on the enclosure of the waste containers and the stability and possible subsidence of the ground's surface. At this stage one must fall back on the laws of salt mechanics.

For example, the mechanical behavior of rock salt, as with all polycrystalline solids, depends on the crystal properties and grain boundary behavior. The extent to which the mechanical behavior is determined more by processes inside the grain or by interaction between the grains depends to a large extent on the effective stress, the temperature and the grain size. Characteristic of salt rocks is their ability of creeping; that means under load the rocks may significantly deform plastically in a time-dependent manner. The plastic and viscous deformation components resulting from the creep cannot be clearly separated experimentally. They are therefore summarized for visco-plastic deformation.

To obtain firmly grounded knowledge about the time-dependent inelastic deformation behavior from the acting stresses and the temperature, requires a considerable number of different laboratory tests and investigations on simple underground structures at the site, since any scratching of the rock (geological barrier) must be largely excluded. The results and the described behavior of the salt type are always site specific [25]. Extensive investigations are available for the ERAM. These are considered in [17].

In salt cavern mining for example, the prediction of the convergence- and load-bearing behaviour of heavily loaded load-bearing elements, depends mainly on taking into account the progressive behaviour of the material as a consequence of the mining operations; deformation resulting from load-transfers is determined by the operating period, varying from a few years to a few decades. For this purpose, material models in the transient area were required that enabled reliable predictions and statements regarding the development of stability over time. However, underground disposal of radioactive waste (as in ERAM or in the planning of repositories) requires forecasts over several millennia, using material models which enable stationary creep behavior to be described. Initially only the undamaged rock body is considered as the geological barrier. However, this can be damaged by mining operations as for example at ERAM and/or by inflow of solutions and may thus show a completely different mechanical behavior, in terms of bearing capacity and consequently the development of pathways. Questions about the impermeability of the saline barrier or the stability of a supporting element cannot therefore be answered conclusively without knowing the state of deformation and damage in addition to the stress state, see ERAM. The possibility of mending existing crack structures should be noted here, see Fig. 6.13.

The complexity of the task also includes the modeling of the geotechnical barriers, here the backfilling (salt grit backfilling) and various construction options for dam structures. The material models for the backfilling are significantly different from those of the rock body. On the basis of the existing model approaches, a continuous development resulting in today's complex material models has occurred. In the course of this development, a growing understanding of the deformation phenomena and -mechanisms is reflected, see Fig. 6.13.

The required material parameters are obtained from laboratory tests and from the recalculation (benchmark calculations) of stress-controlled creep tests and deformation-controlled strength tests from the laboratory as well as simple underground structures, see Figs. 6.12 and 6.13. They are initially valid for a specific type of salt.

The following phenomena, described by most of the material laws for a salt type with a uniform variable set, have been included in this comparison, also see requirements for the rock salt geological barrier,

- transient creep after stress increases and -reductions,
- stationary creep,
- development of volumetric deformation (loosening/dilatancy),
- development of damage,
- fracture and creep fracture,
- post peak behaviour and residual strength,
- influence of different differential stresses (loads),
- Influence of the confining pressure (in laboratory tests)/the minimum principal stress (underground), as well as
- Influence of moisture on the deformation.

The functional requirements of rock salt and of the backfilling material, as derived and described in [26, 27], are summarized as follows, see also [28]:

- mechanically stabilize the natural geological barrier,
- transmission of the decay heat of the heat-generating waste into the host rock (not applicable for the ERAM),
- reduction of the cavity volume in the mine workings,
- have a high initial density and be installable flush to the gallery roof,
- take over a long-term sealing function like the host rock.

The theoretical basis for the description of the mechanical behavior is described in [29, 30]. The starting point for the material behavior contained in this chapter is the additive decomposition of the rate of the strain tensor, see Eq. (6.1).

$$\underline{\dot{\varepsilon}}(T, \hat{\sigma}, \sigma_0, \eta) = \underline{\dot{\varepsilon}}_{el}(T, \hat{\sigma}, \sigma_0, \eta) + \underline{\dot{\varepsilon}}_{vpl}(T, \hat{\sigma}, \sigma_0, \eta) + \underline{\dot{\varepsilon}}_{th}(T) \qquad (6.1)$$

$\underline{\dot{\varepsilon}}(,\hat{\sigma},_0,)$:   Rate of total strain tensor;
$\underline{\dot{\varepsilon}}_{el}(T, \hat{\sigma}, \sigma_0, \eta)$:   Rate of elastic part of total strain tensor;
$\dot{\varepsilon}_{vpl}(T, \hat{\sigma}, \sigma_0, \eta)$:   Rate of the viscoplastic part of total strain tensor;
$\dot{\varepsilon}_{th}(T)$:   Rate of thermal part of total strain tensor; : Temperature;
$\hat{\sigma}$:   von Mises-equivalent stress; σ_0: medium stress; η: Porosity.

Within the general context according to Eq. (6.1), the strain rate of the respective barrier (rock body, backfilling, dam structure) can be determined under the respective site conditions, which also applies to rock salt—carnallitite as in the ERAM.

At present, there is no physical reasoned modeling of rock salt creep that would be scientifically substantiated and accepted worldwide. Nevertheless, a phenomenological model proved from an early stage to be very successful, and that is widely used as a Norton model. To specify this approach for the respective conditions encountered, requires considerable adaptations in order to develop a material- and site-specific material model; this consists of a multitude of different tests and post calculations (benchmark calculations)to be derived, in order to determine the required material parameters under the existing boundary conditions and the existing time dependence. The Norton approach proves to be limited as it is obviously unsuitable for correctly detecting a very large temperature and voltage range. For the site conditions at ERAM (rock salt, carnallitite), a large number of triaxial experiments have been run at higher hydrostatic pressure and higher temperatures. To fit the experimental data, a double Norton approach was used, see also Hunsche and Schulze [31]. With the model propagated by the Federal Institute for Geosciences and Natural Resources (BGR) as a power approach for stationary creep with dislocation climb as decisive deformation mechanism, see context Eq. (6.2), the conditions of the ERAM are described adequately.

$$\dot{\varepsilon}_{cr} = \left(A_{S,1} \cdot \exp\left[\frac{-Q_1}{R \cdot T}\right] + A_{S,2} \exp\left[\frac{-Q_2}{R \cdot T}\right] \right) \cdot \left(\frac{\sigma_{eff}}{\sigma_0}\right)^{n_s} \qquad (6.2)$$

$\dot{\varepsilon}_{cr}$ – creep rate, $A_{S,i}$—preexponential factor in power approach; $Q_1 = 42 \frac{kJ}{mol}$; $Q_2 = 113 \frac{kJ}{mol}$, $n_S = 5$; $\sigma_0 = 1{,}0$ MPa

σ_{eff}—effective stress, see Fig. 6.10, here $\sigma_{\text{eff}} = \sqrt{3} \cdot \Delta\sigma$; Q—activation energy; n_S—stress exponent in power approach; R—general gas constant; R $=$ 83,144,621 $\frac{\text{kJ}}{\text{mol K}}$ [5]

From the creep behavior Eq. (6.2) one can determine the convergence of a cavity filled with salt grit in the rock salt. The salt grit blown into the cavity initially has an initial porosity:

$n = \frac{V_p}{V}$ with n—porosity; V_p—pore volume; V—total volume (e.g. of backfilling).

This is usually set to $n_0 > 0.4$. Over time, this becomes smaller and decreases within the cavity from the top to the bottom. Due to the convergence of the rock salt (cavity), the salt grit is finally compacted so that the porosity of the salt grit coincides with that of the surrounding rock formation and the containers located in the emplacement areas are surrounded by rock salt. The detection of the relative temporal change of the cavity volume takes place with the aid of the convergence rate, which takes place as temporal change of the void volume via the differential Eq. (6.3).

$$\frac{d}{dt}V(t) = -\dot{K} \cdot V(t) \tag{6.3}$$

with K—convergence:

$$K(t) = \frac{V_0 - V(t)}{V_0}$$

In addition to the general description of the convergence behavior of rock salt, two dependencies of the convergence rate must be pointed out here:

- The convergence rate depends on the cavity geometry. Therefore, the convergence approaches require different parameter values when used on drifts and chambers. For the individual emplacement areas of the ERAM, see Fig. 6.9, these were to be determined, and
- Low moisture contents of about 0.4% by weight, -% [32] can drastically reduce the supporting effect of the backfilling. This is also done with special reference to the Asse II mine, see Chap. 6.3.4.

The influence of temperature on the rate of convergence is given by Eq. (6.2).

---

[5]Uncertainty is 9.1×10^{-7}, to be taken into account in probabilistic long-term safety proof.

6.3.4 Closure of a Nuclear Facility According to § 9a Atomic Energy Act (AtG)—Asse II Mine

6.3.4.1 Description of the Location Situation

The Asse II mine, see Fig. 6.16, in the municipality of Remlingen-Semmenstedt in the district of Wolfenbüttel in Lower Saxony is a 100 year old potash and salt mine at the southern tip of the Asse- mountain range. The mine was mined in the Zechstein salt of the Leash (Na3) and Staßfurtserie (Na2) in an approximately 8 km long saddle, which trends from NW to SE. The vertical section in Fig. 6.17 shows in the center of the mine, that the south flank with an angle of approximately 70 degrees is tipping more strongly than the northern flank. The rock salt field on the southern flank was excavated in the immediate vicinity to the southern overburden. On the upper levels, the rock salt barrier to the overburden has only a minimal thickness of about 10 m. It is clear that the mine workings were directly influenced by the construction of the Buntsandstein and Muschelkalk mines within the Triassic rocks; whose tectonic disturbances can also be seen in Fig. 6.14 (Fig. 6.15).

The salt table is about 300 m below the terrain surface. At the salt table, easily soluble minerals are dissolved (sub-corrosion) leaving behind insoluble and poorly soluble constituents and transformation products, such as clay and gypsum. These form the cap rock.

Not discussed here are the disputes that are often fueled by political and with tendentious "scientific" statements and presentations about the condition of the Asse mine. Rather we consider the mine's location and its suitability for long-term safe, long-term stable disposal of the LAW and MAW. The long-term safety assessment is set out at Chap. 8. Additional statements on salt mechanics can be found in chapter 6.3.3.3 (ERA Morsleben).

The Asse mine is classified as an installation according to § 9a AtG, but it is not a repository. The costs for the entire process are borne by the federal government.

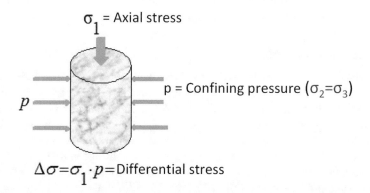

$$\sigma_1 = \text{Axial stress}$$

$$p = \text{Confining pressure } (\sigma_2 = \sigma_3)$$

$$\Delta\sigma = \sigma_1 \cdot p = \text{Differential stress}$$

Fig. 6.14 Stress-controlled creep test. *Source* According to the BGR, [24]

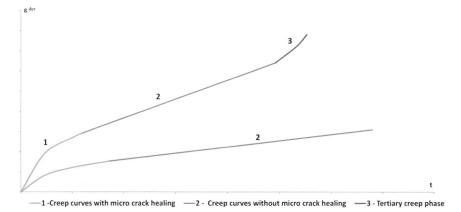

—1 -Creep curves with micro crack healing   —2 - Creep curves without micro crack healing   —3 - Tertiary creep phase

Fig. 6.15 Behavior of the deviatoric creep strain εdev with time: primary creep phase (1); secondary; stationary creep (2); tertiary creep phase (3); final failure by breakage

Fig. 6.16 Asse II mine, winding machine house and winding tower *Source* M. Lersow

According to reputable estimates, these may add up to 6 billion euros. The final project costs will depend on the final option for the waste disposal that is selected for implementation.

- **Geomechanical situation**

The excavation of the panels on the southern flank (Fig. 6.17) was drilled on 13 levels with a total of 131 excavation chambers. In most cases panels on one level are arranged side by side in direction of strike and there are 9 mine workings. A standard mine working is 60 m long (strikewise), 40 m wide (crosswise) and 15 m high. Between the mine workings pillars exist that are normally 12 m wide, with the exception of the 20 m wide pillar between the mine workings in rows 4 and 5.

The vertical roof thicknesses[6] between the mine workings above the 700 m level amount to about 6 m, between the 700 and 725 m level about 8.5 m and between

---

[6]Roof—mining term for the false ceiling between two chambers in a mine.

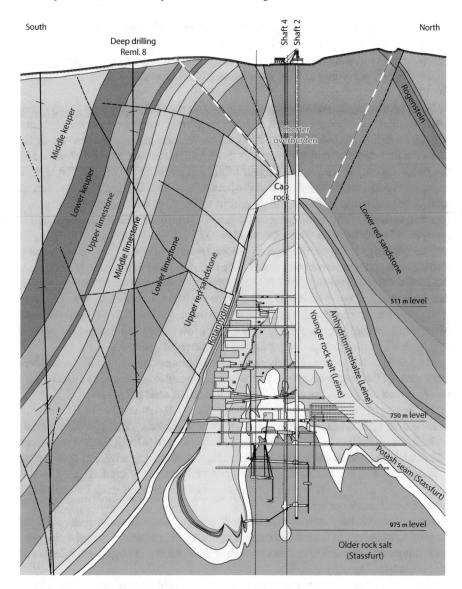

Fig. 6.17 Vertical cross section of Asse II in the center of the mine with the panels on the southern flank, in the core of the salt anticline and on the northern flank/legend next to it. *Source* BfS 2010

the 725 and 750 m level about 14 m. The panel has a dimension of approximately 650 m in the strike direction and 275 m in the vertical direction, see Fig. 6.15. The panels in rock salt and carnallitite are, for the present scheduled to be operating as a nuclear facility according to § 9a, without additional support measures (backfilling), IfG [33].

Such bearing systems are referred to as yielding. They react with creep deformations, dilatant softening as well as local fracture processes on the acted ground stresses, see Eqs. (6.1–6.3). From these processes, the pillar and contours of the roofs are first detected and the excess of the bearable stresses must be absorbed by the better supported core zones (higher minimum pressure fixing). If, due to softening and local cracks, the pillar and roof cores lose load-bearing capacity, further load is transferred to adjacent pillars and in the end, to the borders of the panels (bearing systems). This geomechanical behavior (bearing systems) can be found in many panels excavated in the evaporite rock.

The ongoing stress rearrangement processes across decades are very complex. Extensive in situ measurements, which have been installed in the Asse II mine since the 1980s, demonstrate the stress rearrangement processes. It could thus be shown, that, as a result of the stresses relocated from the mine, excesses of strength occur with subsequent repercussions on the southern flank, see Fig. 6.18. The upper levels of the southern flank are too close to the southern, low-strength and strongly fractured overburden and thus pose a danger for the formation of pathways.

Especially in the region of the horizontal displacement maximum of the southern flank most of the roof is completely broken (only outer rings of roofs still exist), see Fig. 6.19 and the pillars are crossed by a multitude of cracks.

Due to model calculations of the Institute of Rock Mechanics GmbH, IfG [33], the load-bearing system of southern flank system is in a limited state of dilatant softening with effects on the immediate overburden, which also lead to local fracture processes. The model calculations, however, cannot replace a comprehensive and coordinated system of in situ measurements, inspections and laboratory studies. This is also shown by the many catastrophic scenarios derived from such model calculations, none of which have occurred.

From the rock mechanics point of view, the ongoing softening process of the supporting elements in the mine workings on the southern flank required a complete backfilling of all remaining open excavations. This process is largely completed and is constantly being supplemented.

Due to the, geologically speaking, very short periods, based on the introduced backfilling, significant backfilling pressures have not yet built up. In order to offer some support, all the roofs should be topped up; due to the relatively high porosity and the beginning of the settlement of the backfilling due to the dead weight, there were relatively large roof clefts. Also, the roof clefts filling is largely completed and can be permanently supplemented if necessary. However, an immediate effect cannot be expected, see Eqs. (6.1–6.3). However, this no-load absorption of the backfilling can still be achieved. The supporting effect is detectable by the digressive tendency of the compression rates of pillars as well as the decreasing microseismic activity in the mine workings.

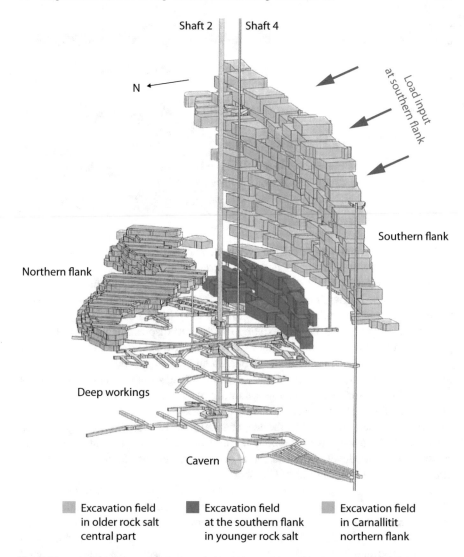

Fig. 6.18 Location of the panels in the mine workings of the Asse II mine. *Source* IfG [33]

Since 2009, the operator of the Asse II mine has initiated a series of preventive measures to stabilize the mine workings and protect the emplacement chambers through residual cavity backfilling and construction of sealing structures that noticeably improve the geomechanical and hydrological situation directly and sustainably. There is currently no immediate danger from the geomechanical situation, which will allow for the time needed to develop the best closure technology for the final disposal solution for the radioactive waste stored in the Asse II mine.

Fig. 6.19 Remains of a broken roof, mine workings row 7 south flank, photograph 2003. *Source* IfG [33]

Since 1988, groundwater from the overburden or dirt bed has been entering the pit. Most of the inflowing waters are collected. Part of it in the so-called "lye handling" which are temporarily stored and after measurement cleared and discharged from the mine to the mine Mariaglück. The accumulated saline solution volume has gradually increased since 1988 and, according to Michel [34], input flow currently stands at around 80 m^3 per week. Due to existing overburden displacements, local fracture processes, the low overburden distance of the upper levels of the southern flank, the low stability and heavily fractured overburden and the large number of pathways, a reliable forecast on the amount and rate of the inflowing groundwater is currently not possible. This is surprising because a number of methods (e.g. tracers) are known that would allow volumes to be measured and tracked.

Part of the water inflow enters the deeper areas of the mine, soaks the emplacement chambers and emerges as contaminated lye at the 750 m level. Contaminated salt solutions are in the level of the 2nd southern pilot drift on the 750 m level and, since 1988, in a water sump in front of chamber 12.

Based on the assumption of only a limited residual capacity in the middle region of the southern flank and an increase in the inflow of groundwater, model calculations have indicated that the shaft will probably become flooded. However, it should be pointed out that reliable statements are not yet available and therefore this thesis should be classified (without clarification of the data) as a scenario with low likelihood of occurrence, as long as the operation of the system according to the approved criteria continues as planned. Should it become necessary, additional water could be recovered through the mine's pumping boreholes and, if necessary, released into the receiving waters via a water treatment plant. For this purpose, pumping boreholes would have to be drilled at a suitable location.

In 2007, the IfS had "shown" in a model calculation (prognosis without any interventions) that, as a result of the softening- and fracture processes that have been going on for decades, there will be an increasing loss of bearing capacity from 2014. Quite apart from the fact that these are model calculations that do not allow such detailed time data, the following reactions from this statement are typical for the approx. past 50 years of Asse II mine. All that is proved is that there can be an increase in displacement rates of overburden. Although the IfS had limited this statement itself Quote: "Immediately after this year, however, no collapse of the mining field is derived, but the calculation shows, acc. the target of a degressive deformation process, the ability to forecast is limited." As the year 2014 began there was a lot of activity related to this question.

However, this process has something positive. According to the view expressed here, the precautionary and emergency planning of the BfS for the Asse II mine combined with transparency measures has contributed significantly to calm the situation among the citizenry, but with no hurry to achieve progress.

This "tangible emergency" is also reflected in the overall planning for emergencies. The BfS defines emergency planning as the totality of all planning with regard to emergencies with the goal:

- the limitation of beyond-design-basis events, that means reduction of the probability of occurrence,
- improving the design of the Asse II mine, d. H. reduction of the probability of occurrence of unforeseen events, and
- minimizing the consequences of beyond-design-basis events—inside and outside the plant.

See BfS: "Evaluation of the fact-finding procedure and the approach for the Retrievability"; As of April 27, 2016.

The emergency planning provided a strong message to the concerned community and is associated with the implemented planning of emergency measures, such as securing the building materials and media supply, provision of the necessary material resources for building material production and backfilling, e.g., the delivery of $MgCl_2$-rich solution for the counterflood and a sufficiently strong pumping system to be able to transfer the water from the mine in the event of a fault. Emergency preparedness, i.e. the ability to immediately implement the emergency measures should be established in the original planning of the BfS in 2016; currently the year 2024 has been adopted.

- **Geo-hydrological situation**

The structure of the Asse is separated from the regional hydraulic system by deep trough structures. On the one hand, the permeability of the regional aquifers decreases significantly with depth. Furthermore, the groundwater has an increasing salt mineralization with depth and is therefore significantly heavier in the deeps than near-surface groundwater which has low mineralization. Both effects hinder the regional groundwater flow, so that the influence of the regional hydrological system on the local conditions of the Asse is very small.

The groundwater moves along discrete pathways with properties of jointed aquifers. The local conditions represent a pronounced alternation of aquifers and aquitards. Significant groundwater-conducting horizons in the immediate vicinity of the mine facility are located in the dumped overburden and in shell limestone. Due to karstification, the Röt anhydrite in the area of the salt table can also be considered as a potential aquifer. Important local groundwater ladders are found in the clayey-marly strata of the pelitic Röt-series 2–4 in the southern flank and the low-permeability strata of the Lower and Middle Buntsandstein in the northern flank, see Fig. 6.20. Disturbances and rock-mechanical stresses increase the integral permeability of the groundwater overhead conductor. Due to the geomechanical deformations in the southern flank, it can be assumed that there is a rock formation (shearing deformation area) which is heavily affected by shear stress and extends from the salt flank to the Lower Muschelkalk. In this rock formation, the integral permeability can be significantly increased, despite the clayey layers, so that pathways exist between the salt flank and the Muschelkalk, see Fig. 6.20.

The permeability of the pathways decreases significantly with depth. The groundwater has an increasing salinity with depth. In the southern flank of the salt anticline, water enters the Muschelkalk and moves predominantly parallel to the structure. Along the shear deformation area or fault zones, saline solutions can penetrate the low-conductivity overlapping pelitic Röt-series and reach the salt flank. From the dumped overburden and the cap rock groundwater flows to reach the salt table. The flow paths at the salt table are in hydraulic contact with the Röt anhydrite, through which the saline solution on the southern flank of the salt anticline can penetrate up to the shear deformation area, see Fig. 6.17.

In summary, it can be said, that the groundwater moves in the area of the Asse predominantly not at depth, but near the surface and parallel to the Asse-mountain range. Water probably enters the mine via the aquifer "dumped overburden", plaster cap and the "Röt anhydrite". The mine itself is practically a deep discharge point. But it is also possible that water flows through the lower Muschelkalk into the Röt anhydrite, because in this layer, which normally dams the water, some places are broken, see Fig. 6.20. Further, see [33, 36, 37].

- Summary of the location description

– Studies have shown that groundwater can penetrate into the crest up to the salt table and into the salt flank up to the southern flank. Other groundwater conducting layers in the Muschelkalk can be hydraulically connected to the flank of the salt anticline via disturbances,
– The salinity of the groundwater increases with the depth, in the depth range of the Asse II mine is NaCl—rich saline solution,
– In the area of the Asse II mine, permeability has increased in tectonically preconditioned and mechanically disturbed rock formations,
– Currently, radioactivity can not escape into the rock against the existing groundwater flow. Danger for humans and environment exists then, if the mine should flood in an uncontrolled manner. As long as the operation of the plant continues as planned in accordance with the approved criteria (precautionary measures), this scenario is unlikely.
– Uncontrolled flooding does not necessarily lead to the mobilization of radioactive substances; which source term can arise in this case and which discharges path this

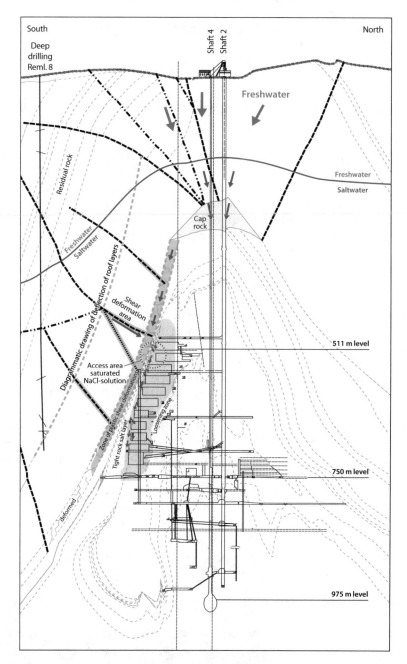

Fig. 6.20 Representative cross section through the Asse-salt anticline in cross cut direction with fault zones and scenario model of saline solution inflows, HZM [35]

may possibly take are currently pure speculation, since no reliable investigations are required to be made.

- **Stored radio-toxic inventory in the Asse II mine**

Between April 7, 1967 and December 31, 1978, a total of 125,787 containers of low-level and medium-level radioactive waste (LAW and MAW) were stored on different levels in various chambers, see Fig. 6.21. This corresponds to a volume of approx. 47,000 m^3.

The radioactive waste is located in rock salt chambers in the core of the salt anticline, in depth levels of 750 and 725 m, and in depth levels in the southern flank at 750 and 511 m. In total there are 13 emplacement chamber—zones: emplacement chamber 8 on the 511 m level; ones: emplacement chamber 7 on the 725 m level and 11 emplacement chambers on the 750 m level. 1293 of the stored containers are declared as medium radioactive waste (MAW) in 200 litre barrels, the remaining containers contain low-level radioactive waste (LAW). According to the admission declaration, the total activity of the stored waste was 2.11E+05 Ci or 7.81E+15 Bq. The mid-level radioactive waste (MAW) accounted for 1.36E+05 Ci (5.03E+15 Bq), the low-level radioactive waste (LAW) and the lost concrete shieldings with 7.52E+04 Ci (2.78E+15 Bq) being the Total activity declared. The total inventory activity listed in Table 6.10 is larger. This is due to replication of some radionuclides. For further information see [36, 37].

According to Gerstmann [38], the total radioactivity of the radioactive wastes deposited in the Asse mine was 8.35E+15 Bq β/γ emitters and 0.12E+15 Bq α emitters as at reporting date: 01.01.1980. As at reporting date: 01.01.2003 this inventory had decreased to 2.84E+15 Bq β/γ emitters, but by replica the inventory of α-emitters had grown to 0.18E+15 Bq. The highest levels of activity were reported as at reporting date: 01/01/2003 for Pu-241 (1.05E+15 Bq), Ni-63 (0.75E+15 Bq), Cs-137 (0.56E+15 Bq) and Sr-90 (0.32E+15 Bq), see Table 6.10. The inventory listed in [38] contained 11.6 kg of plutonium, 102 tonnes of uranium, 87 tonnes of thorium, and about 5.5 g of radium-226. See the decay curve Fig. 6.22.

This radioactive inventory must be isolated. Currently, radioactivity can not escape from the emplacement chambers of the Asse II mine into the rock formation against

Fig. 6.21 MAW-barrel cone, storage on the 511 m level (emplacement chamber 8). *Source* Michel [34]

Table 6.10 Radionuclide vector at various reporting dates in the Asse II mine, acc. [38] Gerstmann (2002) and [16] K-MAT 14 (2015)

Nuklide	Half life [a]	Asse II (01.01.1980) Ai [Bq]	Asse II (01.01.2003) Ai [Bq]	Asse II (01.01.2005) Ai [Bq]	Asse II (31.12.2013) Ai [Bq]
C-14	5700	3.86E+12	3.85E+12	3.82E+12	2.60E+12
Co-60	5.3	2.66E+15	1.29E+14	0.99E+14	1.10E+13
Ni-63	100	8.82E+14	7.52E+14	7.40E+14	2.60E+14
Sr-90	28.5	5.67E+15	3.25E+14	3.10E+14	2.00E+14
Cs-137	30.2	9.45E+14	5.56E+14	5.30E+14	3.60E+14
Ra-226	1600	2.02E+11	2.00E+11	2.00E+11	2.00E+11
U-234	250,000	1.33E+12	1.33E+12	1.33E+12	1.40E+12
U-238	4.40E+09	1.26E+12	1.26E+12	1.26E+12	1.30E+12
Pu-238	87.7	4.38E+13	3.65E+13	3.63E+13	9.20E+12
Pu-239	24,000	1.92E+13	1.92E+13	1.90E+13	4.50E+13
Pu-240	6600	2.23E+13	2.22E+13	2.20E+13	5.10E+13
Pu-241	14.4	3.21E+15	1.05E+15	9.69E+14	1.30E+15
Am-241	432.6	2.69E+13	9.58E+13	9.94E+13	2.40E+14
Cm-244	18.1	4.56E+12	1.89E+12	1.75E+12	8.00E+11
Sum α		1.20E+14	1.78E+14	1.81E+14	3.49E+14
Sum β/γ		1.34E+16	2.82E+15	2.65E+15	2.13E+15
Sum		1.35E+16	2.99E+15	2.83E+15	2.48E+15

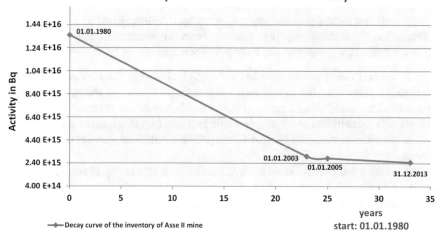

Fig. 6.22   Decay curve of the Asse inventory acc. Table 6.10

the existing groundwater flow (mine workings). It can be assumed that an activity of approximately 3.0E+13 will still be present after an observation period of 10,000 years.

6.3.4.2   Site Concept of the Long-Term Safety of the Asse II Mine

In 2007, Helmholtz Zentrum München (HMGU), a 100% subsidiary of the federal government, presented a closure concept for the Asse II mine as the operator after a plan for the final closure was requested. The closure concept was politically contentious. However, the decision was under some time pressure, since the geomechanical stability of the mine structure seemed assured for only a few years. From today's perspective, this approach was completely unacceptable. Initially, the HMGU had considered three "concepts":

1. Closure of the mine without the introduction of a protective fluid,
2. Closure of the mine with introduction of a protective fluid,
3. Retrieval of radioactive waste.

In its closure concept, the HMGU came to the conclusion that "only with the introduction of a protective fluid can the safe retention of of radioactive waste against the biosphere be ensured" HMGU [35]. The HMGU claimed that the closure concept met the basic requirements of the state of science and technology and had no alternative. The closure of the plant should be carried out as soon as possible, since only a limited residual bearing capacity of the southern flank was given.

The closure concept of the HMGU formulated the basic requirements for the closure of the facility:

– Sufficient and fastest possible support of the load-bearing system
– Protection of carnallitite against solutions from the overburden and
– Secure boundary conditions for long-term forecasts.

A long-term safety assessment has not been included in the closure concept of the HMGU, other than the hypothesis: Long-term protection of the carnallitite can be achieved by introducing a saline solution that becomes almost in chemical equilibrium with carnallitites and the hydrostatic pressure (equilibrium) of the protective fluid in the mine workings at the same time provides sufficient support of the overall system.

The missing long-term safety evidence and a lack of a safety check of the mine facility, considering beyond-design solution inflows into the mine workings are to be expected, This led to reasonable doubts that the protection goal is achievable with the presented closure concept, both in limiting the external dose rate and preventing large-scale contamination of the groundwater body.

On 1 January 2009, the BfS took over the operator function of the Asse II mine and rejected the closure concept of the HMGU. For its part, the BfS awarded contracts for the development of closure options for the Asse mine. Since then three closure options have been developed and presented:

– **Option: relocation of waste into deeper stratas of Asse**

The radioactive waste will be kept in the mine structure, in deeper, newly excavated cavities. The barrels are to be removed and safely packed for transport. Subsequently, the cavities are sealed for the long-term.

– **Option Retrievability**

Materials in the underground area prepared for final disposal as radioactive waste, are taken from the Asse II and put into a prepared and approved repository. The long-term closure of the repository is part of the permit.

– **Option complete backfilling of the mine facility**

The radioactive waste remains at the current location. The mine facility is completely backfilled and secured and is to be closed safely in the long term.

After submission and presentation of the closure options, the BfS has examined the options in an options comparison, BfS [36], stating that only the option for retrievability can be demonstrated in a long-term safety assessment. The option retrievability was considered a preferred option. Thus, the BfS has subsequently submitted 3 location concepts for the options, see Fig. 6.20.

The German Bundestag (DBT) dealt with the Asse and the comparison of options while maintaining a possible spontaneous failure of the mine workings and came to the conclusion that the option of retrievability is without alternative.

It also enacted an "Act to Accelerate the Retrieval of Radioactive Waste and the Decommissioning of the Asse II Mine", which came into force on April 24, 2013. In it, the Asse mine is classified as a nuclear facility according to AtG § 9a. The Act amends § 57b AtG "Operation and decommissioning of the Asse II mine" in which: (2) "The mine facility must be shut down immediately. For further operation until decommissioning, including the retrieval of radioactive waste and related measures, a plan of procedure for approval according to § 9b is not required. The decommissioning should preferably take place after the retrieval of the radioactive waste. The retrieval must be discontinued if its implementation is not justifiable to the population and employees for radiological or other safety-related reasons."

In this way the DBT has selected the option "retrieval and the fastest possible decommissioning". It is interesting that for the retrieval option no planning approval procedure should be necessary and exceptions are provided for radiation protection regulations, with the justification of the time pressure! There should be no time pressure, but there is an absolute need for comprehensive transparency.

However, due to the extensive precautionary measures introduced, combined with an emergency plan and a considerably expanded public participation, some realism seems to be coming into play. Thus, the federal environment ministry states that there will be no retrieval before 2030. In the meantime, it has also been determined that the Schacht Konrad repository will not be available to take the radioactive waste of the Asse, and the planned repository for HAW as a possible location will also be excluded. Thus, the year 2030 for the retrieval is completely unrealistic. Unless one would prefer a transitional storage, this is proposed in Chapter 7. However, this includes the transport of these radioactive wastes. Extensive transport should not be planned or at the worst minimized in repository projects. In any case, the retrieval poses a high radiological risk for the operating personnel and possibly also for the population in the vicinity of the Asse II mine.

In the meantime, expert opinions on the evaluation of the comparison of options are available, according to the ESK, ESK [37] and the Working Group "Options Comparison", AGO [39]. These come to the following statements (excerpt):

- The comparison of options is based on a partially insufficient level of knowledge;
- Lack of reliable data for assessing the radiological and chemotoxic consequences of uncontrollable groundwater ingress (environmental impact of unmanageable solution inflow);
- For the option retrieval, there is currently no final disposal possibility. There is only the principle of hope that a suitable final disposal option can be found in Germany;
- The time requirement assumed for the options in each case appears to be of little resilience in the view of the working group "comparison of options";
- Preliminary long-term safety estimates only assume that the long-term safety proof for the Schacht Konrad repository is not significantly affected by the additional Asse waste; the storage in Schacht Konrad is meanwhile rejected, an alternative is not shown;
- Safety in the operating phase has priority;
- The retrieval option did not take into account the radiation expositions of the employees.

Quotation ESK [37]: "Even though the radiation exposure of personnel and the population arising from a retrieval of the Asse II mine can not currently be estimated as reliable; the ESK points out that in order avoidance of hypothetical dose values as conservative calculated in the future, if the waste were to remain in the Asse II mine, real radiation exposure to a considerable extent in the coming decades would have to be accepted for the operating personnel if the waste were to be retrieved from the Asse II mine. Also for people around the facility, additional real radiation exposure would result from direct radiation and higher emissions."

As a result of these assessments, the BfS is now carrying out a fact finding study, see Fig. 6.23, which was launched in 2012 and where three holes 750-m deep were drilled into chamber 7, see Fig. 6.17, with the following result:

- No risk of explosion due to gases produced;
- Radioactivity levels in both chamber 7 and in its environment correspond to the current state of knowledge;
- It was possible to determine the exact position of the closure structure, the chamber ceiling and the walls;
- Fact-finding in the current form takes too much time.

Furthermore, no radioactive discharges from the inventory of wastes were detected in the environmental monitoring. There is no increased radioactive contamination in the vicinity of the Asse II mine in the Wolfenbüttel county. This emerges from

Fig. 6.23 Fact-finding, drilling site with preventer. *Source* M. Lersow

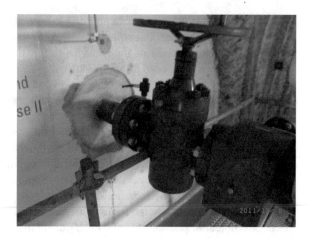

an interim report submitted by the Agricultural Exploration- and Research Institute (LUFA) North-West and the Braunschweig district office to the Lower Saxony Chamber of Agriculture. The environmental monitoring and the permanent disclosure of the determined data is an essential project of public participation and therefore mandatory.

6.3.4.3   Scenarios/Outlook

When and how the long-term safe storage of radioactive waste takes place is, from the local point of view, currently unclear. It seems clear, however, that the available data on the inventory can no longer be doubted and can be used for work on conceptual plans. Asse GmbH currently keeps the mine workings open and carries out safety measures for the stability of the mine facility and of averting a danger due to mining. Permanent surveillance, precautionary measures and emergency plans exclude, from the view point of the author, spontaneous failure and worsening of the rock-mechanical and hydrological conditions. This is also due to the fact that in the vicinity of the mine two more shafts exist, Asse I (the mine had to be abandoned in July 1906 due to flooding) and Asse III (the shaft was drilled from 1911 to 1921, and the workings was finished in 1924). Both shafts have been "flooded" for decades - filled with water there is a stable state of hydrostatic tension. Convergence will not take effect under such conditions. Under these conditions, the likelihood of an irresponsible increase in the present risk situation is very low. A mechanical spontaneous failure is to be expected to have very low probability.

Based on Michel [34] 2013, the following description of the condition of the Asse II mine can be given:

- The 2010 comparison of options is without foundation;
- A flooding of the Asse II mine facility is the probable development; flooding should be assumed as a worst case scenario in a review;

- Upon entry of an uncontrollable solution inflow, which results in an evacuation from the mine, the introduction (counter-flooding) of $MgCl_2$-rich solution, the pneumatic pressurization of the mine workings during the counterflood of the mine workings is proposed as an option;
- The submitted assessment of the potential radiation exposure in the vicinity of the Asse II mine, is approximate only, interspersed with technical errors and is too conservative, see Küppers et al. [40].
- The entry of uncontrollable solution inflows corresponds to the scenario "beyond-design-basis" inflow rates of brine solutions from the over burden during the operational phase, see Chap. 8. There must be a robust estimation of the potential radiation exposure in the operational and post-operational phases. A modeling of the overburden has not been made. The proposed propagation paths are speculative.
- Based on current knowledge robust assessments of workers' real exposures and real and potential population exposures during the retrieval are not possible. An estimation of the real exposures of the workers and the real and potential exposures of the population is also missing in the presented option: retrieval. Such data are fundamental for its evaluation;
- We recommend an iterative procedure to model the radiological consequences of a possible Asse II mine flooding in the various options.

Following the assessment of R. Michel, all previously existing site specific models have to be re-evaluated, based on the information resulting from the fact check data. It can be stated that a careful consideration of the options for handling the radioactive waste stored in the Asse, see Table 6.10, based on robust data, the technologies available, the burden on operating staff and possibly the outgoing dangers to the environment, allows the possibility of finding the best solution for the existing conditions. There is no cause for extreme hurry, even if the integrity of the mine workings is limited, see [41, 42].

Despite the assessment that the retrieval option is devoid of any specialist basis, the considerations on the option to return are further advanced by the Asse GmbH and facts created.

Since there is currently no repository site for the waste from the Asse II mine and it is not yet clear where and when such a facility can be found, one has to deal with the problem of the radioactive waste to be transported to the surface. A planning contract for the underground tasks was awarded in 2015. When dealing with the radioactive waste transported to the surface, the problem of intermediate storage is touched on. An interim storage facility at the site with a conditioning system is favored by the authorities, but not by the affected population. The discussion was conducted without a robust assessment of the real exposures of the workers and the real and potential exposures of the population. It is recommended to connect this discussion with the transitional storages for HAW-HGW. This creates transparency and takes the suspicion of a one-sided determination. Such a discussion pre-supposes that an assessment of the real exposures of the workers and the real and potential exposures of the population would be presented.

For the retrieval and transport, including the entire supply of fresh air (mine ventilation) and the material transport, a separate recovery shaft is required. For this purpose, exploratory drilling from the surface and from the underground side have already been undertaken to establish the possible connection of various levels to the recovery shaft. It is recommended to include the drillings in the emergency planning in such a way that these can also be expanded as production wells should there be an emergency of "inrush of water". In the case of uncontrollable solution additions, these production wells would also have the possibility of introducing (counterflushing) $MgCl_2$ -rich solution, optionally also the pneumatic pressurization of the mine workings. If core drillings are carried out, then a model can be constructed using the data obtained.

There is a comprehensive environmental monitoring network set up which includes the following sampling activities: gamma local dose rate (ODL); Gamma dose; Aerosol activity; Soil samples; Grass samples; Water samples.

The Agricultural Exploration- and Research Institute (LUFA) North-West and the Braunschweig district office of the Lower Saxony Chamber of Agriculture are presenting the results of additional environmental monitoring in annual reports. The results of the additional environmental monitoring in the area of the Asse II mine confirm that an activity contribution of the Asse II mine is not recognizable or demonstrable. Together with the fact check, the data situation improves noticeably, so that this can allow a robust assessment of the radiological consequences of a scenario "beyond-design-basis inflow rates" of brine solutions from the overburden, see [43].

The option of retrieval must be further detailed. Thus, the consequences of a retrieval, in particular of the radiological situation, for the staff and the population have not yet been demonstrated enough. The feasibility of this option is thereby decisively determined, in addition by the final disposal site and the costs. The current "Perspectives" for the Asse II mine are summarized in Fig. 6.24, supplemented by a non-time limited operation for retaining the mine, with production wells and water treatment.

From the point of view of the author, the opportunity should be seized to gather as much data as possible, which can be made available when considering final disposal in Germany as a whole. The creation of a comprehensive database in which all available, evaluated and verified data, e.g. for final disposal in the host rock "rock salt" (carnallitite), can be put together and made available to the public is long overdue.

This also includes, if possible, opening an emplacement chamber and examining the condition of the containers. In order to gain reliable data, the Asse II mine should be instrumented for the entire time of "operation for keeping the mine open" and possibly beyond so that reliable data on the behavior of the repository after decommissioning can be collected and thus scientifically evaluated.

Because a scenario "beyond-design-basis inflow rates" of the of brine solutions from the overburden for the mine facility Asse II cannot be ruled out, there is a particular need for research on solution transport in the mine workings. Chapter 8 deals with solution transport in the mine workings and radionuclide transport into the

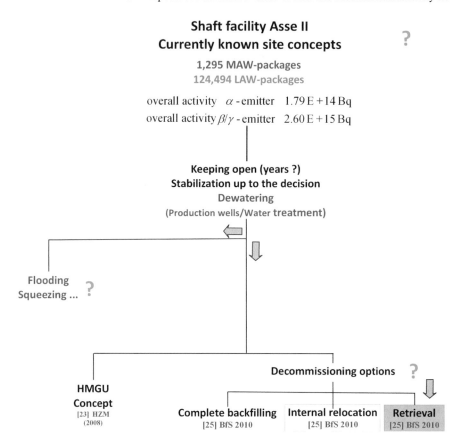

Fig. 6.24 Asse II mine facility; currently known site specific concepts

biosphere and its implications for a long-term safety assessment. However, an investigation of this site specific problem has so far been almost completely omitted for the Asse II mine, including the generating of necessary parameters. The description of solution transport in the mine workings also requires a description of the geochemical environment likely to develop in the storage areas and in the mine structure generally, taking into consideration solubility limits, retardation and sorption.

In Küppers e al. [40] in 2011, is a recalculation of the effects of a "beyond-design-basis inflow rates" in the Asse II mine which was made after the flooding concept of the HZMU was available. Because of the short comings and the unrealistic scenario description, Michel [34] has already dealt with it. It should be noted here that a "squeezing out" of solution from the mine workings over a long period of time is not to be expected, which is why this scenario currently has no priority.

The scenario of the escape of brine solutions from the overburden and the necessary instructions for remedial action are compiled below as an example, see also Chap. 8.5.2.

6.3.4.4 Szenario "Beyond-Design-Basis Inflow Rates" of the of Brine Solutions from the Overburden

- **Formation of a radionuclide source term in the mine workings**

In Chap. 8, it is shown how in general the dissolved radionuclides in the solution can be transported in the mine workings by three different effects. These are solution transport through pressure gradients, exchange processes, dispersion and diffusion. In addition, a diffuse radionuclide transport takes place, which is driven by existing concentration gradients of the radionuclides, see Chap. 8 and Eqs. (6.4), (6.5), (6.6). Diffusive radionuclide stream J_D:

$$J_D = -D(T) \cdot A \cdot \varphi \cdot \nabla c \tag{6.4}$$

Convective solution transport; solution exchange stream:

$$Q_{T,C} = \frac{g \cdot k \cdot \beta}{8 \cdot v} H^2 B \frac{\Delta T}{\Delta L} - \frac{g \cdot k}{8 \mu} H^2 B \frac{\Delta p_C}{\Delta L} \tag{6.5}$$

Advective solution stream:

$$Q = \frac{k \cdot A}{\mu} \nabla p \tag{6.6}$$

In order to be able to realistically describe this solution transport for the Asse II mine, the following parameters must be determined:

C—Molar concentration in mol m^{-3}; ∇c—concentration gradient of radionuclide
Unit for J_D for example $[J_D] = \text{mol m}^{-2} \text{ s}^{-1}$
k—Permeability of backfilling
μ—dynamical viskosity of solution
v—kinematic viskosity
β—coefficient of thermal expansion of backfilling
D(T)—Diffusion coefficient at the temperature T in the solution
Φ—Porosity of backfilling
B, H, L—width, height, length of transport pathway

The entire radionuclide transport results from the superimposition of all partial processes: transport by forced solution movement, convective exchange processes and diffusion, Eqs. (6.4), (6.5), (6.6). These processes are very different from each other, so that an analytical treatment of the super imposition is not possible. Extensive experiments can be found in Klaus-Peter Kröhn (GRS) et al. [44]. However, a sensitivity analysis and uncertainty considerations cannot be dispensed with here either. In order to be able to describe the solution transport realistically, the following investigations need to be carried out:

– Release of radionuclides from the waste container,
– Retention of radionuclides in the mine workings; sorption and retardation,
– Solubility limits of the solution,
– Dissolution and reprecipitates of sealing structures,
– For gas formation due to stored biological residues; Measures to limit the formation
  of gas, possibilities of removing gas-forming substances (metal, wood, plastics).

● **Mass transport modeling for the overburden**

To be successful the transport modeling in the overburden requires a sufficiently accurate description of the groundwater flow. An overburden map is currently not available and must be constructed.

The radionuclide stream which passes from the mine workings to the overburden forms the basis for consideration. The modeling on the basis of the equations quoted in Chap. 8 Eqs. (8.33–8.35) can be considered sufficiently secured. The groundwater flow is assumed to be laminar. For modeling, the fluid mass is balanced locally (i.e., within a volume element) and for the entire model area, see Eq. (8.33). For density-driven flows, the mass of the dissolved salt, Eq. (8.34) and/or the amount of heat are also simultaneously balanced. Equation (8.34), is omitted in the case of a non-density-driven flow, see Chap. 8.6.3. The structure and the properties of the overburden must be described in detail. In addition to the dilution of the radionuclide current, the sorption properties of the flow through overburden play an essential role in achieving a robust assessment of the radionuclide transport into the biosphere.

● **Possible radiological effects in the biosphere**

In order to determine the radiological effects of an uncontrollable solution release to the biosphere, the transfer of radionuclides within the biosphere has to be modeled. Also, a person or a group of people is to be defined, including their living habits, for which the effects are to be determined (critical group). For examples, see Gellermann [45]. There are substantial research achievements for the evaluation of the different location models. However, these require robust assessments of the radionuclide currents in the model areas.

However, a different assessment should be mentioned here, which should be considered in connection with the evaluation of possible radiation exposures in the vicinity of repository structures. This concerns the radiation exposure of reference persons of the general population (critical group) shown in Fig. 6.25.

The calculated radiation exposure of the critical group is correlated with figures from the Epidemiological Cancer Registry of Lower Saxony (EKN), which showed that in the "joint municipality" of Asse between 2002 and 2009 about 10 additional cases of leukemia and about 8 additional cases of thyroid cancer were observed. The rating alone speaks for itself. Although these are model calculations and there is no statistical evidence, a connection to the Asse II mine is practically produced directly here. The discharges of radioactive substances with the exhaust air from the Asse II mine and the environmental monitoring has resulted in no measurable concentrations in the environment. Even with an extremely conservative modeling of

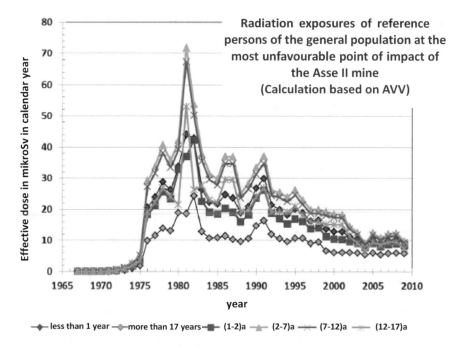

Fig. 6.25 Radiation exposure according to General Administrative Regulation to § 47 of German Radiation Protection Ordinance (AVV) (With the introduction of the new radiation protection legislation in Germany in 2019, the AVV will also be revised.) determined in 2013, see Michel [34]

the potential radiation exposure of the population around the Asse II mine, negligible annual doses below the region's natural radioactivity are reported. Of course, there must be a comprehensive, monitoring and evaluation of the radiological situation in the vicinity of the Asse II mine, but there is no cause for concern at this time.

It is also a task of these presentations here to handle carefully the radiological consequences of a final disposal of radioactive material. Therefore, a connection to the residues from uranium ore extraction and treatment in Saxony and Thuringia should be produced and checked for any unequal treatment of similar situations of hazards under the same legal regulations.

In the territory of Saxony and Thuringia, after the Second World War, uranium ores were extensively mined in the regions of Aue and Koenigstein in Saxony and Ronneburg in Thuringia; operations were undertaken by the reparation enterprise called SAG Wismut and later by SDAG Wismut. SDAG Wismut was at the time the largest uranium producer in the world. And the GDR, as a country, historically ranked third place in the list of uranium-producing countries and still occupies the 5th place in today's ranking. During the entire operating period about 230,000 tons of uranium ore were mined. The Federal Office for Radiation Protection has records of the radiation exposures of the workforce of the SDAG Wismut in the Wismut uranium miners cohort study, Schnelzer et al. [46]. The, the radiation exposures of the professional radon loaded miners of the SDAG Wismut have been scientifically evaluated and analyzed. Based on a stratified random sample of all employees using existing wage- and salary-payroll records from the register of the SDAG Wismut, data were extracted for all cohort members. It has been possible to extract information on personal data and on the work records of the Central Reception Office Wismut (ZeBWis) of the German Social Accident Insurance (DGUV), which then become the basis of scientific evaluation. A cohort was formed with about 59,000 members available. The applicable documents were from the residents' registration offices and public health offices or the former pathology archive of the SDAG Wismut. In summary: In the early years of SAG Wismut virtually every radiation and occupational protection procedure was missing, also the radon and quartz fine dust exposures were very high. The working conditions gradually improved from 1954 with the founding of SDAG Wismut and by about 1971 had reached international standards. Up to 5492 lung cancers and 14,531 silicosis cases of former Wismut employees were recognized as occupational diseases. After the reunification of Germany between 2008 and 2011, a further 3696 cases of lung cancer and 2720 cases of silicosis were added to the statutory accident insurance, see Schnelzer et al. [47]. Figure 6.26 shows the cumulative exposure of exposed Wismut uranium miner cohort members. From the underlying figures it can be seen that the cumulative exposure between 1989 and 1991 increased slightly (about 2.2 mSv/a). In December 1991, the state-owned redevelopment company Wismut GmbH was founded, which has to eliminate the legacies of the SDAG Wismut in such a way that the endangerment

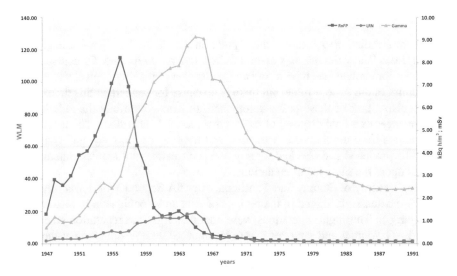

Fig. 6.26 Average annual cumulative exposure to radon daughters (RnFP) in WLM (The Working Level Month (WLM) is a historical unit specifically used in uranium mining. One WLM corresponds to an exposure of 1.3×10^3 mega electron volts of potential alpha energy per litre of air over a period of one working month (170 h).), external gamma radiation in mSv and long-lived radionuclides (LRN) in kBqh/m^3. *Source* [47]

of the affected population is largely excluded. It is inadmissible and misleading to attribute all the 6416 registered cases of lung cancer and silicosis cases registered by the German Social Accident Insurance between 1991 and 2011 to Wismut GmbH.

However, what must be ascertained is that, especially in the first years of Wismut GmbH, there was an increased risk of cancer for exposed persons, but this had decreased significantly by 2017. Figures 6.27 and 6.28 document the remediation success. In particular, the long term safety guarantee of the uranium mill tailings ponds of the SDAG Wismut must be based on a guardian-principle (stewardship) because of the lack of base sealing in the containment.

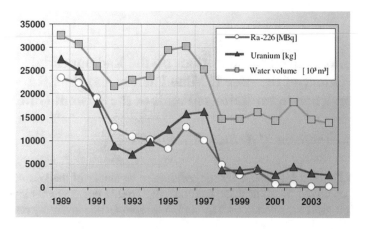

Fig. 6.27 Evolution of radioactivity in leaked water at the water measuring points

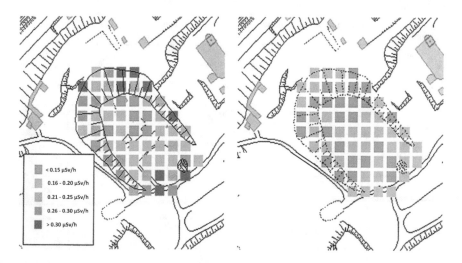

Fig. 6.28 Gamma dose rates on a remediated area before and after remediation. *Source* Second Joint Convention 2006/Lersow, M.

Extensive long-term environmental monitoring ensures that escaping contaminants and the radionuclide flux via the water pathway are collected and sent to a water treatment plant and finally discharged into the receiving waters, see also Figs. 6.27 and 6.28.

As a guideline for radiation exposure for remediated areas, Wismut GmbH has established a gamma dose rate ODL < 0.15 μSv/h.[7]

The conclusion to be drawn here is that a comparison of the radioactive inventories of the uranium mill tailings ponds of SDAG Wismut and the final deposition sites for radioactive waste in a plant acc. § 9a AtG also enables a real risk assessment and comparison to be made between the locations and in particular the representations relative to the Asse II mine.

6.4 Comparative Considerations on the Radionuclide Inventory of the Asse II Mine Facility and the ERA Morsleben to the Tailings Ponds of the Wismut GmbH

6.4.1 Comparison of Activity Inventories

In Table 6.11 some parameters for the radioactive inventory of the Wismut tailings ponds are compiled. In all of these facilities, the Th-232 decay chain plays no role in the radiological considerations. The complete radionuclide inventory results if the

Table 6.11 Radionuclide vector of the Wismut remediation project "Tailings Ponds" (the inventory of pore water is neglected here) see also [48]

Tailings Impoundment	Culmitzsch A	Culmitzsch B	Trünzig A	Trünzig B	Helmsdorf	Dänkritz I
U_{nat} in solids (t)	4800	2200	1500	700	5000	1000
U_{nat} in solids (TBq)	122	56	38	18	127	25
Ra-226 in solids (TBq)	790	240	130	50	550	40
Ra-226 im Porenwasser (mBq/l)	… 5000	… 2300	… 630	N.A.	500 … 2000	N.A.
U_{nat} im Porenwasser (mg/l)	0.3 … 3.9	1.0 … 16.5	1 … 19	1 … 20	2 … 30	10 … 85
$\sum$ (TBq) in solids	912	296	168	68	677	65

---

[7]0.15 μSv/h is approximately 1.31 mSv/a, if a person were permanently one year at one place. This is completely unrealistic, so this conversion should not be done.

daughter nuclides of the U-238 and U-235 decay series are taken into account. As a result, the physical activity inventory is about 10 times larger than the radionuclide-related inventories.

The tailing ponds of SDAG Wismut will be closed to be stable in the long term. The main barrier is a multifunctional surface cover, see also Chap. 5. Neglected in the compilation of Table 6.11 is the radioactivity contained in the pore water of the uranium mill tailings mass. This share of total activity is only low. The hazard assessment assumes that the pore water exits via the base floor of the uranium mill tailings ponds. The effects of the lack of base sealing are extensively discussed in Chap. 5.

The physical activity inventory of the uranium mill tailings ponds of SDAG Wismut is compiled in Table 6.12. Here reference is made to the exemption limits (FGi) acc. Table 1, Annex 4 StrlSchV.

For the existing plants in Germany acc. § 9a AtG, where low- and medium-active (LAW and MAW) waste is stored in geologically deep layers, the inventories for the respective key data relating to the FGi are summarized in Table 6.13. On the one hand for the radioactive waste repository Morsleben—ERAM in Saxony-Anhalt and on the other for the Asse II mine facility near Remlingen in Lower Saxony. The inventory and the radionuclide composition (radionuclide vector) in the two plants according to § 9a AtG are sufficiently well documented, K-MAT 14 [16].

The activity potential of the ERAM is significantly lower than that of the Asse II mine. If the number of nuclides where the half-life—is less than 100 years, the radiotoxicity quotient Ai/FGi of the activity inventory of the ERAM has decreased to approx. 1.0E+08.

In view of the very different radionuclide composition, the radioactive inventory of the ERAM and the Asse II mine are not directly comparable to that of the Wismut Tailings ponds. In order to be able to compare the activity inventories of the tailings ponds and the facilities according to § 9a AtG, the activity inventories for each radionuclide have been related according to StrlSchV (Annex 4, Table 1, Column 2) and on the exemption limit of the total activity. The sum of these ratios across all radionuclides gives a dimensionless activity index, which characterizes the radiotoxicity of the total inventory, since the radiological hazards of single nuclides above the respective exemption limit are taken into account.

The results of such an evaluation are summarized in Tables 6.10 and 6.11. It has been taken into account that U_{nat} compiles in terms of activities as follows: 238U/234U = 1.0, 235U/238U = 0.046. In addition, in terms of the 230Th activity, it is noted that for the daughter nuclides of uranium, 230Th/226Ra is about 1 and 231 Pa/227Ac is about 0.046. Tables 6.10 and 6.11 make it clear that the inventories of Wismut tailings and the Asse, when evaluated according to this simple radiological indicator, are comparable. However, there are differences in the mobility of radionuclides. The mobility of radionuclides in the tailings ponds inventory is significantly higher than in the repositories for radioactive waste with negligible heat generation. In addition, the pore water contained in each tailings mass provides a transport medium, which in nuclear facilities acc. § 9 AtG is largely kept away from the stored radionuclides. Without a transport medium (water, air) radionuclides can

Table 6.12 Radionuclide inventories of Wismut uranium mill tailings ponds normalized to the exemption limits (FGi), Lersow and Gellermann [49]

Tailings pond/nuclide[a]	Half-life period [a]	FGi [Bq]	Culmitzsch A	Culmitzsch B	Trünzig A	Trünzig B	Helmsdorf	Dänkritz I
			Ai/FGi	Ai/FGi	Ai/FGi	Ai/FGi	Ai/FGi	Ai/FGi
U-238+ (α)	4.468E+09	1.00E+04	1.22E+10	5.60E+09	3.80E+09	1.80E+09	1.27E+10	2.50E+09
U-234 (α)	245.500	1.00E+04	1.22E+10	5.60E+09	3.80E+09	1.80E+09	1.27E+10	2.50E+09
Th-230 (α)	7.538E+04	1.00E+04	7.90E+10	2.40E+10	1.30E+10	5.00E+09	5.50E+10	4.00E+09
Ra-226++ (α)	1602	1.00E+04	7.90E+10	2.40E+10	1.30E+10	5.00E+09	5.50E+10	4.00E+09
U-235+ (α)	7.038E+08	1.00E+04	5.56E+08	2.55E+08	1.73E+08	8.21E+07	5.79E+08	1.14E+08
Pa-231 (α)	3.276E+04	1.00E+03	3.60E+10	1.09E+10	5.93E+09	2.28E+09	2.51E+10	1.82E+09
Ac-227+ (β^-)	21.773	1.00E+03	3.60E+10	1.09E+10	5.93E+09	2.28E+09	2.51E+10	1.82E+09
Ra-223+ (α)	11.435 d	1.00E+05	3.60E+08	1.09E+08	5.93E+07	2.28E+07	2.51E+08	1.82E+07
Sum			2.55E+11	8.15E+10	4.57E+10	1.83E+10	1.86E+11	1.68E+10

[a]"+", "++" or "sec" are parent nuclides which are in equilibrium with the daughter nuclides specified in Table 2, Appendix 4 StrlSchV

Table 6.13 Compilation of the total activity Ai and the values of the activity indicator Ai/FGi (FG—Exemption limit according to Annex 4 Table 1 Sp. 2 StrlSchV) for the radiologically most important radionuclides in the inventory of the Asse II mine facility and in ERA Morsleben

Disposal plant/nuclide	Half-life period [a]	FGi [Bq]	Asse II (31.12.2013)		ERAM (31.12.2013)	
			Ai [Bq]	Ai/FGi [–]	Ai [Bq]	Ai/FGi [–]
C-14 (β^-)	5730	1.00E+07	2.6E+12	2.6E+05	3.20E+12	3.20E+05
Co-60 (β^-)	5271	1.00E+05	1.1E+13	1.1E+08	5.40E+12	5.40E+07
Ni-63 (β^-)	100.1	1.00E+08	2.6E+14	2.6E+06	1.40E+13	1.40E+05
Sr-90 (β^-)	28.79	1.00E+04	2.0E+14	2.0E+10	4.80E+12	4.80E+08
Cs-137 (β^-)	30.17	1.00E+04	3.6E+14	3.6E+10	6.30E+13	6.30E+09
Ra-226 (α)	1602	1.00E+04	2.0E+11	2.0E+07	2.30E+10	2.30E+06
U-234 (α)	245,500	1.00E+04	1.4E+12	1.4E+08	1.10E+09	1.10E+05
U-238 (α)	4.468E+09	1.00E+04	1.3E+12	1.3E+08	4.30E+08	4.30E+04
Pu-238 (α)	87.74	1.00E+04	9.2E+12	9.2E+08	7.80E+10	7.80E+08
Pu-239 (α)	24,110	1.00E+04	4.5E+13	4.5E+09	6.90E+10	6.90E+06
Pu-240 (α)	6563	1.00E+03	5.1E+13	5.1E+10	6.60E+10	6.60E+07
Pu-241 (β^-)	14.35	1.00E+05	1.3E+15	1.3E+10	9.00E+11	9.00E+06
Am-241 (α)	432.2	1.00E+04	2.4E+14	4.4E+09	2.30E+11	2.30E+07
Cm-244 (α)	18.10	1.00E+04	8.0E+11	8.0E+07	4.80E+09	4.80E+05
Sum α			3.49E+14	6.12E+10	4.72E+11	8.88E+08
Sum β/γ			2.13E+15	6.91E+10	9.13E+13	6.84E+09
Sum			2.48E+15	1.30E+11	9.18E+13	7.72E+09

Sources K-MAT 14 [16]

not escape from the containment into the biosphere. This also applies to the solid tailings mass, which is protected by a multifunctional surface sealing especially against seepage water, see Chap. 5.

For long-term predictions, it is also important to note that the activity of uranium tailings hardly changes over a period of 1000 years (the radionuclide Th-230 has a half-life of about 80,000 years). However, the activity index of the Asse inventory amounts to about 4.4E+10 in 100 years and in about 1000 years decreases it to about 2.4E+10. The different requirements for geotechnical environmental structures—repository structures—for long-term safe storage of radioactive waste and residues from uranium ore processing in Germany cannot be justified from the activity inventories.

6.4.2   Comparison of the Long-Term Safety Concepts of ERA Morsleben with Residue Storage Facilities from Uranium Ore Processing

6.4.2.1   Summary of the Long-Term Safety Closure Concept of Tailing Ponds from Uranium Ore Processing

The concept for the long-term safe and stable custody of the tailings ponds of SDAG Wismut is set out extensively in Chap. 5 and is listed here only as a summary for comparison. The basis is a multi-barrier concept consisting of the following main elements:

• Dam construction, cover (encapsulation), base sealing as geological barrier and/or as technical barrier, water collection and water treatment as well as the diversion of purified water into the receiving water (if discharge values of pollutant concentrations are maintained), planting and vegetation, monitoring system; (Multi-barrier system with long-term monitoring).

Both the dam structures and the multi-functional covers are constructed as technogenic loose rock bodies whose classification is possible over the entire loose rock body. Nevertheless, the material parameters of the functional elements (mineral, geosynthetic, etc.) of the technogenic loose rock bodies are stated variables and thus time-dependent. Also, aging during the period of observation cannot be ruled out.

The geotechnical environmental structure of the tailings ponds of SDAG Wismut must be operated with the missing base seal and according to the guardian-principle for a long time. The escape of radioactive and toxic contaminants can enter the biosphere via the soil-, water- and air pathways (gas- fluid- and solid phases of the loose rock body). A discharge via the water pathway is collected in a controlled manner and fed to the water treatment plant, cleaned and delivered in compliance with water-legal exemptions in the Vorflut. Water treatment is part of the security system for the long-term safety of tailing ponds from uranium ore processing. For the monitoring of the water pathway, in addition to a corresponding groundwater monitoring system, a network of wells and groundwater measuring points, with production wells to lift groundwater is required.

In order to prevent uncontrolled escapes of contamination from the tailings body and considering the groundwater conditions at the site, the groundwater level, can be lowered in the area of the tailings pond by means of production wells. This will establish a defined safety distance between the tailings body and the highest groundwater level, and help to prevent or/and limit contamination transfers into the groundwater body.

The proof of the long-term stability of the multi-functional cover of a tailings pond should be designed for a period of 1000 years. An assessment value for the radiation exposure of ODL $\leq 0.15\ \mu$Sv/h, has been established by Wismut GmbH for their own use.

There may be a risk of subsequent human intervention in the encapsulation. Such "human intrusion scenarios", must be prevented, this includes securing the surrounding area. A follow-up remediation and after-care provision is possible at any time and should form part of the long-term security system. Damaged points will be indicated in particular by increasing radon exhalation rates. The scenario of an: aircraft crash is not considered but should also remain manageable here, because only a small part of the total area would be likely to be affected and it is expected there would only be low spontaneous emissions of radioactivity.

The record (final documentation) of the abandoned mining site with all relevant data remains in the register of the relevant mining authority. The time of removal of the site from supervision is part of the long-term safety system for the geotechnical environmental structure and is approved or denied by the competent mining authority on application.

The parameters of the geotechnical environmental structures for long-term safe, long-term stable storage of the tailings ponds of SDAG Wismut have been summarized in Table 6.14.

6.4.2.2 Summary of the Long-Term Security Concept for Long-Term Storage of LAW and MAW

The long-term safe storage of this radioactive waste is extensively covered in this chapter. Table 6.15 summarizes the main parameters of the geotechnical environmental structures, as installations according to § 9 a AtG for the long-term safekeeping of LAW and MAW in Germany (according to the current situation). A comparison of Tables 6.14 and 6.15 shows the main differences for the safekeeping of uranium tailing ponds of SDAG Wismut and LAW and MAW in Germany, with the appropriate applicable laws and regulations. This is also reflected in the risk evaluation and assessment of outgoing hazards.

The current decommissioning scenario in Germany for repositories for radioactive waste subject to monitoring stipulates that "after completion of all decommissioning measures no control and monitoring program is required". However, for the uranium mill tailings ponds with the guardian principle their long-term security will be guaranteed.

With the specification of an extremely long observation period for long-term safety in the final disposal of radioactive waste subject to supervision, the impression of a high level of safety is to be conveyed. However, long-term safety is independent of the given observation period and depends solely on the parameters of the Geotechnical Environmental Structure as an installation according to § 9 a AtG.

The "geological" observation period for the long-term safety of a repository encounters scientific limits, insofar as it has to involve variables of the geotechnical environmental structure. Since these variables are time-dependent, extrapolation of the measurement results from time series of a few decades to more than 100,000 years is generally not permitted. However, one effect may be achieved by such an approach. Potentially it may fuel fears and uncertainties with the side effect that long-term safe

Table 6.14 Parameters of the geotechnical environmental structures for the long-term safe, long-term stable storage of the tailings ponds of the SDAG Wismut

Multifunctional cover	Dam construction/hollow form	Activity of inventory Tailings pond Culmitzsch A	Type of storage deposit	Period of observation	Proofs	Radionuclide transport	Gathering of contaminated water
Total cover thickness > 6 m	Hight ca. 72 m; 159 ha; 61 Mio m³	Ai/FGi = 2.55E+11	Hollow form: former uranium open pit	1.000 a	**Dam construction:** mechanical stability/extraordinary events—earthquakes	Soil-, Water- and air path	Seepage water collection, production wells
Design	*Design*	*Activity of inventory*	*Base sealing*	*Assessment value for radiation exposure*	*Multi-functional cover*	*Long term monitoring*	*Water treatment*
Work platform, intermediate cover, contouring layer, erosion-proof end cover, revegetation layer, instrumentation as a part of encapsulation	Contouring, profiling; support dumps, erosion-proof end cover vegetation layer, planting, instrumentation as a component of encapsulation	1.94E+15 Bq	Natural and/or technical barrier: None	≤0.15 µSv/h	Internal and external Erosion—consolidation	Soil samples; groundwater monitoring points network; Exhalation rate (enrichment box), [50] Tracer tests (SF6) **Total: Local dose rate**	Technical and biological water treatment plants

Table 6.15 Parameters of the geotechnical environmental structures as installations according to § 9 a AtG for the long-term storage of LAW and MAW in Germany according to the current situation

Emplacement areas	Format of delivery	Inventory of activity of ERAM and of Asse II mine	Disposal	Period of observation	Proofs	Radionuclide transport	Reversibility
Former production mines; Morsleben in Lower Saxony Remlingen in Lower Saxony	Containers, usually 200 l—barrels; Container not designed as a technical barrier	Ai/FGi = **ERAM** 6.95E+09 **Asse II** 1.30E+11	In deep geological formations: **ERAM:** rock salt **Asse:** Rock salt, in each case Emplacement chambers	1.000.000 a	Integrity of the repository construction/long-term safety assessment Prevention of radio-nuclide exit into the biosphere	Water- and air pathway	**ERAM:** None **Asse II mine:** Option of orderly recovery and transfer of the radioactive inventory to a authorisable repository site

(continued)

Table 6.15 (continued)

Emplacement areas	Format of delivery	Inventory of activity of ERAM and of Asse II mine	Disposal	Period of observation	Proofs	Radionuclide transport	Reversibility
Multi-barrier-system	*Emplacement chambers*	*Inventory of activity*	*Defects/packages*	*Assessment value for radiation exposure*	*Incident*	*Environmental Monitoring*	*Long term monitoring*
Dam structures and closure construction	**ERAM:** Backfilling of the Emplacement chambers, possible placing of concrete, optimisation during the operating phase until the plan approval decision is issued. **Asse II:** 3 options: retrieval; transfer to deeper levels, safekeeping on site; See description of Options. Currently securing of the mine workings, fact finding, emergency planning	**ERAM** 9.18E+13 Bq **Asse II** 2.48E+15 Bq	**ERAM:** nothing important known **Asse II:** Restricted integrity, uncontrolled entrance of water possible	$\leq 0{,}3$ mSv/a **ERAM:** Exposition proof of lower expositions **Asse II mine:** currently no reliable proof of exposition is available	Incident: "beyond-design inflow" of brine solution In the Asse II mine an emergency planning is active with the scenario: Give up of the underground mine	Determination of the transfer of radionuclides within the biosphere and assignment to a person or a group of people with their living habits. Sampling of spread media **Total: Local dose rate**	None intended

storage of this waste is not controllable. The experiences show something else and indicate to the community, for social acceptance, that the responsible and proper handling and long-term safe storage of radioactive residues is achievable.

A quick decommissioning is not recommended in any case. The immediate closure of a final disposal on the principle of "Close your eyes and hope for the best!" also carries the message that parties want to "get rid" of the radioactive waste and shift the problem to subsequent generations as soon as possible. Subsequent generations should be given the opportunity to intervene as they are affected anyway, because of the longevity of the radioactive waste.

References

1. Vieno, T; Nordman, H; Taivassalo, V; Nykyri, M: Post-closure safety analysis of a rock cavern repository for low and medium level waste. Radioactive Waste Management and the Nuclear Fuel Cycle 1993, Vol. 17(2), pp. 139–159.
2. André Rübel; Ingo Müller-Lyda; Richard Storck: „Die Klassifizierung radioaktiver Abfälle hinsichtlich der Endlagerung"; GRS – 203 ISBN 3-931995-70-4; Dezember 2004.
3. U.S. Government: Code of Federal Regulations No. 10 CFR 61 § 55: Waste classification. U.S. Government, Washington 2001.
4. Glombitza, F. et. al.: Verfahren zur Fassung, Ableitung und Reinigung bergbaulich kontaminierter Grundwässer; Herausgeber: Landesamt für Umwelt, Landwirtschaft und Geologie - Freistaat Sachsen; EU-Ziel 3 Projekt VODAMIN.
5. M. Paul et al: Water Management – a Core Task of the Wismut Remediation Programme; Mining Report 151 (2015) No. 6; p. 5008–518.
6. Bundesministerium für Wirtschaft und Technologie; 20 Jahre Wismut GmbH, Sanieren für die Zukunft, 2011.
7. Märten, H: Neueste Trends zur aktiven Wasserbehandlung und Anwendungsbeispiele; 57. Berg- und Hüttenmännischer Tag 22. und 23. Juni 2006; Technische Universität Bergakademie Freiberg Institut für Geologie; Wissenschaftliche Mitteilungen 31 (proceedings), S. 14 – 22; Freiberg 2006; Hrsg: B. Merkel, H. Schaeben, Ch. Wolkersdorfer, A. Hasche-Berger.
8. Gesetz zur Neuordnung der Organisationsstruktur im Bereich der Endlagerung, vom 26. Juli 2016; Bundesgesetzblatt Jahrgang 2016 Teil I Nr. 37; S. 1843 ff.
9. Karin Kugel, Kai Möller (Hrsg.): Anforderungen an endzulagernde radioaktive Abfälle (Endlagerungsbedingungen, Stand: Februar 2017) - Endlager Konrad - Bundesamt für Strahlenschutz, BfS-Bericht SE-IB-29/08-REV-3, Salzgitter, 10.02.2017.
10. https://www.endlager-konrad.de/Konrad/DE/themen/endlager/eignung/geologie/geologie.html.
11. Bundesministerium des Innern (1983): Sicherheitskriterien für die Endlagerung radioaktiver Abfälle in einem Bergwerk; GMBl. 1983.
12. Niedersächsisches Umweltministerium: Planfeststellungsbeschluss für die Errichtung und den Betrieb des Bergwerkes Konrad in Salzgitter als Anlage zur Endlagerung fester oder verfestigter radioaktiver Abfälle mit vernachlässigbarer Wärmeentwicklung vom 22. Mai 2002.
13. Brennecke, P.; Steyer, St.: Endlager Konrad - Bilanzierungsvorschrift für Radionuklide/Radionuklidgruppen und nichtradioaktive schädliche Stoffe; BfS; SE-IB-33/09-REV-1; Stand: 07. Dezember 2010.
14. Gehobene wasserrechtliche Erlaubnis zur Einleitung von Niederschlagswasser, Grundwasser und Abwasser aus dem Endlager für radioaktive Abfälle, Schacht Konrad 2, in Oberflächengewässer, Niedersächsisches Umweltministerium, Planfeststellungsbeschluss Konrad, Anhang 3.

15. BfS: Radionuklidvektor der relevanten Radionuklide Stichtag 30.06.2005 – ERAM.
16. K-MAT 14; Gemeinsames Übereinkommen über die Sicherheit der Behandlung abgebrannter Brennelemente und über die Sicherheit der Behandlung radioaktiver Abfälle- Bericht der Bundesrepublik Deutschland für die fünfte Überprüfungskonferenz, Mai 2015.
17. BfS: Stilllegung ERA Morsleben, Plan zur Stilllegung des Endlagers für Radioaktive Abfälle Morsleben, Salzgitter, 15.09.2009.
18. Geologische Bearbeitung des Hutgesteins über der Allertal-Salzstruktur, BGR, http://www.bgr.bund.de/DE/Themen/Geotechnik_vor20090101/Geologische_Bearbeitung_Hutgestein_Allertalstruktur.html.
19. BMUB: Programm für eine verantwortungsvolle und sichere Entsorgung bestrahlter Brennelemente und radioaktiver Abfälle (Nationales Entsorgungsprogramm) August 2015.
20. BfS 2009 - Stilllegung ERA Morsleben, Plan zur Stilllegung des Endlagers für Radioaktive Abfälle Morsleben, Bundesamt für Strahlenschutz, Salzgitter, 15.09.2009.
21. Projekt ERA Morsleben. Hydrogeologische Standortbeschreibung und Modellgrundlagen. Teile 1 - 3.- Bearbeiter: Langkutsch, U., Käbel, H., Margane, A. & Schwamm, G., Verfahrensunterlage P 070, 1998.
22. Langzeitsicherheitsnachweis für das Endlager für radioaktive Abfälle Morsleben (ERAM); Stellungnahme der Entsorgungskommission vom 31.01.2013.
23. Brasser, T.; Droste, J.: Endlagerung wärmeentwickelnder radioaktiver Abfälle in Deutschland - Anhang Endlagerstandorte -; Gesellschaft für Anlagen- und Reaktorsicherheit (GRS) mbH; Anhang zu GRS-247; ISBN 978-3-939355-22-9; Braunschweig/Darmstadt; September 2008.
24. BGR: https://www.bgr.bund.de/DE/Themen/Endlagerung/Methoden/Labor/salzmechan_fortgeschrittene.html.
25. Günther, R.-M.: Erweiterter Dehnungs - Verfestigungs – Ansatz, Phänomenologisches Stoffmodell für duktile Salzgesteine zur Beschreibung primären, sekundären und tertiären Kriechens, Dissertation zum Dr.-Ing., TU Bergakademie Freiberg, Fakultät für Geowissenschaften, Geotechnik und Bergbau, 10-2009.
26. Leuger, B., Staudtmeister, K., Zapf, D.: The thermo-mechanical behavior of a gas storage cavern during high frequency loading. In: Mechanical Behavior of Salt VII, Tijani, M., Berest, P., Ghoreychi, M., Hadj-Hassen, F. (Editoren), S. 363–369, ISBN Print 978-0-415-62122-9, 2012.
27. Popp, T., Salzer, K., Schulze, O., Stührenberg, D.: Hydro-mechanische Eigenschaften von Salzgrusversatz - Synoptisches Prozessverständnis und Datenbasis. Memorandum, Institut für Gebirgsmechanik (IFG), Bundesanstalt für Geowissenschaften und Rohstoffe (BGR): Leipzig, 30.05.2012.
28. Peiffer, F., McStocker, B., Gründler, D., Ewig, F., Thomauske, B.,: Havenith, A., Kettler, J.: Abfallspezifikation und Mengengerüst. Basis Ausstieg aus der Kernenergienutzung (Juli 2011). Bericht zum Arbeitspaket 3, Vorläufige Sicherheitsanalyse für den Standort Gorleben, GRS-278, ISBN 978-3-939355-54-0, Gesellschaft für Anlagen- und Reaktorsicherheit (GRS) mbH: Köln, September 2011.
29. ITASCA: FLAC3D: Fast Lagrangian Analysis of Continua in 3 Dimensions - Manuals. Users Guide. 3. Edition, Version 3.1, ITASCA Consulting Group Inc.: Minneapolis, Minnesota, USA, 2006.
30. Backhaus, G. (1983), Deformationsgesetze, Akademie-Verlag, Berlin
31. Hunsche, U., Schulze, O.: Das Kriechverhalten von Steinsalz. In: Kali und Steinsalz, Band 11, Heft 8/9, Dezember 1994.
32. Salzer, K., Popp, T., Böhnel, H.: Investigation of the Mechanical Behaviour of Precompacted Crushed Salt in Contact to the Host Rock. NF-PRO Deliverable 3.5.6, 2007.
33. Gebirgsmechanische Zustandsanalyse des Tragsystems der Schachtanlage Asse II, Institut für Gebirgsmechanik GmbH, 2007.
34. Rolf Michel; Strahlenexpositionen bei der Stilllegung der Schachtanlage Asse II, 18. Sommerschule für Strahlenschutz der LPS, Berlin 2013.
35. Entwicklung und Beschreibung des Konzepts zur Schließung des Schachtanlage Asse, Helmholtz-Zentrum – München, 2008.

36. BfS 2010; Optionenvergleich Asse Fachliche Bewertung der Stilllegungsoptionen für die Schachtanlage Asse II; urn:nbn:de:0221-201004141430, Jan. 2010.

37. Entsorgungskommission (ESK) 2010, Stellungnahme zu Fragen des BMU zur möglichen Rückholung und Konditionierung von radioaktiven Abfällen aus der Schachtanlage Asse II; Bonn, 07.04.2010.

38. Dr. U. Gerstmann (2002), H. Meyer, M. Tholen; Abschlussbericht: Bestimmung des nuklidspezifischen Aktivitätsinventars der Schachtanlage Asse, August 2002, GSF – Forschungszentrum für Umwelt und Gesundheit, see also http://www.greenpeace.de/fileadmin/gpd/user_upload/themen/atomkraft/GSF_Bericht_nuklidspezifisches_Aktivitaetsinventar.pdf.

39. Arbeitsgruppe „Optionenvergleich"; Projektträger Karlsruhe – Wassertechnologie und Entsorgung (PTKA-WTE) am Karlsruher Institut für Technologie (KIT); Stand: 29.04.2010; http://docplayer.org/4705847-Bundesamt-fuer-strahlenschutz-bfs-arbeitsgruppe-Optionenvergleich.html.

40. Küppers, Ch., et. al.: Neuberechnungen zu den Auswirkungen eines auslegungsüberschreitenden Lösungszutritts in der Schachtanlage Asse II- Weiterentwicklung der radioökologischen Modellierung; Darmstadt, 2011.

41. Kock, I.: Integritätsanalyse der geologischen Barriere Bericht zum Arbeitspaket 9.1 - Vorläufige Sicherheitsanalyse für den Standort Gorleben; Gesellschaft für Anlagen- und Reaktorsicherheit (GRS) mbH; GRS – 286; ISBN 978-3-939355-62-5; Dezember 2012.

42. Minkley, W.: Integrität von Salzgesteinen und praktische Relevanz für die Verwahrung von Salzkavernen; Twente University, Enschede, 27. Oktober 2015.

43. Evaluierung der Faktenerhebung und der Vorgehensweise zur Rückholung; Bundesamt für Strahlenschutz, Asse-GmbH; Salzgitter, den 27. April 2016; http://www.bfs.de/SharedDocs/Downloads/Asse/DE/IP/stellungnahmen/150415-evaluierung-faktenerhebung-vorgehensweise-rueckholung.html.

44. Klaus-Peter Kröhn (GRS), et. al.: Restporosität und -permeabilität von kompaktierendem Salzgrus-Versatz, REPOPERM - Phase 1; September 2009.

45. Gellermann, R.: Schachtanlage Asse II Radioökologisches Modell zur Berücksichtigung der gekoppelten Migration von Tochternukliden; FUGRO-HGN GmbH – 2010.

46. M. Schnelzer et. al., The German Uranium Miners Cohort Study (Wismut cohort), 1946–2003; Technical Report; BfS - February 2011.

47. M. Schnelzer et. al., Berufliche Exposition und Mortalität in der deutschen Uranbergarbeiterkohorte, ASU Arbeitsmed Sozialmed Umweltmed 49 | 10.2014.

48. M. Lersow (2006), P. Schmidt; The Wismut Remediation Project, Proceedings of First International Seminar on Mine Closure, Sept. 2006, Perth, Australia, p. 181 – 190.

49. Lersow, M.; Gellermann, R.: „Sichere Verwahrung von Rückständen aus der Erzaufbereitung, insbesondere aus der Uranerzaufbereitung – Ein Überblick", Geotechnik, Sonderheft 2013, S. 67–79, Herausgeber: Deutsche Gesellschaft für Geotechnik e.V., Essen; ISBN 978-3-943683-18-9.

50. Schulz, H. et. al.: Entwicklung einer Messmethodik zur Bestimmung der Radonquellstärke großer Flächen und Bewertung der Radondämmwirkung von Abdeckschichten; Bundesministerium für Umwelt, Naturschutz und Reaktorsicherheit; Bonn – 2003.

Chapter 7
Final Disposal of Radioactive Waste with High Radioactivity (Heat-Generating Radioactive Waste)

The radioactive wastes with high radioactivity (HAW) are those which are relatively low in volume worldwide, but represent over 90% of the total radioactivity held in wastes. The decay of radionuclides produces significant quantities of heat, which makes direct transport into a safe deposit impossible. High-level radioactive waste is characterized according to the IAEA classification, [1], see Fig. 2.1, by having a specific activity of $A > 10^{14}$ Bq/m^3, typically: 5×10^{16} to 5×10^{17} Bq/m^3, with a specific decay heat of 2 $KW/m^3 \le Q \le 20$ kW/m^3.

In order to enable a comparison of repository sites, it is first necessary to refer to a legal and regulatory framework, as this is the only way to make geotechnical environmental structures comparable. Even if the authors refer in the following to the German regulations as a matter of priority, it is in any case necessary to make a comparison with the legal and regulatory framework of the USA. The authors are keen to emphasise this point because, on the one hand, the USA has the world's greatest experience in dealing with radioactivity and, on the other hand, it is in the USA that by far the largest quantities of radioactive waste and residues are generated annually. These materials arise from the peaceful and military uses of nuclear energy. Thirdly, because the USA uses a different classification of radioactive waste to Germany. This is due to tradition and because in the USA the need for classification arose earlier, before the IAEA announced a classification, see Table 7.1.

According to the World Nuclear Association [3], the IAEA and the US Nuclear Regulatory Commission (NRC), approximately 12,000 tonnes of high active waste (HAW) are produced annually, primarily from commercial energy production. According to the US Department of Energy (DOE), it is estimated that approximately 390,000 tonnes of HAW are currently awaiting disposal worldwide, of which about 90,000 tonnes are HAW in the USA. The US government's nuclear weapons program has produced 14,000 tonnes of spent nuclear fuel and high-level radioactive waste, C [4]. Also according to the DOE approximately 3000 tons are added annually. However, this corresponds to less than 0.5% of the total radioactive waste and residues, see Chap. 1, Fig. 1.2. A detailed overview is not available from the IAEA, as the reports of the individual member states are not standardised and often

© Springer Nature Switzerland AG 2020
M. Lersow and P. Waggitt, *Disposal of All Forms of Radioactive Waste and Residues*, https://doi.org/10.1007/978-3-030-32910-5_7

Table 7.1 Subclasses of low-level waste according to the US NRC and definitions of material designations that qualify waste classifications, according to [2]

High-Level Waste (HLW)	Definition
Spent Nuclear Fuel (SNF)[a]	**Nuclear Spent Fuel (SNF)**: irradiated commercial reactor fuel **Reprocessing Waste**: liquid waste from solvent extraction cycles in reprocessing. Also the solids into which liquid wastes may have been converted **Note**: The Department of Energy (DOE) defines HLW as reprocessing waste only, while the Nuclear Regulatory Commission (NRC) defines HLW as spent fuel and reprocessing waste
Low-Level Waste (LLW) Class	Definition
Class A	Low levels of radiation and heat, no shielding required to protect workers or public, rule of thumb states that it should decay to acceptable levels within 100 years
Class B	Has higher concentrations of radioactivity than Class A and requires greater isolation and packaging (and shielding for operations) than Class A waste
Class C	Requires isolation from the biosphere for 500 years. Must be buried at least 5 m below the surface and must have an engineered barrier (container and grouting)
Greater than Class C	This is the LLW that does not qualify for near-surface burial. This includes commercial transuranics (TRUs) that have half-lives >5y and activity >100nCi/g (3.7 kBq)
Material	Designation definition
Transuranic Material (TRU)	**(TRU)** Material containing or contaminated with elements that have an atomic number greater than 92
Contact Handled (CH)	**(CH)** Materials or packages with a surface exposure rate <200 mR/h (2 mSv/h) may be handled without shielding for radiation workers
Remote Handled (RH)	**(RH)** Materials or packages with a surface exposure rate >200 mR/h (2 mSv/h) must be handled remotely for protection of radiation workers. Individual sites may have upper limits, as well
Hazardous Waste (Mixed Waste, MW)	**(Mixed Waste, MW)[a]** Waste that contains both hazardous material, regulated under RCRA by the EPA, and radioactive material, regulated under the AEA and its by the NRC or DOE, is called mixed waste. There are high-level mixed wastes, low-level mixed wastes, and TRU mixed wastes (DOE treats all of its TRU waste as mixed waste. EPA has not yet determined whether SNF will be designated as mixed waste

[a]US EPA has not yet determined whether SNF will be designated as mixed waste

Table 7.2 Basis of storage of HLW in USA acc. [5], Table 1.5.1-1. Summary of Repository Inventory

Type of waste	Estimated number of canisters	Tonnes of heavy metal
Commercial SNF and HLW (West Valley)	~221,000 assemblies ~7500 TAD canisters 275 HLW canisters	63,000
HLW	~9300 canisters	4667
DOE SNF	~2500 to ~5000 canisters	2268
Naval SNF	~400 canisters	65
Total	–	70,000

incomplete. Table 7.2 gives an overview of the composition of the HAW and the associated activities.

As there is only one classification for Low Level Waste and High Level Waste (HLW) in the USA, see [6], High Level Waste would include all materials above Low Level Waste, class C. In the USA, the High Level Wastes are closely related to the physical characteristics of the IAEA classification. However, this is not sufficient for practicable use for HLW. Therefore, in the USA, a definition has been enforced, and is quoted from US NRC https://www.nrc.gov/waste/high-level-waste.html as follows:

High-level radioactive wastes are the **highly radioactive materials produced as a byproduct** of the reactions that occur inside nuclear reactors. **High Level Wastes take one of two forms**:

- Spent (used) reactor fuel when it is accepted for disposal
- Waste materials remaining after spent fuel is reprocessed legally implemented and thus applies as a basis for HLW management:

  Spent Nuclear Fuel (SNF) is used fuel from a reactor that is no longer efficient in creating electricity, because its fission process has slowed. However, it is still thermally hot, highly radioactive, and potentially harmful. Until a permanent disposal repository for spent nuclear fuel is built, licensees must safely store this fuel at their reactors.

  Reprocessing extracts those isotopes from spent fuel that can be used again as reactor fuel. Commercial reprocessing is currently not practiced in the United States, although it has been allowed in the past. However, significant quantities of **High-Level radioactive Waste (HLW)** are produced by the defense reprocessing programs at Department of Energy (DOE) facilities, such as Hanford, Washington, and Savannah River, South Carolina, and by commercial reprocessing operations at West Valley, New York. These wastes, which are generally managed by DOE, are not regulated by NRC. However, they must be included in any **High-Level radioactive Waste disposal plans**, along with all high-level waste from spent reactor fuel, see Fig. 1.2.

  Because of their highly radioactive fission products, high-level wastes and spent fuel must be handled and stored with care. **Since the only way radioactive waste finally becomes harmless is through decay which for high-level wastes can take hundreds of thousands of years; thus the wastes must be stored and finally disposed of in a way that provides adequate protection of the public for a very long time.**

For radioactive wastes containing mixtures of radionuclides, in [6] it is necessary to use a calculation rule, in which the total concentration is to be determined as the sum of fractions. The results are used for the classification of the radioactive waste, see Table 7.1.

The waste forms to be disposed of are categorized as follows:

- Commercial SNF—Spent Nuclear Fuel (SNF)
- HLW—High-Level Radioactive Waste (HLW)
- DOE SNF
- Naval SNF.

This also shows the difference to the HAW accruing in Germany, because in the USA about 17% of the total high-level waste arises from the nuclear weapons program, i.e. from military purposes. These High Level Wastes are assigned to the DOE, while those from civilian use are under the supervision of the US NRC. The above-mentioned High Level Waste from civilian and military use are to be stored together in a repository—Yucca Mountain—for long term safety.

The construction of a repository for High Level Waste is based on the Nuclear Waste Policy Act of 1982. In this context, the amount of high level waste to be disposed of is also stipulated. The deposit capacity was given as 70,000 tons. Among this total are 63,000 tonnes of spent fuel from commercial nuclear power plants, the remainder is spent fuel assemblies from military used reactors and vitrified high-level radioactive waste from military used reprocessing facilities of the Department of Energy (DOE), see Table 7.2.

The planning for the Yucca Mountain repository in Nevada, USA, provides for a storage capacity of approximately 77,000 tonnes of High Level Waste, see Sect. 7.4.1. Whether or not Yucca Mountain will be designed to accommodate the ongoing High Level Waste cannot be predicted at this time. There is currently no need for action because of the necessary cooldown period for the waste.

In the US, military use of radioactivity generates waste, containing the so-called transuranic elements (TRUW), which does not fall into the classification of High Level Waste. These are listed as a special category of transuranic elements, internationally known as long-lived intermediate level waste. These wastes, also known as "mixed waste[1]", are jointly overseen by the US EPA Environmental Protection Agency and the New Mexico Environment Department.

The series of transuranic elements, having an atomic number (AN) > 92, begins with neptunium (AN 93). In addition to the element plutonium (94), which is important for nuclear fission, Americium (95), Curium (96), Berkelium (97), Californium (98), Einsteinium (99), Fermium (100), Mendelevium (101), Nobelium (102) and Lawrencium (103) as well as all other heavy elements also belong to the transuranic elements.

According to [7], transuranic elements are waste (TRUW) with more than 100 nCi alpha-emitting transuranic isotopes per gram of waste (3700 Bq/g) with a half-life of more than 20 years.In 1979, following successful surface exploration work, the US Congress approved the Waste Isolation Pilot Plant (WIPP) as an underground research and development laboratory to demonstrate the safe disposal of transuranic

---

[1]The DOE designates all TRU waste as mixed waste.

Table 7.3 Final form radionuclides. Site: Argonne National Laboratory, according [7, 8, page 51]

Isotope	Typical concentration (Ci/m^3)	Isotope	Typical concentration (Ci/m^3)
Am-241	7.59E+09	Pu-244	7.33E+03
Am-243	2.72E+08	Sr-90	5.03E+09
Cm-244	3.67E+10	Th-229	2.12E+04
Cs-137	6.29E+09	Th-230	1.40E+05
Np-237	1.09E+07	Th-232	1.09E+05
Pu-238	3.47E+08	U-233	5.62E+06
Pu-239	4.03E+09	U-234	9.18E+07
Pu-240	1.87E+09	U-235	1.02E+05
Pu-241	7.96E+09	U-236	9.88E+02
Pu-242	2.23E+07	U-238	2.74E+06

waste (TRUW) and mixed waste (MW) from US nuclear weapons production, see Sect. 7.4.1. The 10 WIPP-tracked radionuclides are: (241Am, 238Pu, 239Pu, 240Pu, 242Pu, 233U, 234U, 238U, 90Sr, and 137Cs). Table 7.3 shows an example of the site—Argonne National Laboratory—accepted waste at the Waste Isolation Pilot Plant (WIPP)/New Mexico/USA repository. This waste is not within the remit of the Nuclear Regulatory Commission (NRC) but within the remit of the DOE. For the WIPP the operating phase is planned until 2033. A possible total of 176,000 m^3 of radioactive waste is to be stored, see [7].

Overall, it can be stated that in the USA there is a very tightly and clearly organised supervisory and licensing system for the produced radioactive waste, which assigns the tasks to the US EPA and the US DOE, with the involvement of the supervisory and licensing authorities of the respective federal state. In the USA, a definition of CPR has been established that can be implemented by law and thus serves as the basis for CPR management, which is subject to the objective: *"Because of their highly radioactive fission products, high-level wastes and spent fuel must be handled and stored with care. Since the only way radioactive waste finally becomes harmless is through decay, which for high-level wastes can take hundreds of thousands of years, the wastes must be stored and finally disposed of in a way that provides adequate protection of the public for a very long time".*

Both the US EPA and the US NRC publicly disclose the details of the handling of the radioactive waste and residues generated, the repository planning and the generated data obtained. The transparency is appropriate to the task and may also provide guidance for other countries.

In addition, the USA has the largest and longest experience in handling radioactive waste and residues, especially with high-level waste. It should therefore be sensible to consider incorporation of the experience gained in the USA into one's own considerations. It is unlikely that there will be an international solution for the storage of radioactive waste, in particular for HLW (HAW), since the waste is subject to provisions of the Treaty on the Non-Proliferation of Nuclear Weapons (Non-Proliferation Treaty—NPT).

Most countries have decided to keep their HLW (High Active Waste—HAW) in deep geological formations and under an arrangement using a multi-barrier system. The search for suitable geological formations depends on the overall geological situation in the individual countries. While in Germany three suitable geological formation are available: Salt rock, crystalline rocks and clay formations. However, the Scandinavian countries can resort only to crystalline formations. This makes it clear that in addition to the different definitions of CPR in the individual countries, it should also be taken into account that the suitability of different host rock formations is linked to their availability in the individual countries.

The earth is not created to be a repository. Thus, the individual host rock formations have various defects, which must be compensated for by a multi-barrier concept. This obviously means that only by the inclusion of geotechnical and technical barriers at a planned repository site can a long-term stable Geotechnical Environmental Structure be established, especially for HLW. There are quite a number of host rocks which are in principle suitable for creating a safe geotechnical environmental structure, but usually only in connection with other barriers, in particular the final disposal containers. The attribute "best possible" in terms of a long-term safe and stable repository may be considered by some to be devoid of all sense. However, the US authorities also use this attribute.

The first to respond to this constraint—existing and appropriate host rock formation(s) and the need to reconcile geotechnical and technical barriers—are the Scandinavian countries of Sweden and Finland. Thus safe repository systems can be developed in different geological formations. The term "sure" is not tied to a repository formation. This clearly shows that the selection of a repository site is by no means only subject to the geological situation. Selection means that one has to decide between at least two repository systems.

It is therefore a question, if a geotechnical environmental structure, taking into account the existing site-specific conditions, is to be developed; that the socially specified protection goal is achieved; the final disposal of radioactive waste and residues is designed in such a way, in particular for HLW (HAW), that harmful effects of the waste on humans and the environment are prevented in the long term. That means, transfers and spread of radionuclides from the radio-toxic inventory of the repository in the biosphere can only be tolerated within socially and legally acceptable limits.

It should be noted that the boundaries accepted by society differ between individual countries. However, the IAEA provides a general orientation.

First of all, is the assessment of Americans to agree: Since the only way radioactive waste finally becomes harmless is through decay, which for high-level wastes can take hundreds of thousands of years.

It is therefore important to find a place where this decay can take place without any impairment, or only to the extent permitted. That means one has to monitor the geotechnical environmental structure and its immediate surroundings, over a very long period of time. The scope and boundaries of the surveillance area are initially set but must be adjusted and updated from time to time based on the interpretation of the available measurement results and observations.

However, what can generally be stated is that when radioactivity escapes from the inventory of the geotechnical environmental structure, only a limited area is influenced around the site. The receiving municipality is thus affected in two ways by a repository built in its area of responsibility:

- that a repository building is built in the district of a municipality first causes damage to the image of the area
- and in the event of a worse case, the receiving municipality and its inhabitants also bear the consequential damage to property, goods and health.

From the authors' point of view, it is imperative that both the affected municipalities and their citizens must be fully involved in all phases of final disposal. In every case one should reach an agreement. Anyone who serves overriding social interests in this way, especially if consequential damage cannot be completely ruled out, must also be adequately compensated for it. This justifies both technical and social long-term monitoring. However, this also answers the question of why, even in states where there is a friendly climate for nuclear energy, there may still be considerable regional resistance to a repository site.

7.1 High-Level Radioactive, Heat-Generating Waste to Be Disposed of in Germany

In addition to the description of the radioactive inventories, this chapter discusses the existing repository solutions for highly radioactive waste, based on the description of the situation in Germany. Alongside the choice of the repository site, the repository concept and the interaction of the geological, geotechnical and technical barriers is also the task of ensuring the long-term security between the various scenarios. Reference is also made to the special significance of monitoring with environmental and health monitoring and this is discussed in context. However, detailed descriptions of long-term safety and monitoring are reserved for Chaps. 8 and 9.

The disposal of radioactive waste is considered to be a major economic and environmental issue worldwide, but even the IAEA cannot say exactly what amount of radioactive waste is stored where. The IAEA-led list of repositories contains nearly all of the world's under construction, in operation, as well as closed and closed repositories for low, medium and high level radioactive wastes. This is a considerable volume of material. However, the completeness of the list cannot currently be guaranteed because the declaration of a repository to the IAEA takes place on a voluntary basis and not all countries have submitted reports on their facilities.

According to the World Nuclear Association (WNA), 12,000 tonnes of high-level radioactive waste are produced worldwide every year. Highly radioactive waste in Germany primarily consists of fuel elements from German nuclear power plants and waste from the reprocessing of fuel elements from German nuclear power plants (for example, vitrification of highly active fission product solutions), which is conditioned in France and England. Internationally, highly radioactive waste from the military use of nuclear energy is still to be added to the list.

Chapter 2, Fig. 2.1 shows in the classification presented there that highly active waste (HAW) develops a decay heat of (2–20) kW/m^3 and a specific radioactivity of >1.0E+14 Bq/m^3, see [9, 10]. Internationally and relevant to the final disposal is the distinction: SL—short lived (short-lived); ML—medium lived (medium) and LL—long lived (durable).

In Germany one is able to state the quantity of HAW sufficiently precisely because of the defined exit from the production of electricity by nuclear fission. The BMU regularly updates the balances of final radioactive waste in the National Waste Management Program. A summary of all available findings and the results of the previous investigations as well as all radiologically relevant data are summarized in the "Preliminary Safety Analysis Gorleben" (VSG) [11].

VSG serves as a planning basis for the final disposal of heat-generating radioactive waste—Heat Generating Waste (HGW). The radioactive waste of approx. 29,030 m^3 reported in this report is distributed as follows in 2075:

Total activity of about 6.2×10^{19} Bq

- spent fuel from power reactors (about 5.3×10^{19} Bq)
- vitrified high-level radioactive fission products and feed clarification sludge [vitrified waste (CSD-V)] from reprocessing (about 9.0×10^{18} Bq)

The total activity of α-emitters anticipated in 2075 is approximately 6.4×10^{18} Bq

- From this spent nuclear fuel from power reactors (about 6.2×10^{18} Bq)
- Of which HAW-canisters (CSD-V) (about 2.0×10^{17} Bq)

The fuel elements are structured differently depending on the reactor type. They consist of individual fuel rods. A fuel rod is a long sturdy metallic shell (tube) surrounding a nuclear fuel. Typically, the nuclear fuel consists of many sintered tablets (pellets) containing enriched uranium dioxide, possibly also plutonium dioxide (MOX fuel = mixed oxide fuel element). The pellets are introduced into the fuel rods. A coil spring presses on top of the pellets and holds them firmly together in a column. At the end, the fuel rods are sealed with an end cap. Typically, a number of fuel rods are bundled together to form a fuel assembly. A fuel rod is considered to be burned if it can no longer be used effectively to generate energy. At this time the fuel should be replaced. The remaining fraction of fissile uranium-235 in spent fuel averages about 0.8–0.9% and is thus not much greater in concentration than natural uranium as obtained from mining (0.71% U-235).

The mined uranium (uranium ore) is first processed into "Yellow Cake", see Chap. 5. The uranium has therein only a technical purity and exists in its natural isotopic composition (about 99.3% U-238 and 0.7% U-235). In order to use the uranium in the reactor, the proportion of U-235 must be increased from 0.7% to between 3 and 5%; at much higher purity.

An increase in burnup to less than 0.5% U-235, as possible today, resulting in less heat-producing HAW. In the past, spent fuel was often shipped to France (WAA La Hague) and England (WAA Sellafield) for reprocessing. However, since July 1 2005, transport to reprocessing plants (WAA) are no longer permitted by law in

Germany. Since then, only direct final disposal has been pursued. In addition, waste from reprocessing which was produced up to 2005 must also be disposed of.

In Table 7.4 summarizes spent fuel data and how they may be used in criticality calculations. The Nuclide Inventory of Spent Fuel elements, Table 7.4, which is ultimately to be stored, depends on:

- the reactor type
- the initial enrichment of U-235 or Pu_{fiss} and
- degree of Burn-up.

Table 7.4 Characteristic data of nuclide inventories used for criticality calculations; from [9]

Fuel element[a]	Enrichment % U-235 e.g. % Pu_{fiss}	Burn-up (GWd/t SM)	Decay time (years)	wt% U-235	wt% Pu_{fiss}	wt% Pu_{total}	wt% U-238	wt% any other
du40a1e2	3.6% U-235	40	100	0.78	0.58	0.95	97.53	0.74
du55a1e2	4.4% U-235	55	100	0.67	0.65	1.14	97.19	1.0

Burn-up: in Gigawatt-days per metric tons Heavy metal (GWd/t HM); Pu_{fiss}—(Pu-239 + Pu-241); Pu_{total}—(Pu-238 + Pu-239 + Pu-240 + Pu-241 + Pu-242 + Am-241)
[a](German) name: du – PWR – fuel (uranium dioxide); 40—Burn-up in GWd/t HM; a1e2—decay time, here 1×10^2 a

Table 7.5 Heat output for irradiated nuclear fuel in dependence from the decay time acc. [11]

Time after discharge	Heat output [W/tHM]			
	Type of reactor/type of spent fuel element			
	PWR/UO_2 Burn-up 55 GWd/tHM	PWR/MOX Burn-up 55 GWd/tHM	BWR/UO_2 Burn-up 50 GWd/tHM	BWR/MOX Burn-up 50 GWd/tHM
0	2.80E+06	2.80E+06	2.00E+06	2.00E+06
5	3.40E+03	7.10E+03	2.90E+03	5.70E+03
10	2.20E+03	5.50E+03	1.90E+03	4.40E+03
15	1.80E+03	5.00E+03	1.60E+03	3.90E+03
20	1.70E+03	4.60E+03	1.40E+03	3.60E+03
30	1.40E+03	3.90E+03	1.20E+03	3.00E+03
40	1.20E+03	3.50E+03	9.90E+02	2.60E+03
50	9.80E+02	3.10E+03	8.30E+02	2.30E+03
60	8.40E+02	2.80E+03	7.10E+02	2.10E+03
70	7.30E+02	2.60E+03	6.10E+02	1.90E+03
80	6.40E+02	2.40E+03	5.40E+02	1.80E+03
90	5.70E+02	2.20E+03	4.70E+02	1.60E+03
100	5.10E+02	2.10E+03	4.20E+02	1.50E+03

The spent fuel elements will be transported to an intermediate storage facility using Castor containers and current technology. However, this can only happen when the spent fuel has cooled to a temperature that makes transport in Castor containers possible. For this reason the materials remain in "cooldown" in the decay tank of the reactor. Decay time (cooldown)—in nuclear technology—refers to the period spent fuel has to stay in the cooling pond before it generates so little heat that it is allowed be transported using the Castor. The cooldown is not described here by a half-life, since it is a mixture of many different radionuclides, see Chap. 2.

All interim storage[2] facilities in Germany are designed as dry storage facilities in which transport and storage containers loaded with irradiated fuel elements or vitrified high-level radioactive waste are stored. In Germany there are 12 intermediate storage sites at the nuclear power plant sites and 4 off site storage facilities (Ahaus, Gorleben, Jülich and Rubenow). The capacity in the interim storage facilities in Germany is sufficient to accommodate all accumulating HAW until 2075. However, it may become critical due to the 40 years operating license for the interim storage facilities, which expires at all interim storage facilities around 2050.

On the basis of the "Act on the Reorganisation of Responsibility in Nuclear Waste Management" [12], the federal BGZ Company for Interim Storage mbH, located in Essen, was founded. From 2019, BGZ mbH will be responsible for the 12 interim storage facilities at the sites of the German nuclear power plants (NPP) and from 2020 also for the twelve interim storage facilities with low and intermediate level radioactive waste from the operation and decommissioning of the NPPs. In Germany, the Federal Government is thus responsible for the interim storage of all radioactive waste subject to monitoring. NPP operators are left with only decommissioning and dismantling. Interim storage facilities should be part of a repository concept.

The spectrum of the decay heat of HAW is given as $(2–20)\,kW/m^3$, see Chap. 2. The heat generation of radioactive waste, see Table 7.6, is relevant for disposal (effective containment zone—ewG), because the heat effect on the host rock leads to changes. Changes occur due to the thermal expansion of the rock and the resulting stresses. During cooling, the reverse process takes place, but the decline is slower than the build-up of tension during heating. As a consequence, cracks may form due to stress build-up and degradation, and new pathways may form, see [13]. New pathways can also arise when heat causes temperature-related changes in rock properties or mineral changes, see [14]. However, the heat development of final-stored HAW must not lead to a reduction in the insulating capacity of the geological barriers of a repository system, e.g. by the thermally induced stresses or by changes in barrier properties. In order to avoid such disadvantages, the heat effect of the waste packages on the host rock and the total heat load introduced into the repository must be limited. This also includes a limitation of the contact temperature based on the host rock in the repository. In [15] a limitation of the contact temperature to 200 °C is recommended where the host geology is rock salt. For the host rocks mudstone and granite, lower contact temperatures (100 °C in each case) must be defined, as these conduct heat less

---

[2]The official designation reads: "Interim storage facility for heat-generating radioactive waste and spent fuel elements".

Table 7.6 Fission products of fuel elements [16]

Fission products of medium term life time (0.3% of inventory)

Fission product	$T_{1/2}$ in a	Q in keV	Decay by
Eu-155	4.76	252	$\beta\gamma$
Kr-85	10.76	687	$\beta\gamma$
Cd-113m	14.10	316	β
Sr-90	28.90	2826	β
Cs-137	30.23	1176	$\beta\gamma$
Sn-121m	43.90	390	$\beta\gamma$
Sm-151	90.00	77	β

Fission products of long term life time (0.1% of inventory)

Fission product	$T_{1/2}$ in 10^6 a	Q in keV	Decay by
Tc-99	0.211	294	β
Sn-126	0.230	4050	$\beta\gamma$
Se-79	0.327	151	β
Zr-93	1.53	91	$\beta\gamma$
Cs-135	2.3	269	β
Pd-107	6.5	33	β
I-129	15.7	194	$\beta\gamma$

4% U-235 Enrichment at time of loading
Burn-up 45 GWd/t HM
Q—released energy

well than rock salt, and changes in properties take place even at lower temperatures. Compliance with these temperature limits can be achieved by an appropriate deposit concept. Influencing factors are above all the decay times of the fuel elements and vitrified HAW, the spatial and temporal deposit patterns as well as the arrangement of the boreholes or the emplacement chambers and their distances between them, see Table 7.5 with Fig. 7.1.

Possible effects of heat generation on geological barriers' insulation capacity, significantly greater protective measures to reduce the heat-induced stress of the rock, and possible impairment of the barrier function of low permeability host rock by gas pressure build-up, will not only affect the site selection. The establishment of temperature limits for contact temperatures of the host rock also has an influence on the design of the repository, such as barriers and the dimensions of emplacement chambers.

The radionuclide vector of spent fuel rods has alpha-emitting nuclides with sometimes very long half-lives (Np-237, Pu-238, Pu-239, Pu-240, Cm-243 and Cm-244). These are included in Table 7.4 under Other. In the case of direct final disposal and for the proof of the safety of storage for very long periods, legally enshrined over one million years, the radionuclides with a long half-life play an important role. With

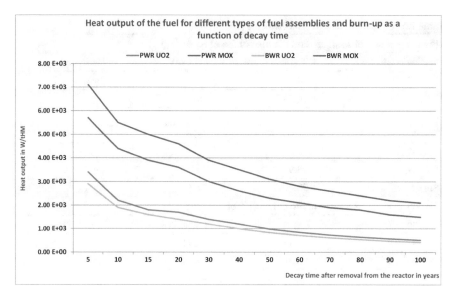

Fig. 7.1 Heat output of fuel for different types of fuel elements and burn-up in dependence of the decay time, according Table 7.5 and [11]

regard to the long-term safety of a repository, in addition to radiotoxicity, the mobility of radionuclides is of particular importance. The proportion of fission products in the inventory is summarized in Table 7.6 and gives an indication of the heat release.

7.2 Comparative Contemplations on Stored Radionuclide Inventories in Repository Constructions (Geotechnical Environmental Structures) in Germany

The activity inventory of repositories of HG HAW differs from that of the previously described inventories of Asse and ERAM not only by the much higher activity, but also in the completely different composition and heat generation resulting from the decay process of the nuclides involved. So for example evidence of subcriticality is required to show that "self-sustaining chain reactions are excluded in both probable and less likely developments." [17].

In addition to the two sites in Asse, ERAM, where low and medium-active (LAW and MAW) waste is stored in geologically deep layers, the Konrad mine is currently being converted into a repository for radioactive waste with low heat generation, see Chap. 6.

In addition, there are large facilities with residues of the former uranium ore processing, in which materials are also kept according to radiation protection aspects.

The task for the future is also to find for LAW and MAW of the Asse mine facility as well as long-term final disposal possibilities of HG HAW from NPPs.

Although the legal framework for these cases varies greatly, it is useful to illustrate the problems, with which to compare the radioactive inventories existing in the aforementioned waste categories. However, such a comparison encounters difficulties since very different radionuclides dominate in the waste categories. Since the physical size of the activity is easily measurable but is not very meaningful in terms of the potentially associated hazards, different parameters are devised to compare inventories. So can for example the computational doses resulting from a complete consumption of the waste ("ingestion") are determined. The doses thus obtained are completely fictitious and can be manipulatively misused by conversion to "cancer deaths". Therefore, another indicator is used here. By the activity inventories Ai for each radionuclide i are related to the exemption limit of the total activity (FGi) of the respective radionuclide according to StrlSchV (Appendix 4 Table 1 column 2), so a dimensionless ratio results is defined. The sum of these ratios across all radionuclides yields a dimensionless activity index capable of characterizing the overall inventory of activity inventories of different composition. The radiological hazard of single nuclides is taken into account as an orientation beyond the respective exemption limit of single nuclides.

In addition, weighting for hydrochemical mobility would be useful for assessing radioactivity with regard to long-term safety during disposal.

In the release of radionuclides, the processes in play include: container corrosion, decomposition of the waste matrix, entry into solution of pollutants, sorption processes and radioactive decay The radionuclides behave differently with mobilities varying according to their chemical properties. All release processes can only occur in aqueous systems, see Chap. 8. However, since there are no binding data sets for weighting according to hydrochemical mobility, indicators that take into account hydrochemical effects are currently subjectively influenced and therefore not very suitable for general comparisons. Thus, the mobility of inventories resulting from the hydrochemical mobility of the individual radionuclides is disregarded. Mobility alone cannot describe the spread of radionuclides into the biosphere. In addition to the isolation capacity of the ECZ, this also includes the pathways in overburden and waste rock, etc.

Table 7.7 shows spent fuel data from pressurized and boiling water reactors (PWR + BWR) and waste from reprocessing (CSD-V and CSD-C) at the end of the operational phase (here defined to start in 2075) of a repository [18] and a calculation of activity indicators to weight the radioactivity with regard to the long-term safety of a repository.

If the repository is put into operation in 2035 and has an operational life of 40 years, the date 2075 represents the earliest possible date for the completion of the storage of all heat-generating radioactive waste in a repository in Germany. The information given in [18] inventories of spent PWR and BWR fuel elements are based on burn-up rates of 55 and 50 GWd/tHM (gigawatt days per metric ton of heavy metal). Inventories of waste inventories are based on burnup and activation bills for burnup of 33 GWd/tHM (uranium oxide fuel with a 3.5% enrichment in

Table 7.7 Half lifes, Exemption limits (FGi), Activity inventories Ai [Bq] and radiotoxicity indicators Ai/FGi [–] of spent fuel elements and waste from reprocessing at end of the operational phase of a German HAW repository (here defined start in 2075) acc. [18]

Isotope	$T_{1/2}$ [a]	PWR + BWR [Bq]	CSD-V/C [Bq]	SUM [Bq]	Exemption limit FGi (Bq)	Sum Ai/FGi
C-14	5.73E+03	3.15E+14	1.22E+14	4.38E+14	1.00E+07	4.37E+07
Cl-36	3.00E+05	9.07E+12	3.92E+12	1.30E+13	1.00E+06	1.30E+07
Se-79	6.50E+04	2.53E+13	8.53E+12	3.38E+13	1.00E+07	3.38E+06
Sr-90	2.91E+01	8.47E+18	1.87E+18	1.03E+19	1.00E+04	1.03E+15
Tc-99	2.13E+05	7.74E+15	2.64E+15	1.04E+16	1.00E+07	1.04E+09
I-129	1.57E+07	1.65E+13	4.97E+12	2.14E+13	1.00E+05	2.15E+08
Cs-135	2.30E+06	2.66E+14	8.21E+13	3.48E+14	1.00E+07	3.48E+07
Cs-137	3.00E+01	1.44E+19	2.89E+18	1.73E+19	1.00E+04	1.73E+15
U-232	7.20E+01	1.82E+13	7.96E+09	1.82E+13	1.00E+03	1.82E+10
U-234	2.45E+05	8.04E+14	9.30E+11	8.05E+14	1.00E+04	8.05E+10
U-238	4.47E+09	1.12E+14	9.11E+10	1.12E+14	1.00E+04	1.12E+10
Np-237	2.14E+06	2.39E+14	6.30E+13	3.02E+14	1.00E+03	3.02E+11
Pu-238	8.78E+01	2.12E+18	1.97E+15	2.12E+18	1.00E+04	2.12E+14
Pu-239	2.41E+04	1.65E+17	4.10E+14	1.66E+17	1.00E+04	1.65E+13
Pu-240	6.54E+03	3.72E+17	1.21E+15	3.73E+17	1.00E+03	3.73E+14
Pu-241	1.44E+01	4.70E+18	2.84E+15	4.70E+18	1.00E+05	4.70E+13
Pu-242	3.87E+05	2.32E+15	2.14E+12	2.33E+15	1.00E+04	2.32E+11
Am-241	4.33E+02	3.05E+18	1.54E+17	3.20E+18	1.00E+04	3.20E+14
Am-242m	1.52E+02	9.80E+15	8.31E+14	1.06E+16	1.00E+04	1.06E+12
Am-243	7.39E+03	2.96E+16	2.76E+15	3.24E+16	1.00E+03	3.24E+13
Cm-244	1.81E+01	5.63E+17	1.16E+16	5.74E+17	1.00E+04	5.75E+13
Cm-245	8.51E+03	1.18E+15	2.41E+13	1.20E+15	1.00E+04	1.20E+11
Cm-246	4.73E+03	2.11E+14	3.41E+12	2.14E+14	1.00E+03	2.14E+11
Sum		3.39E+19	4.94E+18	3.88E+19		3.82E+15

uranium-235) and assumed uranium separation factors of 0.998 and Plutonium of 0.994 within the reprocessing process. Recent studies on measured inventories of vitrified waste from the reprocessing plant (WAA) in La Hague, [18] show that the real and calculated inventories do not always coincide. In particular, for the volatile elements such as iodine and chlorine, the real values are considerably lower.

The data from the 20 radionuclides with the highest values of the activity indicator are listed in Table 7.7. Table 7.7 shows that the highest levels of activity indicated in all activity inventories considered here are from Cs-137 and Sr-90. However, both radionuclides decay relatively quickly due to their half-life of about 30 years and are less important for long-term considerations. Of the long-lived radionuclides, Pu-239,

Pu-240 and Am-241 are among the most important 10 radionuclides of LAW/MAW and HAW.

If the radionuclides of the HAWs that have less than 100 years of half-life are also deleted from consideration, then 11% of the activity inventory of the fuel elements and 3% of the activity inventory of the reprocessing residues remain, but which get: 22% of the activity indicator of fuel rods and 4% of the activity indicator of waste of reprocessing. The activity index relevant for long-term considerations in Table 7.4 is reduced to 7.44E+14.

The following considerations are intended to show that due to the radiotoxicity of the various types of radioactive waste and residues on the one hand, very different requirements for the Geotechnical environmental structures—repository construc- tions derived, on the other hand, the site-specific disposal solutions presented here cannot be justified with the radioactive inventories deposited there.

For the Konrad mine facility, the maximum storage activities can be used for comparison, see Table 7.8. If one also ignores the nuclides H-3 and Pu-241, which are not relevant for long-term safety of the repository, the activity index of shaft Konrad changes to 2.01E+11.

The balancing of the activity values for radionuclides and radionuclide groups, total α-emitters and total β/γ-emitters is stipulated in the plan approval decision for the shaft "Konrad" repository. The deviations in the activity data in Table 7.8, see also Chap. 6, between the sum values (total α-emitter and total β/γ-emitter) and the individual values are due to the fact that the radionuclides from the Th-232 decay

Table 7.8 Maximum deposited activities (radiotoxicities) of relevant radionuclides and radionu- clide groups at the end of the operational phase of the shaft "Konrad"-repository, see [19] Brennecke et al.

Radionuclide/Radionuclide group	Half life [a]	FGi [Bq]	Activity [Bq]	Radiotoxicity [–]
H-3	12.3	1.00E+09	6.0E+17	6.00E+08
C-14	5700	1.00E+07	4.0E+14	4.00E+07
I-129	1.60E+07	1.00E+05	7.0E+11	7.00E+05
Ra-226	1600	1.00E+04	4.0E+12	4.00E+08
Th-232	1.40E+10	1.00E+04	5.0E+11	5.00E+07
U-235	7.00E+08	1.00E+04	2.0E+11	2.00E+07
U-236	2.30E+07	1.00E+04	1.0E+12	1.00E+08
U-238	4.40E+09	1.00E+04	1.9E+12	1.90E+08
Pu-239	24,000	1.00E+04	2.0E+15	2.00E+11
Pu-241	14.4	1.00E+05	2.0E+17	2.00E+12
Total α-emitters			1.5E+17	2.20E+12
Total β/γ-emitters			5.0E+18	6.41E+08
Summe	–	–	5.15E+18	2.20E+12

Source [20] Brennecke/Steyer (SE-IB-33/09-REV-1)

series and of the U-238 decay series in total α-emitter and total β/γ-emitter, see Chap. 2, Fig. 2.2. The influence on the radiotoxicity index (right column, Table 7.8) is low.

The comparison to the nuclear installation Asse II shows that the activity indices are of the same order of magnitude, see Table 7.9.

Table 7.9 shows activity indicators for the Asse and the ERAM based on the published data on the radionuclide inventory of Asse [22] and ERAM: [21]. The data show that the Asse II mine has an overall higher inventory of radionuclides than the approved Morsleben repository. Based on the activity index calculated here, the Asse' inventory of 1.50E+11 is almost 20 times larger than that of the ERAM (7.72E+09). If one eliminates the radionuclides from the balance sheets, which have less than 100 years half-life (see column half life ($T_{1/2}$) in Table 7.9) and which make only a negligible contribution to the inventory after 1000 years, then the reduced Long-term, relevant inventory activity index of the Asse becomes 7.98E+10 and that for the ERAM becomes 9.87E+07. A calculation of activity indicators with the above called approach for the inventories of Wismut tailings ponds (see Table 5.5, Chap. 5) is shown in Table 7.10. In order to derive a standardised activity index from the data

Table 7.9 Compilation of the overall activity Ai and the values of the activity indicator Ai/FGi (FG—Exemption limit according to Annex 4 Table 1 column 2 StrlSchV) for the radiologically most important radionuclides in the inventory of the Asse II mine and in the ERA Morsleben

Nuclide	Half life [a]	FGi [Bq]	Asse II (31.12.2013)		ERAM (31.12.2013)	
			Ai [Bq]	Ai/FGi [–]	Ai [Bq]	Ai/FGi [–]
C-14	5700	1.00E+07	2.60E+12	2.60E+05	3.20E+12	3.20E+05
Co-60	5.3	1.00E+05	1.10E+13	1.10E+08	5.40E+12	5.40E+07
Ni-63	100	1.00E+08	2.60E+14	2.60E+06	1.40E+13	1.40E+05
Sr-90	28.5	1.00E+04	2.00E+14	2.00E+10	4.80E+12	4.80E+08
Cs-137	30.2	1.00E+04	3.60E+14	3.60E+10	6.30E+13	6.30E+09
Ra-226	1600	1.00E+04	2.00E+11	2.00E+07	2.30E+10	2.30E+06
U-234	250,000	1.00E+04	1.40E+12	1.40E+08	1.10E+09	1.10E+05
U-238	4.40E+09	1.00E+04	1.30E+12	1.30E+08	4.30E+08	4.30E+04
Pu-238	87.7	1.00E+04	9.20E+12	9.20E+08	7.80E+10	7.80E+08
Pu-239	24.000	1.00E+04	4.50E+13	4.50E+09	6.90E+10	6.90E+06
Pu-240	6.600	1.00E+03	5.10E+13	5.10E+10	6.60E+10	6.60E+07
Pu-241	14.4	1.00E+05	1.30E+15	1.30E+10	9.00E+11	9.00E+06
Am-241	432.6	1.00E+04	2.40E+14	2.40E+10	2.30E+11	2.30E+07
Cm-244	18.1	1.00E+04	8.00E+11	8.00E+07	4.80E+09	4.80E+05
Sum α			3.49E+14	8.08E+10	4.72E+11	8.88E+08
Sum β/γ			2.13E+15	6.91E+10	9.13E+13	6.84E+09
Sum			2.48E+15	1.50E+11	9.18E+13	7.72E+09

Sources K-MAT 14 [21] and Gerstmann [22]

Table 7.10 Radionuclide inventories of Wismut Tailings ponds, normalized to the exemption limits (FGi), Lersow and Gellermann [23]

Tailings Pond	FGi [Bq]	Culmitzsch A	Culmitzsch B	Trünzig A	Trünzig B	Helmsdorf	Dänkritz I
		Ai/FGi	Ai/FGi	Ai/FGi	Ai/FGi	Ai/FGi	Ai/FGi
U-238+	1.00E+04	1.22E+10	5.60E+09	3.80E+09	1.80E+09	1.27E+10	2.50E+09
U-234	1.00E+04	1.22E+10	5.60E+09	3.80E+09	1.80E+09	1.27E+10	2.50E+09
Th-230	1.00E+04	7.90E+10	2.40E+10	1.30E+10	5.00E+09	5.50E+10	4.00E+09
Ra-226++	1.00E+04	7.90E+10	2.40E+10	1.30E+10	5.00E+09	5.50E+10	4.00E+09
U-235+	1.00E+04	5.56E+08	2.55E+08	1.73E+08	8.21E+07	5.79E+08	1.14E+08
Pa-231	1.00E+03	3.60E+10	1.09E+10	5.93E+09	2.28E+09	2.51E+10	1.82E+09
Ac-227+	1.00E+03	3.60E+10	1.09E+10	5.93E+09	2.28E+09	2.51E+10	1.82E+09
Ra-223+	1.00E+05	3.60E+08	1.09E+08	5.93E+07	2.28E+07	2.51E+08	1.82E+07
Sum		2.55E+11	8.15E+10	4.57E+10	1.83E+10	1.86E+11	1.68E+10

in Table 5.6, Chap. 5, it was taken into account that Unat is composed as follows with regard to activities: 238U/234U = 1.0, 235U/238U = 0.046. In addition, for the daughter nuclides of uranium, 230Th/226Ra is about 1 and 231 Pa and 227Ac have about 0.46 of 230Th activity. The standardized radionuclide inventories in Table 7.10 result from these data. Table 7.10 makes it clear that the inventories of Wismut tailing ponds and Asse assessed by this simple radiological indicator are comparable. However, there are differences in the mobility of radionuclides. The hydrochemical mobility of radionuclides in the tailings ponds inventory is significantly higher than in radioactive waste repositories with negligible heat generation, see Chap. 5.

Mobility is not the only factor that describes the spread of radionuclides into the biosphere, but also the insulation capacity and the pathways that develop in the geotechnical environmental structure must be considered.

It is also important for long-term statements that the activity of tailings hardly changes over a period of 1000 years (the radionuclide Th-230 has a half-life of 80,000 years). In Chap. 5, it was further demonstrated that even with the escape of all pore water from the tailings body, this has little effect on its radiotoxicity. This makes it clear that the existing site-specific disposal solutions cannot be justified with the respective radioactive inventory deposited there.

A summary of the activity inventories of all waste or residues considered here is given in Table 7.11. In addition, the activity indicator per 1 million Bq (MBq) of the respective radionuclide mixture was given in this table.

However, the decisive difference between the inventories of LAW/MAW and HAW is the radioactivity and radiotoxicity per unit volume (see Table 7.12). Only at high concentration of activity in a small volume, the radiation energy of the radioactive decays at the HAW leads to a significant increase in temperature. The hazard potential is highlighted when the comparison is made with the natural radioactivity of a soil.

Table 7.11 Long-term safely to be stored activity inventories and associated radiotoxicity, measured as an activity indicator for various inventories of radioactive waste or residues

Final disposal facility	Reporting date	Activity (Bq)	Activity indicators (Radiotoxicity)	Activity indicators per MBq
HAW (Table 7.4)	Prognosis 2075	3.88E+19	3.82E+15	3.82E+09
Konrad (Table 7.5)	Prognosis 2080	5.15E+18	2.20E+12	2.20E+06
Asse II (Table 7.6)	31.12.2013	2.48E+15	1.50E+11	1.50E+05
ERAM (Table 7.6)	31.12.2013	9.18E+13	7.72E+09	7.72E+03
Culmitzsch A (Table 5.6)	Adopted 31.12.2000	1.94E+15	2.55E+11	2.55E+05

Table 7.12 Comparison of volume-related activities and activity indicators for different inventories of radioactive waste or residues

	Volume (m^3)	Activity per volume (Bq/m^3)	Activity indicator per m^3	Remarks
HAW (2075)	29,000	1.34E+15	1.32E+11	Volume acc. [11]
Konrad (2080	300,000	1.71E+13	6.67E+06	Volume acc. [24]
Asse (2013)	47,000	5.28E+10	3.19E+06	Volume specification = package gross volume
ERAM (2013)	Appr. 40,000	2.30E+09	1.93E+05	36,753 m^3 solid or solidified waste and 6,621 sealed sources
IAA Culmitzsch A adopted 2000	6.1E+07	3.18E+07	4.18E+03	Volume acc. [23]
Soil/Earth crust		1.57E+06	134	Presumation: U-238 s = Th-232 = 40 Bq/kg; density 1.6 t/m^3

7.3 Stand of the Long-Term Safe and Long-Term Stable Disposal of HAW in Germany

For a long time in Germany, the prevailing opinion regarding the location search for a repository for high-level radioactive waste was that it should be stored in deep geological strata, and that particularly suitable host rocks would be saline, although these have high solubility in water and low sorption capacity. By sorption is meant the

Table 7.13 Country-specific repository projects HAW (heat-generating radioactive waste, OECD-NEA—as of July 2015)

Country	Reprocessing	Dir. repository	Program description	Underground laboratories	Candidates for repository sites	Geological formation	Comment
Belgium	X		Investigation of the Boom-clay formation on suitability for all types of radioactive waste	Underground Research Laboratory (URL) HADES near Mol	Mol (2035–MAW 2050–HAW)	(Boom-)clay	Site for low-, medium- and high radioactive waste
Germany	X[a]	X[b]	Investigation of the Gorleben site since 1979. 2013 open-ended site search; according to (StandAG). Gorleben remains a candidate	Originally URL Asse II, closed, (storage tests and research not completed)	open-ended with Gorleben site	Rock salt, clay (mudstone) or granite	**See further description**
Finland		X	Location approved by Parliament. Approval procedure since the end of 2012. Start of construction for 2015, operation planned from 2022	Onkalo	Olkiluoto	Granite	**See further description**

(continued)

Table 7.13 (continued)

Country	Reprocessing	Dir. repository	Program description	Underground laboratories	Candidates for repository sites	Geological formation	Comment
France	X		Reference concept is geological deep-storage with retrievability. Approval procedure since 2015, operation planned from 2025	Bure (Lothringen)	Bure	Clay (mudstone)	**See further description** (P&T of long-lived radionuclides)
Great Britain	X		2006 Basic decision for geological deep storage for high- and intermediate level radioactive waste. Searching for interested communities for repository site	–	–	–	The responsibility for carrying out the disposal lies with the waste producers
Russia	X		Site decision planned for 2025. Underground lab planned in granite	Schelesnogorsk (near Krasnojarsk)	Selection procedure for a repository site in the immediate vicinity of existing nuclear installations	Granite	German-Russian scientific and technical cooperation (**Rosatom—BGR**)

(continued)

Table 7.13 (continued)

Country	Reprocessing	Dir. repository	Program description	Underground laboratories	Candidates for repository sites	Geological formation	Comment
Sweden	X		Location Östhammar near Forsmark decided in 2009. Approval procedure since 2011. Start of construction for 2020, operation expected from 2030	Stripa (up to 1992) URL Äspö	Site Forsmark	Granite	**See further description** Operator of the nuclear power plants are responsible for the disposal
Switzerland	X	X	Technical feasibility of a repository by government 2006 confirmed. Nagra[c] had identified six locations in a first stage	Grimsel (Canton of Bern/Granite) Mont Terri (Canton of Jura/Clay)	Proposal of the Nagra two location regions: Jura East and Zurich Northeast	Granite Opalinus Clay	Organization of a regional participation
Spain	X		Examination of possible geological formations completed. In the medium term only pursuit of foreign activities	–	–	granite, clay (mudstone) (Salt stone)	For low- and intermediate-level radioactive waste since 1992 above ground repository in EL Cabril

(continued)

Table 7.13 (continued)

Country	Reprocessing	Dir. repository	Program description	Underground laboratories	Candidates for repository sites	Geological formation	Comment
Japan	X		Two underground laboratories in operation; Site identification in three phases, application, selection for explore the sites, site exploration; so far no applications	Mitzunami (on Honshu) Horonobe (on Hokkaido)	–	granite sedimentary rock	Burned fuel rods are stored at the nuclear power plant sites
Canada		X	Concept demonstration at the Whiteshell URL in Lac du Bonnet, Manitoba (a.o. 2 large-scale sealing tests)	Lac du Bonnet, Manitoba (2010 closed)	–	granite	Atomic Energy of Canada's underground research laboratory (URL)
USA	X		Decision on Yucca Mountain Waste (Civil Use of NE) adopted by President and Congress in 2002 was lifted in 2009. Basic decision on the suitability of Yucca Mountain is available. Work continued under President Trump	Yucca Mountain (Nevada)	Open	Open??? (Ignimbrite is a variety of hardened tuff—called Topopah Spring Tuff)	**See further description** (WIPP and Yucca Mountain)

[a]Reprocessing permitted until mid-2005

[b]Direct final disposal permitted since 1994

[c]National Cooperative for the Disposal of Radioactive Waste (Nagra)

Bold means, there is in the following text a description of the repositories in this country and Rosatom ist a special case in connection with the Germans (BGR)

Table 7.14 December 1995 estimates of inventory of actinide waste[a], according to [25]; https://www.nap.edu/read/5269/chapter/3#9

Isotope	Half-life[c] (years)	FGi (Bq)	Total CH-TRU (Ci)	Total CH-TRU (Bq)	Total CH-TRU Ai/FGi	Total RH-TRU (Ci)	Total RH-TRU (Bq)	Total RH-TRU Ai/FGi
Th-232[b]	1.41E+10	1.0E+04	9.11E-01	9.11E-01	3.37E+06	9.24E-02	3.42E+09	3.42E+05
U-233[b]	1.59E+05	1.0E+04	2.00E+03	2.00E+03	7.40E+09	3.18E-01	1.18E+12	1.18E+08
U-234[b]	2.45E+05	1.0E+04	5.53E+02	5.53E+02	2.05E+09	3.93E-01	1.45E+12	1.45E+08
U-235[b]	7.00E+08	1.0E+04	1.28E+01	1.28E+01	4.74E+07	5.20E+00	1.92E+11	1.92E+07
U-238[b]	4.47E+09	1.0E+04	3.96E+01	3.96E+01	1.47E+08	1.44E+00	5.33E+10	5.33E+06
Np-237	2.14E+06	1.0E+03	5.49E+01	5.49E+01	2.03E+09	4.85E-02	1.79E+09	1.79E+06
Pu-238	8.77E+01	1.0E+04	3.80E+06	3.80E+06	1.41E+13	1.45E+03	5.37E+13	5.37E+09
Pu-239	2.4E+04	1.0E+04	7.82E+05	7.82E+05	2.89E+12	1.03E+04	3.81E+14	3.81E+10
Pu-240	6.5E+03	1.0E+03	2.08E+05	2.08E+05	7.70E+12	5.07E+03	1.88E+14	1.88E+11
Pu-241[b]	1.44E+01	1.0E+05	2.61E+06	2.61E+06	9.66E+11	1.42E+05	5.25E+15	5.25E+10
Pu-242	3.70E+05	1.0E+04	1.17E+03	1.17E+03	4.33E+09	1.50E-01	5.55E+09	5.55E+05
Am-241	4.33E+02	1.0E+04	4.39E+05	4.39E+05	1.62E+12	5.96E+03	2.21E+14	2.21E+10
Sum			7.84E+06	2.90E+17	2.73E+13	1.65E+05	6.10E+15	3.06E+11

[a]Not shown are minor concentrations of many other isotopes

[b]Because the half-life of Pu-241 is less than 20 years, it falls outside the EPA 40 CFR 191 definition of TRU waste, although it is a transuranic isotope. The thorium and uranium isotopes are likewise actinides shown here that are excluded from classification as TRU waste, since they are not transuranic isotopes

[c]The level of radioactivity decreases exponentially with time, with the half-life denoting the time at which the initial amount of any radionuclide is halved. After ten half lives, approximately 0.1% of the initial radioactivity remains. For long times, the WIPP inventory is dominated by Pu-239, ten half lives of which is a quarter of a million years

Table 7.15 Characteristics of WIPP according to [25]; https://www.nap.edu/read/5269/chapter/3#9

Characteristic[a]	CH-TRU	RH-TRU
Total anticipated activity (Ci), all isotopes	7,840,000[d]	165,000[d]
Design capacity of WIPP (m^3)	168,500	7080
Total anticipated volume of WIPP (m^3)	110,000	27,000
Average activity WIPP TRU waste (Ci/m^3)[b]	46.8	143
Average activity of anticipated volume (Ci/m^3)[c]	71.6	37.8

[a]Not shown are minor concentrations of many other isotopes
[b]The Ci/m^3 values shown here derive from the total activity (Ci) and the design capacity of the WIPP repository (m^3). By comparison, the average activity of U.S. spent fuel high level waste is 140×10^3 Ci/m^3 (10-year-old waste) and 140×10^2 Ci/m^3 (100-year-old waste). See Roddy et al. [26]
[c]The Ci/m^3 values shown here indicate the ratio of total activity in curies (Ci) to the total anticipated volume (m^3) of waste to be shipped to WIPP
[d]Value see Table 7.14

Table 7.16 Total thermal power of DOE spent nuclear fuel at a specified time; according to [5]

2010		2030	
Heat output (W)		Heat output (W)	
Nominal	Bounding	Nominal	Bounding
7.18×10^5	1.25×10^6	4.67×10^5	8.54×10^5

retention, here of radionuclides. Clsy in particular has a high sorption property, the ability to bind radionuclides to the surrounding rock and thus to delay and mitigate a release.

The characteristics of salt domes that stand out for the disposal of HAW are:

- Salt domes can have a very homogeneous structure, have an extremely low porosity and thus develop a very good barrier effect,
- Salt domes have a high specific thermal conductivity,
- Due to the convergence of excavated storage areas, the stored containers are completely enclosed in the salt dome over time, which is assessed as an additional securing element.

For repository-relevant properties of potential host rocks, see Table 7.21.

The authors consider the accentuation of the host rock salt rock critical because these areas are particularly vulnerable to accidental/deliberate human intrusion in a repository, see Chap. 8; FEP—Features, Events, Processes. In Germany, salt domes are prospective areas of deposits in which human activities: drilling activities; Mining or other underground activities, such as leaching of caverns, cannot be ruled out over long observation periods, see also [32].

Since salt rock is almost impermeable for fluids, hydrocarbons (petroleum, natural gas) can accumulate below rock salt layers or salt structures, so-called petroleum and natural gas traps. Thus oil and gas deposits are also often associated with this host rock.

Table 7.17 Half-lives, exemption limits (FGi), activity inventories Ai [Bq] and activity indicators Ai/FGi [−] of the planned HLW emplacement volume for the proposed repository site Yucca Mountain

Isotope	$T_{1/2}$ [a]	Exemption limit FGi (Bq)	Commercial (US EPA)[e]		Defense (US DOE)[e]						Naval SNF[d] Ai [Bq]	Naval SNF[d] Ai/FGi
			SNF[a] Ai [Bq]	SNF[a] Ai/FGi	HLW[b] Ai [Bq]	HLW[b] Ai/FGi	Nominal SNF[c] Ai [Bq]	Nominal SNF[c] Ai/FGi	Bounding SNF[c] Ai [Bq]	Bounding SNF[c] Ai/FGi		
C-14	5.73E+03	1.00E+07	4.81E+10	4.81E+03	5.07E+12	5.07E+05	6.77E+14	1.03E+08	1.03E+15	1.03E+08	2.37E+11	2.37E+04
Cl-36	3.00E+05	1.00E+06	3.68E+08	3.68E+02	0.00E+00	0.00E+00	1.10E+13	1.73E+07	1.73E+13	1.73E+07	5.03E+09	5.03E+03
Se-79	6.50E+04	1.00E+07	1.58E+10	1.58E+03	1.40E+14	1.40E+07	1.08E+13	1.99E+06	1.99E+13	1.99E+06	9.88E+09	9.88E+02
Sr-90	2.91E+01	1.00E+04	1.61E+15	1.61E+11	8.33E+18	8.33E+14	1.15E+18	2.12E+14	2.12E+18	2.12E+14	1.13E+16	1.13E+12
Tc-99	2.13E+05	1.00E+07	5.00E+11	5.00E+04	3.58E+15	3.58E+08	3.27E+14	6.03E+07	6.03E+14	6.03E+07	1.89E+12	1.89E+05
I-129	1.57E+07	1.00E+05	1.13E+09	1.13E+04	2.07E+12	2.07E+07	7.22E+11	1.34E+07	1.34E+12	1.34E+07	2.97E+12	2.97E+07
Cs-135	2.30E+06	1.00E+07	1.71E+10	1.71E+03	6.59E+13	6.59E+06	1.16E+13	2.14E+06	2.14E+13	2.14E+06	1.36E+11	1.36E+04
Cs-137	3.00E+01	1.00E+04	2.22E+15	2.22E+11	1.25E+19	1.25E+15	1.41E+18	2.60E+14	2.60E+18	2.60E+14	1.15E+16	1.15E+12
U-232	7.20E+01	1.00E+03	1.18E+09	1.18E+06	1.66E+12	1.66E+09	8.95E+14	1.82E+12	1.82E+15	1.82E+12	1.96E+10	1.96E+07
U-234	2.45E+05	1.00E+04	5.03E+10	5.03E+06	3.01E+13	3.01E+09	2.68E+14	3.77E+10	3.77E+14	3.77E+10	6.88E+11	6.88E+07
U-238	4.47E+09	1.00E+04	1.05E+10	1.05E+06	1.93E+13	1.93E+09	2.87E+13	2.92E+09	2.92E+13	2.92E+09	3.40E+07	3.40E+03
Np-237	2.14E+06	1.00E+03	1.28E+10	1.28E+07	1.38E+13	1.38E+10	7.47E+12	1.39E+10	1.39E+13	1.39E+10	4.33E+10	4.33E+07
Pu-238	8.78E+01	1.00E+04	8.44E+13	8.44E+09	2.31E+17	2.31E+13	3.60E+16	6.62E+12	6.62E+16	6.62E+12	2.89E+14	2.89E+10
Pu-239	2.41E+04	1.00E+04	1.11E+13	1.11E+09	7.07E+15	7.07E+11	1.76E+16	2.85E+12	2.85E+16	2.85E+12	3.65E+11	3.65E+07
Pu-240	6.54E+03	1.00E+03	1.70E+13	1.70E+10	2.76E+15	2.76E+12	1.35E+16	2.30E+13	2.30E+16	2.30E+13	3.85E+11	3.85E+08
Pu-241	1.44E+01	1.00E+05	1.36E+15	1.36E+10	1.66E+16	1.66E+11	5.70E+17	1.19E+13	1.19E+18	1.19E+13	9.47E+13	9.47E+08
Pu-242	3.87E+05	1.00E+04	5.81E+10	5.81E+06	5.55E+12	5.55E+08	1.87E+13	3.10E+09	3.10E+13	3.10E+09	2.09E+09	2.09E+05

(continued)

Table 7.17 (continued)

Isotope	$T_{1/2}$ [a]	Exemption limit FGi (Bq)	Commercial (US EPA)[e] SNF[a] Ai [Bq]	SNF[a] Ai/FGi	Defense (US DOE)[e] HLW[b] Ai [Bq]	HLW[b] Ai/FGi	Nominal SNF[c] Ai [Bq]	Nominal SNF[c] Ai/FGi	Bounding SNF[c] Ai [Bq]	Bounding SNF[c] Ai/FGi	Naval SNF[d] Ai [Bq]	Naval SNF[d] Ai/FGi
Am-241	4.33E+02	1.00E+04	1.12E+14	1.12E+10	8.99E+16	8.99E+12	7.81E+16	1.44E+13	1.44E+17	1.44E+13	1.32E+12	1.32E+08
Am-242m	1.52E+02	1.00E+04	3.21E+11	3.21E+07	2.82E+13	2.82E+09	1.90E+14	3.53E+10	3.53E+14	3.53E+10	1.43E+10	1.43E+06
Am-243	7.39E+03	1.00E+03	5.70E+11	5.70E+08	3.55E+14	3.55E+11	1.50E+14	2.81E+11	2.81E+14	2.81E+11	1.72E+10	1.72E+07
Cm-244	1.81E+01	1.00E+04	2.55E+13	2.55E+09	7.40E+16	7.40E+12	8.58E+15	1.65E+12	1.65E+16	1.65E+12	1.63E+12	1.63E+08
Cm-245	8.51E+03	1.00E+04	5.48E+09	5.48E+05	6.07E+12	6.07E+08	2.64E+12	5.14E+08	5.14E+12	5.14E+08	1.42E+08	1.42E+04
Cm-246	4.73E+03	1.00E+03	1.07E+09	1.07E+06	7.25E+12	7.25E+09	4.07E+11	7.99E+08	7.99E+11	7.99E+08	4.44E+07	4.44E+04
Sum	[Bq]/[–]	[Bq]/[–]	5.44E+15	4.38E+11	2.13E+19	2.13E+15	3.29E+18	5.34E+14	6.18E+18	5.34E+14	2.32E+16	2.31E+12
Total	[Bq]/[–]	[Bq]/[–]	3.43E+20	2.76E+16	4.18E+19	2.23E+15	3.29E+18	5.34E+14	6.18E+18	5.34E+14	9.27E+18	9.24E+14
Radioactivity total Yucca Mountain	Calculated total activity Ai: 4.03E+20 Bq (1.04E+10 Ci); Radiotoxicity Ai/FGi: 3.16E+16											

See Table 7.2—commercial civil waste (US NRC) and military waste (USA) Exemption limits from German StrlSchV, Annex 4, Table 1, column 2

[a]Combined Weighted Average, HLW included— Inventories for spent BWR and PWR fuel are in curies per initial metric ton of heavy metal. Inventories for HLW are for estimated equivalent metric tonnes of heavy metal. Values are based on burnup and cooling histories assumed by DOE

[b]DOE HLW glasses, total

[c]Inventories for DOE spent nuclear fuel (DSNF) are in curies per initial metric ton of heavy metal

[d]Representative Naval SNF Canister 5 Years after Reactor Shutdown

[e]For all basic information and data see: [7, 8, 27]

Table 7.18 Estimate of radioactivity (reference end of 2020) of the HAW and MAW-LL to be deposited in the Cigéo repository, according to [28]

Type of waste	Alpha	Beta-Gamma short life	Beta-Gamma long life
Radioactivity of waste at 31 December 2004			
HAW	1.53E+18	6.4E+19	2.17E+17
MAW-LL[a]	1.34E+16	5.6E+18	3.08E+17
Sum	1.5434E+18	6.96E+19	5.25E+17
Estimate of radioactivity by the end of 2020			
HAW	3.9E+18	1.25 E+20	4.24E+17
MAW-LL	2.66E+16	3.39E+18	4.48E+17
Sum	3.9266E+18	1.2839E+20	8.72E+17

[a]Long Life

A considering of the suitability of different host rocks for erection a repository and thus a specific site selection has not yet taken place in Germany and should now be made in accordance with [33]. From the local point of view, the three previously named host rock types are basically suitable for the site selection. On this basis the site selection is not primarily decisive but reflects the repository concept as a whole and includes the coordination of natural (geological) and geotechnical—technical barriers. During the site search claystone and granite are also to be examined. To improve the sorption behavior of the storage areas, bentonites can be used.

Bentonite is a clay stone formed by mineralogical transformation of (mostly) volcanic ash. The swelling mixture of various clay minerals, quartz, mica and feldspar can be used as a geotechnical barrier in a repository in clay or granite located application.

In 1973, as part of the process to develop a repository for the Federal Republic of Germany, a location search procedure was set in motion which is strongly criticized today, because it selected as a site the salt dome at Gorleben. This was demonstrably not the best possible location and, according to current criteria, would certainly not be shortlisted for the site selection procedure, as only one of the preferred host rocks is used as the basis for the evaluation. At the time of the decision for the Gorleben site for an exploratory mine, the Lower Saxony Ministry of Economic Affairs identified 19 salt domes in Lower Saxony, in addition to the three original locations, Wahn, Lutterloh, Lichtenhorst[3] and Gorleben; in addition there was a 3 × 4 km settlement-free area, with no existing competing land use. On 22 February 1977 the state government of Lower Saxony decided to use the Gorleben salt dome as an exploration site and a provisional location for a possible facility for the disposal of the waste from German nuclear power plants, see [34]. Today it is understood that not only technical reasons were behind the decision. This was not very helpful for the site suitability test that was put in place.

The exploratory mine Gorleben has today been developed via two shafts. These are located in the center of the approximately 14 km long and 4 km wide Gorleben salt

---

[3]Wahn, Lutterloh and Lichtenhorst are salt domes in Lower Saxony.

Table 7.19 The total inventories (Bq) from thirteen radionuclides that are of importance for the decay heat, radiotoxicity and for the calculated long-term risk from the reference scenario, related to the fuel elements from the nuclear power plants and thus for all containers (6000) at the time of encapsulation (penultimate column) in comparison to the total inventory in 2045 (calculated—last column) [29]

Isotope	$T_{1/2}$ [a]	BWR I	BWR II	BWR III	BWR-Mox	PWR I	PWR II	PWR III	PWR-MOX	Total in all type canisters	Aktivität Ai (Bq)
Am-241	4.33E+02	6.6E+17	1.1E+17	3.9E+17	1.2E+17	3.3E+17	1.5E+16	1.5E+17	1.2E+16	1.8 E+18	1.78E+18
C-14	5.73E+03	2.2E+14	3.7E+13	1.5E+14	2.6E+13	7.3E+13	3.3E+12	3.6E+13	2.2E+12	5.5E+14	5.45E+14
Cl-36	3.00E+05	1.0E+12	1.9E+11	7.3E+11	1.1E+11	3.4E+11	1.7E+10	1.8E+11	9.6E-09	2.6E+12	2.57E+12
Cs-137	3.00E+01	9.3E+18	1.2E+18	6.9E+18	9.3E+17	4.2E+18	1.3E+17	2.1E+18	1.3E+17	2.5E+19	2.48E+19
I-129	1.57E+07	5.6E+12	9.8E+11	3.8E+12	6.7E+11	2.6E+12	1.2E+11	1.4E+12	7.5E+10	1.5E+13	1.52E+13
Nb-94	2.03E+04	2.0E+13	3.4E+13	1.3E+13	2.3E+12	6.1E+14	2.6E+13	2.8E+14	1.9E+13	9.8E+14	9.55E+14
Pu-238	8.78E+01	5.3E+17	1.0E+17	4.4E+17	6.9E+16	2.5E+17	1.3E+16	1.7E+17	8.0E+15	1.6E+18	1.57E+18
Pu-239	2.41E+04	5.6E+16	8.0E+15	3.1E+16	8.8E+15	2.6E+16	9.6E+14	1.1E+16	8.5E+14	1.4E+17	1.42E+17
Pu-240	6.54E+03	1.0E+17	1.7E+16	6.6E+16	1.7E+16	4.2E+16	1.9E+15	2.1E+15	1.6E+15	2.7E+17	2.46E+17
Pu-241	1.44E+01	4.0E+18	3.7E+17	3.2E+18	4.5E+17	1.9E+18	3.5E+16	1.0E+18	7.2E+16	1.1E+19	1.10E+19
Sr-90	2.91E+01	6.2E+18	7.6E+17	4.4E+18	6.0E+17	2.8E+18	8.0E+16	1.3E+18	8.2E+16	1.6E+19	1.61E+19
U-234	2.45E+05	2.2E+14	3.7E+13	1.2E+14	3.0E+13	1.3E+14	5.5E+12	5.4E+13	3.4E+12	6.1E+14	5.97E+14
U-238	4.47E+09	5.4E+13	7.8E+12	3.0E+13	6.5E+12	2.2E+13	8.0E+11	8.8E+12	6.3E+11	1.3E+14	1.30E+14
Summe											5.56E+19
Summe											
α-emitter											3.74E+18

Table 7.20 The reference inventory of the potentially safety-relevant radionuclides in 9000 tons of uranium (tU) at 30 years after discharge from the reactor, nach [30]

Isotope	Half-life $T_{1/2}$ [a]	Total inventory at 30 years cooling time: Ai [Bq]
Am-241	4.33E+02	1.74E+18
C-14	5.73E+03	1.45E+15
Cl-36	3.00E+05	2.37E+13
Cs-137	3.00E+01	3.11E+19
I-129	1.57E+07	1.72E+13
Nb-94	2.03E+04	6.77E+15
Pu-238	8.78E+01	2.38E+18
Pu-239	2.41E+04	1.28E+17
Pu-240	6.54E+03	2.81E+17
Pu-241	1.44E+01	1.58E+19
Sr-90	2.91E+01	2.01E+19
U-234	2.45E+05	4.98E+14
U-238	4.47E+09	1.05E+14
Summe		7.15E+19
Summe α-emitter		9.06E+18

Table 7.21 Valuation matrix of repository-relevant properties of host rocks according to Bollinger-fehr [31]

		Aliniferous strata (e.g. salt rock)	Clay/mudstone	Crystalline rock (e.g. Granite)
repository-relevant properties of host rocks	Thermal diffusivity	high	low	medium
	Permeability	practically impermeable	Very low up to low	very low (unfractured) up to permeable (heavily fractured)
	Strength	medium	Low up to medium	high
	Deformation behavior	viscous (creeping)	plastic up to brittle	brittle
	Cavity stability	inherent stability	Lining necessary	high (unfractured) bis low (heavily fractured)
	Residual stress condition	lithostatic isotropic	anisotropic	anisotropic
	Solution behaviour	high	very low	very low
	Sorption behavior	very low	very high	medium up to high
	Temperature resilience	high	low	high

◉ favorable property; ● favorable property; ○ property medium

dome with depths of 933 m and 840 m respectively. The project owner is the Federal Office for Radiation Protection. The reconnaissance mine is operated by the German Society for the Construction and Operation of Waste Disposal Facilities (DBE) on behalf of the Federal Office for Radiation Protection (BfS), see [35]. The Gorleben mine was used to explore the salt dome, see Fig. 7.2, to ascertain its suitability as a possible repository for high-level radioactive waste.

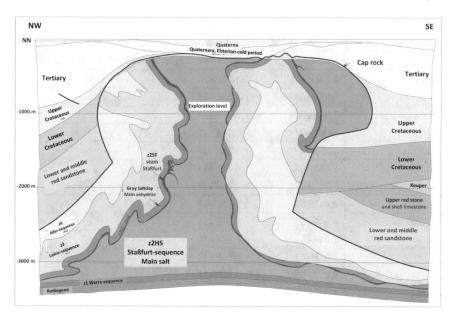

Fig. 7.2 Simplified profile section of the Gorleben salt dome. *Source* © BGR Hannover (Bornemann 1991)

What may be considered the world's most extensive exploration program was then launched. From April 1979 to the 1983, 477 km of seismic profiles were recorded to determine the layer structure and the tectonics and 322 level wells (about 80 m deep) as groundwater measuring point. 44 salt-wash surface boreholes (about 260 m deep), 4 deep boreholes (approx. 2000 m deep) and two shaft pilot holes (just under 1000 m deep) were also installed.

Following the publication of the interim report (1983) of the Physikalisch-Technische Bundesanstalt (PTB) on the results of the site investigation and the sinking of shafts, the Federal Government gave its approval for the start of work to sink shafts and thus approved the beginning of underground exploration work. For this purpose, "Safety criteria for the disposal of radioactive waste in a mine" were published. In 1986, work began on the shaft of Gorleben 1 and the start of underground exploration work. The final depth of 933 m was reached in 1997. The beginning of the sinking of shaft Gorleben 2 was 1989 and in 1995 the final depth of 840 m was reached. The break-through between shafts 1 and 2 took place in 1986, see Fig. 7.2.

The exploration work has been accompanied by constant protests and demonstrations against the repository project. The main reasons for these protests seemed to be the lack of transparency, lack of conflict resolution strategies and opportunities for participation, lack of confidence-building measures and lack of criticism and communication skills of the authorities involved and the lack of incentive for the

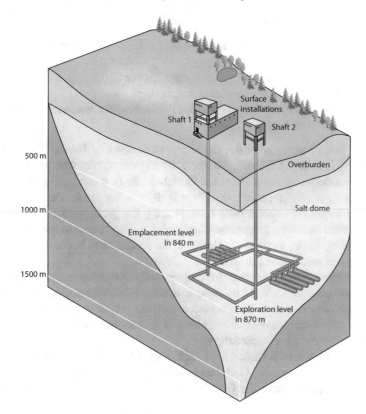

Fig. 7.3 Principle Gorleben exploratory mine. *Source* INFORUM Verlags- und Verwaltungsgesellschaft mbH, Information Group "Nuclear Energy"

host community to participate appropriately in the solution of the national problem (Fig. 7.3).

For example, a 10-year exploration moratorium was first ordered on October 1, 2000 by the then federal government, which ended in October 2010. On 07.11.2012 the exploration work was initially interrupted again, but was not resumed. Thus the repository project can be considered as having failed.

So far, only a small part of the salt dome—the exploration area 1—has been examined. In addition to the historical-political flaw of the site selection, even the low depth of exploration cannot justify a final site selection at present, but it cannot reject it either. Added to this is the expansion of the search area due to the accession of the former GDR.

The Site Selection Act [33], ended the mining exploration of the Gorleben salt dome in July 2013 and stipulated that the salt dome would be included in a new site selection procedure as for any other site in Germany. The previous investigations on the Gorleben site are summarized in [36], "Preliminary Safety Analysis Gorleben (VSG)" (2013), for further use.

In the VSG are summarized:

- The safety requirements for the final disposal of heat-generating radioactive waste of the BMU from September 2010;
- The available data from previous investigations for the Gorleben site, i.e. geoscientific site description and long-term forecasting;
- All inventory of relevant waste expected for to require a repository for heat generating waste in Germany.

On this basis, a safety and detection concept was developed in the VSG and made available for the site search, which includes the examination of whether the disposal concepts developed in the VSG at the Gorleben site or a comparable location are from today's perspective, able to fulfil the BMU's safety requirements. The safety and detection concept shows how, by exploiting the geology of the location and other technical measures (geotechnical barriers), a long-term safe final disposal for 1 million years can be guaranteed. However, the geotechnical barriers, including the containers, are matched to the host rock "salt rock".

Here, the proof of a so-called "zero emission" (completely sealed off) result is sought. This condition should be present if no continuous solution path from the waste to the biosphere is to be expected during the period under consideration. At most, very minor releases of radionuclides, which account for a fraction of the natural radiation exposure, are accepted therein.

A site search has now started [33], including the Gorleben site, which is open-ended. The exploratory mine Gorleben has been transferred to a stand by operation. It must be kept open until the Gorleben site is not excluded in the site selection procedure.

The question of the cost of disposal is clarified so that, according to the polluter pays principle, the waste producers bear the costs. The responsibility to create repositories according to the state of science and technology lies with the federal government.

According to the "Law on the reorganization of the organizational structure in the field of final disposal" [37], the Federal Office for the safety of Nuclear Waste Management (BfE), based in Berlin, was founded. It carries out regulatory, licensing and supervisory tasks in the fields of disposal and intermediate storage as well as the handling and transport of radioactive waste. The BfE is the leader in the site selection procedure for a radioactive waste repository and supervises the execution of the procedure.

According to § 9a (3) of the Atomic Energy Act, it is the task of the Federal Republic of Germany to set up repositories for radioactive waste. This task and the related sovereign powers have been delegated by the Federal Government to a federally-owned company for final disposal (BGE) mbH, with headquarters in Peine/Lower Saxony. Furthermore the federal government takes responsibility for the supervision. According to the Atomic Energy Act, the BGE is responsible for the tasks of Asse-GmbH, the German Society for the Construction and Operation of Waste Management Deposits (DBE) as well as the operational tasks of the Federal Office for Radiation Protection (BfS), including the operating tasks of the Gorleben exploratory mine. It is the procedure leader for the operational tasks of the site search, in which the exploration mine Gorleben is also to be included.

The "Law on the Reassignment of Responsibility in Nuclear Waste Disposal", [37] has re-regulated the financing of waste disposal and set up a public fund for financing, see Sect. 7.7, which is essentially supplied by the utilities. The utilities will then be responsible for the decommissioning and dismantling of the nuclear power plants and the packaging of the radioactive waste. The Federal Republic of Germany assumes responsibility for the interim and final disposal of highly radioactive waste from German nuclear power plants. This fund, which consists of the provisions previously set aside for the storage and disposal of radioactive waste from nuclear power plant operators and a risk premium, finally settles the nuclear payment obligations of the operators in these areas. Thus, the responsibility for execution and financing of the disposal tasks now lie with one organisation.

This task has been completed with the payment by the operators of the nuclear power plants of the amount of 24.1 billion euros and with the takeover of the interim storage facilities by the federal government.

Safe, responsible and environmentally sound dismantling, including decommissioning of nuclear power plants and other nuclear facilities, is not a new task in Germany. The energy supply companies, including the Energiewerke Nord GmbH, have the appropriate know how and have developed different relevant technologies. Nevertheless, decommissioning and dismantling of each individual nuclear power plant is a technically and organisationally demanding major project and a mammoth task for the nuclear phase-out. Of the nuclear power plants that lost the right to continue power operation with the 13th amendment of the Atomic Energy Act in 2011, five plants have received their decommissioning and dismantling permits: Isar 1, Neckarwestheim 1, Philippsburg 1, Biblis A and B. Larger commercial nuclear power plants at Stade, Obrigheim, Mülheim-Kärlich and Greifswald are undergoing decommissioning, while the Würgassen site has already completed its nuclear decommissioning.

In contrast to that stands the construction of a repository structure for HG—HAW. This explains why the focus is on the permanent safe disposal of high-level radioactive waste. Site search, long-term safety proof etc. are described in Chap. 8.

7.4 International Projects (Solutions) for the Long-Term Safe Disposal of Highly Radioactive, Heat-Generating Waste (Selection)

Worldwide final disposal activities (including planning and research) are currently taking place in almost 40 countries. A compilation is shown in [38]. The "Worldwide Review of Geologic Challenges in Radioactive Waste Isolation" process involves an international exchange on repository activities. The International Atomic Energy

Agency (IAEA) has created the "Joint Convention on the Safety of Spent Fuel Man-agement and on the Safety of Radioactive Waste Management" a legal instrument for the management of nuclear fuel and radioactive waste, which, as of 3 July 2017 had been ratified by 76 countries and by EURATOM. Germany is actively involved in this process.

However, results from repository sites can only be transferred to each other in a limited extent because each solution is unique and the geological conditions encoun-tered are often very different. For near-surface solutions where the technical and geotechnical barriers are dominant, the possibility of taking over entire structures is not excluded. In particular, in the production of containers and in the use of immo-bilizates in some solutions, after in situ tests, can be adopted in Germany, as far as possible. An international comparison aims precisely at identifying the structures that are projected on the conditions in Germany and so could also be applied there.

There is a particularly close cooperation in the design of the long-term safety analysis. In doing so, an FEP list is generated on the "NEA-FEP-Database" for the long-term safety analyses, see Chap. 8, of the location, so that the risk that relevant FEPs are disregarded is greatly reduced. In the VSG, a specified FEP list has been prepared for the host rock salt and the Gorleben site, see [39], see Table 7.12.

In the following sections, some international repository sites will be presented in greater detail, allowing a comparison with the repository activities in Germany and between each other.

7.4.1 Repository Projects in the USA

In the US, two independent site search procedures were conducted. One for HAW in the military sector, the Waste Isolation Pilot Plant (WIPP) in New Mexico and since 1982 with a legal mandate to find a suitable repository site for a capacity of 70,000 tonnes of heat-generating waste from civilian use, especially from nuclear power plants. Currently the location studied is at Yucca Mountain in Nevada, see Fig. 7.4, see also Fig. 1.2.

(A) Waste Isolation Pilot Plant (WIPP)

As early as 1955, the US Atomic Energy Commission (AEC), the predecessor of the US Department of Energy (DOE), commissioned the National Academy of Sciences (NAS) to investigate which host rocks could be used for the disposal of radioactive materials from the military use of nuclear energy. The NAS recommended disposal in salt formations, whereupon in 1957 a nationwide "screening" was carried out to find rock salt deposits potentially suitable for the disposal of radioactive waste. After the political failure and safety concerns for a repository in a disused salt mine and detailed investigations of four sites in New Mexico by the US Geological Survey, in 1974, the DOE looked at the salt formation of the Chihuahua Desert, located in southeastern New Mexico, about 40 km east of Carlsbad, as the site for the Waste Isolation Pilot Plant ("WIPP"), see Fig. 7.5. In 1979, the operating license was granted for in situ

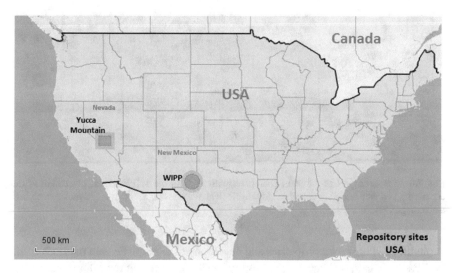

Fig. 7.4 Repositories for HAW in the USA; according to: Witherspoon and Bodvarsson [38]

Fig. 7.5 Location of the salt dome in the Delaware Basin, in the Salado Formation. *Source* Snow [40]

Fig. 7.6 left: Transport of the steel containers with TRU to the WIPP; **right**: placement of the containers in a emplacement chamber. *Source* Nuclear Watch New Mexico; https://nukewatch.org/activemap/NWC-WIPP.html

exploration of host rock properties and development of emplacement methods at the WIPP site. The WIPP began operation in 1999 and is still in the pilot phase today. Since the procedure for the construction of military installations differs from that for civilian facilities in terms of procedure, transparency and the involvement of the public, WIPP has a special position in finding a final disposal location.

This makes it the world's first repository for high-level radioactive waste (TRUW), see Table 7.1. It is not intended for heat-generating high-level radioactive waste, but for transuranics (strong alpha emitters) from research as well as from the production of nuclear weapons. The US defines transuranic waste as containing radionuclides with atomic numbers greater than 92, half lives longer than 20 years, and specific activities greater than 3700 Bq/g. They are subdivided into so-called contact-handled waste (CH) with a permissible dose rate at the container surface of up to 2 mSv/h and remote-handled waste (RH), with a dose rate of between 2 mSv/h and 10 Sv/h, see Table 7.1. The RH waste accounts for only 4% of the storage capacity of the WIPP. About 300 million GBq are to be stored, but the activity is not limited to transuranic radionuclides, see Table 7.14.

In Table 7.14, the total activity of the CH-TRU radionuclides is 290 million GBq and the RH-TRU is 6.10 million GBq. Table 7.1 further summarizes the radiotoxicity as a dimensionless activity indicator A_i/FG_i with reference to the exemption limits according to Annex III Table 1 column 2 of the German Radiation Protection Ordinance (StrlSchV). The planned radiotoxicity of the WIPP total inventory of the CH-TRU radionuclides is 27,300 Giga units and that of the RH-TRU 306 Giga units. The WIPP has been set up solely for the purpose of final disposal. The final deposit area covers 0.5 km² and consists of eight fields each with seven chambers. The authorized storage capacity is approx. 176,000 m³, see Fig. 7.7. Of these, approx. 7100 m³ can be used for Remote Handled (RH) TRU mixed waste, see Table 7.1. The storage chambers are located at a depth of approximately 650 m (2150 ft) below the surface of the terrain in an undisturbed, stratified, 600 m thick Permian rock salt formation. The location and the salt dome are described in detail in [40], see also Fig. 7.7.

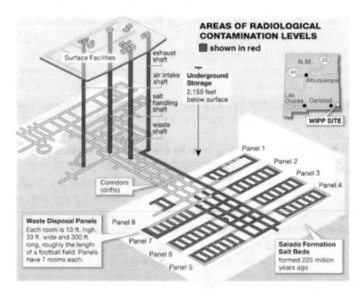

Fig. 7.7 Waste Isolation Plant (WIPP), repository cross section. *Source* US Nuclear Regulatory Commission (NRC)

The FG-exemption limits according to Annex 4 Table 1 column 2 German StrlSchV in Table 7.14 were therefore chosen to make the repositories comparable worldwide. If one extrapolates the total activity to the planned total volume of the final repository "shaft Konrad", then a total activity of approximately 510 million GBq or 0.51 billion GBq would have to be expected, compared to 5.15 billion GBq of the "shaft Konrad" repository.

By 2018, 93,500 m³ of the planned total volume had been used. Under current planning, the repository, once fully occupied and after a one-hundred-year cooldown, is to be sealed under US Department of Energy control. This expected to occur from 2133 onwards, see Table 7.15.

The transuranic waste originates from nine US facilities which are involved in military tasks. The TRU waste is usually placed in steel drums, transported pallet by pallet and deposited in the emplacement chambers, see Fig. 7.6. The emplacement chambers are backfilled with salt grit and permanently closed, so that a long-term safe and stable insulation is created in accordance with the implementing law. A program of underground long-term monitoring is not planned.

A radiological environmental monitoring system of air based particulate sampling has been set up. The results of the air sampling are used to determine the ambient radiological values and to determine if there is any deviation from the established background radiological values. This monitoring takes place at 17 locations. In addition, exposure measurements are taken. The WIPP system has two exhaust air measuring stations. Following a radiological event in 2014, the ventilation system is operating in filtration mode where all the air flows through a series of highly efficient particulate filters. Other radiological environmental samples include groundwater, surface water, soils, sediments, vegetation and wildlife. The purpose of this monitoring is to determine if the local ecosystem has been or will be affected by the operation of WIPP facilities, and if so, to assess the geographical extent and environmental impact. The Carlsbad Environmental Monitoring & Research Center (CEMRC) supports the WIPP primarily through site and environmental monitoring. The results of the random sampling are published annually in the "annual environmental report" of the site (ASER).

In addition, a large-scale facility is to be erected on the surface above the repository. The design envisages 32 monoliths forming a square, inside which 3 km of earthen ramparts enclose 16 more monoliths. These are inscribed in English, Spanish, Russian, French, Chinese, Arabic and Navajo: "This is a dangerous radioactive waste. Never dig or drill ", see Human Intrusion Scenario. In the centre more detailed information is provided together with graphic illustrations, one set above the earth, one set in an underground chamber. The warning system will be installed later. It is today therefore completely unclear whether these plans will be implemented.

(B)   US Central Repository—Yucca Mountain

Parallel to the work on WIPP, a site search procedure for disposal of highly radioactive waste from the civilian use of nuclear power was started in the mid-1970s. Due to widespread criticism of the selection process and the lack of transparency in the selection process for WIPP, the Department of Energy (DOE) was commissioned in 1982 to develop general guidelines for site selection for a repository with a capacity of 70,000 tonnes of heat-generating waste. In 1983, the DOE selected nine sites in six states for preliminary investigations. In 1985, following preliminary investigations, three sites were selected for further in-depth scientific research: Hanford, Washington State, Deaf Smith County, Texas, and Yucca Mountain, Nevada. In 1987, the Congress amended the Nuclear Waste Policy Act and mandated the DOE to focus on the potential Yucca Mountain site.

The site proposed for the repository in a mountain range (Yucca Mountain) consists of volcanic Ignimbrite, a variety of hardened tuff—called Topopah Spring Tuff. Emplacement areas are located about halfway between the top of Yucca Mountain and the deep 300 m—groundwater level, see Fig. 7.8. For insulation, 3 special types of waste packages should be used as technical barriers, see Fig. 7.9, depending on the type of waste, which are automatically driven into the emplacement areas via two access tunnels (north and south portal), see Figs. 7.5 and 7.6. The waste packages are to be emplaced in horizontal sections, which have a total length of 35 miles (56 km). The repositories should then be actively monitored for 300 years. About five miles (8 km) of the emplacement drifts for experimental purposes have been drilled so far.

In 2004, the court imposed an obligation to provide a safety assessment for 1 million years instead of the original 10,000 years. The final storage concept provides for the retrievable emplacement of the final storage containers from the horizontal sections of a mine at about 200–425 m below ground level. In June 2008, the DOE then submitted a construction application for Yucca Mountain to the Nuclear Regulatory Commission (NRC). The permit application was limited to the legally prescribed

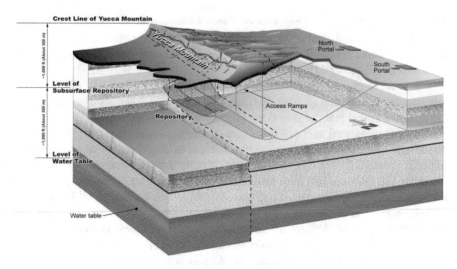

Fig. 7.8 Cross-section of the planned repository for HAW (heat-generating) Yucca Mountain/Nevada. *Source* US Department of Energy (DOE)

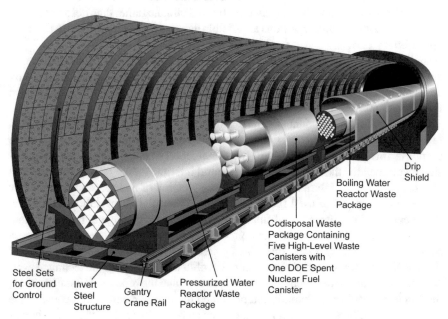

Fig. 7.9 Types of waste packages depending on the type of waste. *Source* US Department of Energy (DOE)

70,000 tonnes, the possible capacity of Yucca Mountain is estimated to be at least 130,000 t. The commissioning of the final repository at Yucca Mountain was initially planned for 2020. By 2040, high-level radioactive waste in the USA is expected to grow to approximately 119,000 tonnes.

The most important arguments in favour of the Yucca Mountain site were the low annual precipitation (approx. 20 mm, of which 95% flows off or evaporates at the earth's surface) and the very deep groundwater level, approx. 300 m below the earth's surface, as well as the closed basin structure, i.e. the planned repository has no influence on the groundwater catchment areas of neighbouring regions. Further important reasons for the selection of Yucca Mountain were ownership and previous use of the site (Nevada Test Site) by the state as well as the weak political position of the state of Nevada. However, it had been overlooked that although all US states—with the exception of Nevada—are in favor of a repository at Yucca Mountain, but for Indian tribe of the Western Shoshone who live there the area is sacred.

The suitability of Yucca Mountain was controversial from the beginning. These include doubts about the appropriateness of ignimbrite (ignimbrite) as a geological barrier, concerns about the imbalance of the selection process and the preference for voluntary commitments of communities and the failure to consider scenarios, in particular possible future climate changes (humid climate rather than desert climate), earthquakes and volcanic eruptions as possible dangers for the repository in the post-operational phase.

Following US government objections Yucca Mountain's budget was significantly cut in March 2009. The US government ordered the NRC to reject the building application. Due to a lack of budget funds, the approval procedure was suspended from September 2011.

In January 2010, President Obama mandated a commission ("Blue Ribbon Commission on America's Nuclear Future") made up of politicians and experts to make proposals for a new disposal strategy. In its final report of January 2012, the Commission recommended the construction of a central interim storage facility for used fuel elements and other high radioactive waste, the creation of an independent authority for the final disposal of used fuel elements and other high-level radioactive waste, the retention of the objective of final disposal in a deep geological repository and a continuation of the debate on the reprocessing of used fuel elements [41]. Yucca Mountain continued to be shortlisted as a potential site. In 2013, a federal court ruled that the Nuclear Regulatory Commission was obliged to complete the technical evaluation, even if the government did not want to continue the process. In January 2015, the Nuclear Regulatory Commission presented the five-part report and came to the conclusion that, from a technical point of view, a repository at Yucca Mountain was suitable according to the design plans. In the meantime, President Donald Trump has taken up $120 million in his 2018 budget to resume the approval process for Yucca Mountain, which was designated by Congress in 1987 as a repository for nuclear waste from power plants. Work continues as of September 2017.

The present plans describe the inventory to be stored so precisely that sufficient security is given for the plans both with regard to the space requirement and the radiological requirements. Table 7.2 lists the various sources of HLW, see also [42]:

The waste forms to be disposed of are categorized as follows:

- Commercial SNF—Spent Nuclear Fuel (SNF)
- CPR—High-Level Radioactive Waste (HLW)
- DOE SNF
- Naval SNF

Information also includes the number of assemblies and canisters, from which the space requirement can be derived.

The plans provided for a emplacement capacity of around 77,000 tonnes of radioactive waste, including 63,000 tonnes of spent fuel elements from commercial nuclear power plants, the remainder being spent fuel elements from military reactors and vitrified high-level radioactive waste from the military reprocessing facilities of the Department of Energy (DOE). The radiological requirements are derived in particular from the Commercial SNF—Spent Nuclear Fuel, see Table 7.17.

The thermal load results from the heat dissipation of the repository containers and in total the emplacement scheme of the final storage containers, the number of final storage containers in the respective storage chamber, the distance to the rock face and the time of storage. Table 7.16 lists the thermal energy. Compare also Sect. 7.1, Table 7.5.

The radioactive inventory which is derived from the planned emplacement volume is summarized in Table 7.17. It is also noted that if the construction of the repository for heat-generating, high-level radioactive waste should be delayed, this will have little effect on the radioactivity of the inventory.

Generally, it can be stated that the total activity will be less than 500 million Terra-Bequerel (about 13.5 GCi), with a radiotoxicity Ai/FGi of 3.5×10^{16}. The exemption limits are also taken from the German Radiation Protection Ordinance, Annex III, Table 1, Column 2. The capacities claimed are similar to those of the authorized HAW (HLW) repository of Bure in France (Lorraine), see Sect. 7.4.2.

It is interesting to note in this context that in the USA a host rock formation was selected for the central repository for HG-HLW, whose retention capacity differs significantly from the other previously favored geological formations such as salt rock and mudstone (clay). The focus of the selection was that the overall migration conditions, i.e. in relation to a multi-barrier system, for radionuclides in the biosphere—in deeper groundwater horizons (water pathway) and to the surface of the earth (air pathway)—are several magnitudes lower than at comparable locations.

The risk potential of the site due to volcanic activity, erosion or other geological processes is therefore extremely low. Due to the positioning of the repository at a depth of approximately 300 m below the earth's surface and 300 m above groundwater level, the emplaced waste is protected against the effects of earthquakes. In Ref. [43] is a discussion of what is currently known about the tectonic setting of the region encompassing the repository site. The data concerning the seismicity of the area and historic earthquake activity are also presented. In this way predictions of future seismic hazards and their potential effects on the repository, as well as the performance of natural barriers, can be made with reasonable certainty, within the limits of the available data.

The arid climate in this region is an important protection criterion. The average annual rainfall is less than 20 mm (for comparison: Federal Republic of Germany approx. 760 mm); the major part (95%) of this rain flows away superficially or evaporates. The seepage velocity of the remaining 5% of the annual precipitation is low (about 1 cm/a). The leachate would therefore take about 31,000 years to cover the distance (in a straight line) from the surface to the repository horizons.

The actual waste repository site will span 1150 acres (this corresponds to 4,653,935.00 m²). From this can be determined a seepage amount of approx. 2500 m³/a, distributed over approx. 470 ha. It can be assumed that the emplacement areas remain completely dry under these conditions.

And because of the closed basin structure, it is not possible that groundwater would flow from the area below the repository into aquifer systems that are used to supply drinking water to the population in the wider vicinity of the repository.

Critics accuse the US NRC that climatic conditions, seen over longer periods of time, may change to a wetter, colder climate. This is the subject of the long-term safety analysis for the repository site, see Chap. 8. With data provided by ice core drilling at the Antarctic station Vostok, see Fig. 8.10, it can be shown that such dramatic climate changes would take more than 125,000 years. This relates to global climate change. Thus, all repository sites worldwide would be affected and forced to include a warm-time—cold-time scenario in the long-term safety analysis. The conclusion then can only be to limit the period of observation to 100,000 years. This consequence has already been reached in the Scandinavian countries.

With the investigation of a repository site located in the host rock: volcanic tuff rocks, the USA has made a significant contribution to repository research. As a long-term radiological target Yucca Mountain has set an emission limit for the post-closure phase: in the first 10,000 years after closure of the repository of 0.15 mSv/a and for the period of 10,000 to 1,000,000 years a limit of 3.5 mSv/a. These limits, as a general radiological criterion, would also apply to the transport of radionuclides by the air path into the biosphere. Current long-term views suggest that the radiological targets at Yucca Mountain are achievable with the proposed multi-barrier system [43, 44].

7.4.2   Selected Repository Projects in Europe

This section presents a brief overview of the state of planning as well as the construction and operation of repositories in selected European countries, all of which are bound by the specifications of the IAEA, Euratom or the OECD NEA, but their national legal implementation show quite considerable differences see also Table 7.13 and from which conclusions for the German waste management concept could also be drawn.

There are currently 15 repository sites in Europe, see also Table 7.13. Some are

– some firmly agreed (Finland, Sweden)
– in some countries, location regions are selected (Switzerland, Russia)

– in some countries neither locations nor regions are selected (Great Britain)

In the following, some locations will be described in more detail, see Fig. 7.10.

(A) Selection of approved European repository sites for HAW—
    Bure/Lorraine/France

As early as the 1970s and 1980s, there were several attempts by the French government to investigate potentially suitable locations for a repository for high level
radioactive waste in claystone, shale, rock salt and crystalline rock. At the beginning
of the search in the 1980s, initially four departments had been in conversation. Three
possibilities of disposal should be tested. The above-ground and underground storage as well as "chemical transmutation". In addition, two laboratory mines should
be set up. However, especially in the southwest of France, resistance was provoked
because winemakers feared for the reputation of their growing areas.

In 1990, the government stopped searching for sites and commissioned a par-
liamentary commission headed by MEP Christian Bataille to come forward with a
proposal for action. This resulted in a unanimously passed law of December 1991,
which postponed the decision on the future repository concept to 2006 and defined
a research program based. After the adoption of the "Disposal Act", municipalities

Fig. 7.10 Locations for repositories for HAW (Heat Generating Radioactive Waste). *Source*
According to OECD-NEC

were sought that generally agree to the establishment of an underground laboratory. A total of 30 municipalities were found ready to receive such a laboratory. Founded in 1979 and since 1991 independent public waste disposal company ANDRA (Agence nationale pour la gestion de déchets radioactifs) has selected three sites from preliminary studies. Two of them in clay, one in granite. There was no decision on a location in a granite formation due to regional resistance. ANDRA's research on granite formations is therefore more general in nature.

In December 1998, the government approved ANDRA to commence the construction of an underground laboratory in a 150 million year old clay formation near Bure (on the border between the departments of Meuse and Haute-Marne).

Among other things, the research in Bure led to a first feasibility and safety study of the ANDRA ("Dossier 2005 Argile"). Between September and December 2005, a public debate on final disposal initiated by the environmental agency ASN took place, a report on which was published in early 2006. As a result, a final repository planning law was passed in June 2006. This regulates the further research in Bure for the location and for the repository concept. Since it is necessary to ensure that the final repository site has geological parameters comparable to those of Bure, a possible area for a repository site of 250 km^2 was initially identified in the Bure region. This was later narrowed by ANDRA to an area extending over 30 km^2.

On 06/08/2015, the French Constitutional Council (similar to the German Federal Constitutional Court) complained about this article not because of substantive concerns, but because it was not passed constitutionally. In July 2016, the French parliament then decided to set up the repository at the Bure site in mudstone, thereby healing the flaw complained of by the French Constitutional Council.

The complex and lengthy consultation and authorisation phase for a Bure repository is still ongoing. However the approval process began in 2017, in which German authorities are also involved. The construction of the plant is expected to start in 2020 and by 2025 with a pilot phase of storage [45]. It plans to place 5% of the inventory in a pilot camp and monitor it for 50 years. Only when this test run is completed successfully, is the final disposal of 95% of the waste which remains in interim storage facilities for planned [28]. It is currently expected to cost about 30 billion euros.

Construction of the Underground Research Laboratory (URL) began in November 1999. The URL consists of two shafts, a test drift at 445 m depth and a drift system with further test drifts at 490 m depth, see Fig. 7.11. There the possibility of a reversible or irreversible disposal of radioactive waste in deep geological formations is investigated at a depth of approximately 490 m. This includes an investigation of the specific properties of the mudstone and the testing of the emplacement concepts.

It is planned to store HAW and medium-level waste with long half-lives (MAW-LL) [28]:

- about 10,000 m^3 HAW (i.e. about 60,000 containers),
- about 73,500 m^3 of long-lived medium-level waste (i.e. about 180,000 containers).

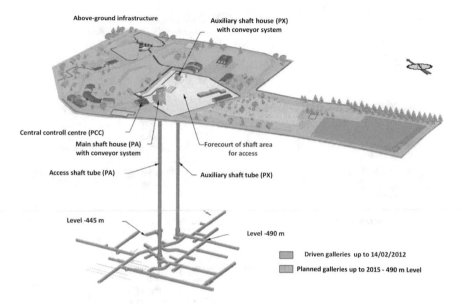

Fig. 7.11 Section through the planned repository Cigéo (Centre Industriel de stockage Géologique) in the district Bure. *Source* ANDRA [25; public domain]

For consideration of the radioactivity of the waste to be stored in Cigéo, see Table 7.18.

Waste management stipulates that transmutation should also be part of the repository concept. As early as 1982, France purposefully examined partitioning and transmutation under the premise of a disposal alternative, see Chap. 2. The French Castaing Report of 1982 concludes that, parallel to the final disposal, the separation and conditioning of plutonium and minor actinides for interim storage and possibly destruction by neutron irradiation should also be studied [46].

France has all the prerequisites for using P & T because the MOX[4] fuel used in the light water reactors is produced there. MOX fuels for P & T scenarios are a further development of the previous MOX fuels, since in addition to plutonium the minor actinides have to be added to the fuel [47]. P & T is part of waste management in the Bure concept, see Fig. 7.12.

The geological repository at Bure is being excavated in the host rock "clay". The bedrock is clays selected from the Upper Jurassic (Callovo-Oxfordia, ca. 160–150 million years) at a depth of 400–550 m. This clay formation has a thickness of about 130 m; it is tectonically practically undisturbed and has only a slight slope to the northwest towards the center of the Paris basin, see Fig. 7.13.

With the adoption of the Law on Economic Growth ("Le Loi Macron"), on 9 July 2015, an article concerning the disposal of French high-level radioactive waste was adopted. The law stipulates in a corresponding passage that the safety of the

---

[4]MOX is the abbreviation for mixed oxide and determines to a substance composed of the oxides uranium dioxide (UO_2) and plutonium dioxide (PuO_2).

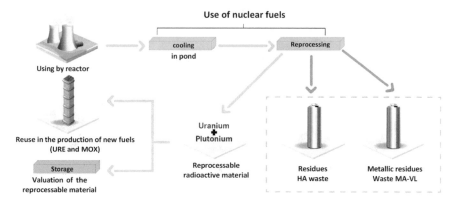

Fig. 7.12 Waste management (fuel cycle) in the Bure conception. *Source* ANDRA (2016); public domain

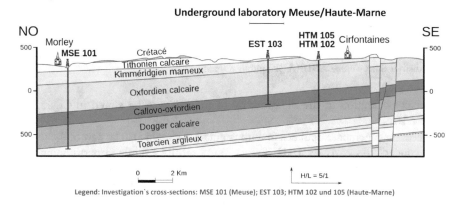

Fig. 7.13 Geological cross section in the repository area Cigéo/Bure. *Source* ANDRA (2016); public domain

repository should be checked before the waste is put into storage during a pilot phase. The observation period is 100,000 years. Furthermore, the waste should be stored in a retrievable way, so that future generations could—if an alternative solution for the disposal of radioactive waste is found - be able to reverse the process of storage at the repository site. Furthermore, it is intended to monitor the development of the repository for at least 100 years. After expiry of the monitoring period (>100 years), the final closure of the repository should take place.

The current status of the work is as follows: ANDRA has begun to set up a research/exploratory mine in the Callovo Oxford clay formation, see also [48]. The shafts have been sunk to a great extent and construction of the underground facilities has begun. In the area of highly radioactive waste, the concept only provides for the storage of waste from reprocessed fuel assemblies. The vitrified reprocessing waste is poured into primary containers of stainless steel and welded watertight with a lid.

Afterwards, they are packed in final disposal containers of non-alloy steel, which are designed to protect against contact with water and can ultimately improve the heat emission. The final disposal containers are intended to secure the waste for a period of about 1000 years over which the activity of the short and medium lived radionuclides are dominant.

The French final disposal concept envisages horizontal or vertical drilling into the clay formation for several vitrified high-level radioactive waste containers. The containers are 1.60 m long, have a diameter of 0.6 m and a wall thickness of 55 mm; because of the return option they are equipped with ceramic gliders. The 40-m-long, 0.7-m-diameter holes are lined with clay buffering material and lined with stainless steel. In the case of vertical drilling, no lining is planned. In the rear storage section, the holes are completely lined with a tight tube. The front of the wellhead is closed with a metal plug as well as a bentonite concrete plug at the end of the operating phase.

Although the reprocessing of spent nuclear fuel is foreseen in France, the disposal concept also includes the storage of spent fuel in steel containers either below the lane of the emplacement drift facility or in horizontal lined boreholes. A final determination has not yet been made. As for vitrified high-level radioactive waste, these lines are to be sealed with clay-based backfill materials. For high-level radioactive waste, these routes will be closed with clay-based offsets.

In order to improve the heat dissipation from the waste products, graphite can be added to the clay-based offsets. By adding silica sand, the swelling power of clay-based materials can be controlled to optimize the pressure load on the barriers due to the swelling processes of the clay. The heat output of the waste and the distances of the storage containers are chosen so that the temperatures in the claystone will be a maximum 100 °C. The determination of this temperature is justified by the fact that in the case of higher temperatures interlamellar water is released and thus significantly worsens the swelling ability of the clays. The distance between the storage cells should be between 8.5 and 13.5 m, depending on the heat output of the containers, all data according to [45].

As part of the ANDRA concept and thus also of the application for authorisation for the Cigéo deep repository planned for 2017, a comprehensive monitoring system should be set up, consisting of monitoring of the repository, monitoring of the environment and health monitoring, see Chaps. 2, 8 and 9.

Supervision of the repository: The purpose of setting up the Cigéo deep geological repository is to set up a monitoring program for the installation, which will monitor all the safety-related parameters of the repository during the entire operating period (speed of the machines, hydrogen concentration, performance of the high-efficiency filters, air in the ventilation system, etc.). In addition to monitoring in the narrower sense, parameters are monitored which may influence the development of the repository in the medium and long term: Ambient air temperature, expansion of the plant (meeting of the galleries), concrete durability, steel corrosion, etc.

For example, individual monitoring of control containers is carried out in separate rooms, which allows easy access to each control container and its regular inspection.

In the first phase of construction of the Cigéo deep storage facility, representative control structures of the various storage components (encapsulation, chambers...) are constructed, which are used only for observation and monitoring. These control structures are extensively instrumented in order to follow their behaviour and evolution in every detail over time. This means that tens of thousands of sensors are installed in the Cigéo deep storage facility. The methods of investigation foreseen use proven sensors already used in the nuclear industry and in construction engineering, for which extensive experience is available (used for several decades in nuclear power plants, dams, etc.), as well as innovative means which are currently being developed by the framework of R&D[5] programs.

The chambers for MAW (SMA chambers), see [10], will be set up during the operating period of the Cigéo deep storage facility. The first chambers will be extensively equipped with instruments for detailed observation. In addition, ANDRA proposes to close one of the control chambers a few years after the storage of the waste containers and then to continue the observations in a closed configuration in order to draw conclusions from these observations for the actual decommissioning of the repository.

A comprehensive environmental monitoring program with health monitoring is planned, which has been included in Chap. 9 and to which reference is to be made here, see [45].

(B)   Selection of approved European repository sites for HAW—Forsmark/Sweden

In Sweden, as in Finland, the operators of nuclear power plants are responsible for disposal and final disposal of wastes. For this purpose, they have set up the joint company SKB (Svensk Kärnbränslehantering AB), which is also responsible for transport and interim storage.

Svensk Kärnbränslehantering AB (SKB) started looking for a repository site as early as 1977. In 1977 SKB also started research work on a repository concept and set up a laboratory for storage technology in the former Stripa iron ore mine site. Later, the research work shifted to the Hard Rock Laboratory Äspö near Oskarshamn at a depth of 460 m; the facility was built between 1990 and 1995.

Because municipalities and local people were not initially involved in the process, many municipalities initially refused to set up a repository on their territory. However, the invitation to apply for a site for the construction of a repository was eventually accepted by several municipalities.

Since 1988, SKB has been operating a near-surface repository for low- and intermediate-level radioactive waste—for the originally twelve, now still ten-nuclear power plants—about 50 m below the Baltic Sea near the Forsmark nuclear power plant. Used fuel elements have been stored since 1985 in the central interim storage facility—Centralt mellanlager för använt kärnbränsle (CLAB)—near the Oskarshamn nuclear power plant.

---

[5]**R**esearch and **D**evelopment.

In 1983, in the report "KBS-3", it published its concept of encapsulating used fuel elements using additional engineering barriers to supplement the natural barriers created by rock formations.

From 1993 to 2000, SKB carried out feasibility studies for eight different sites. The basic prerequisite for a potential site is the approval of the local population. For this reason, the site selection procedure is characterized by close dialogue between the operator SKB and the residents of the municipalities or the provincial government itself. Two of the sites (Storuman and Malä) were ruled out due to negative municipal referenda in 1995 and 1997 respectively. Of the remaining six (Östhammar, Nyköping, Tierp, Oskarshamn, Hultsfred and Älvkarleby), five seemed suitable. Of these, SKB shortlisted the locations Östhammar (near Forsmark), Oskarshamn and Tierp, also because Östhammar and Oskarshamn already have the necessary infrastructure as far as possible—both municipalities also have nuclear power plants. The municipal councils of Östhammar and Oskarshamn approved the drilling of exploration wells. Tierp rejected by a narrow majority (25–23 votes) and thus was not included. Exploratory drilling began in 2002. In June 2009, SKB opted for the Forsmark site because the rock there has a higher thermal conductivity than in Oskarshamn. This results in a better dissipation of the decay heat, resulting in both a smaller area expansion of the repository structure and lower costs. The municipality of Oskarshamn received financial compensation for rejecting "its" repository site.

In March 2011, SKB submitted an application to the Swedish regulatory authorities for the construction of a repository at the Forsmark site. Pilot operation is expected to start by 2020. The construction of a fuel element conditioning plant was also planned for the Oskarshamn site in 2006 (planning application 2006). Due to the in situ storage conditions, the fuel elements will be repacked in copper containers (5 m long, weighing approx. 27 tons). According to SKB, the Forsmark repository for HAW (heat-generating) should be ready for operation in 2030.

The transport of conditioned radioactive waste is to be carried out mainly by sea with a "Specialized Cargo ship" (SKB owned) exclusively developed for this purpose and between ports owned by SKB, see Fig. 7.14. The Baltic Sea is used both by the final storage concept for HAW and by the transport routes. A cross-border EIA was initiated because the Baltic Sea was identified as the study area.

- Within the framework of the transnational participation, the German authorities submitted comments to Sweden.
- The German public had the opportunity to submit comments to the Swedish authority by 15 April 2016.

SKB has also already submitted the long-term safety assessment to the approval authority. The IAEA and the OECD-NEA were also involved in its examination. In the Swedish repository concept, only spent fuel elements (BE) are taken into account; reprocessing is fundamentally not planned. The total amount of spent fuel assemblies from PWR and BWR nuclear power plants is estimated around 12,000 tonnes (heavy metal—initial weight), corresponding to around 54,000 spent fuel assemblies [29]. This corresponds to roughly 6000 canisters in the repository. These figures are based on assumed reactor operational times of 50–60 years. The SR-Site assessment

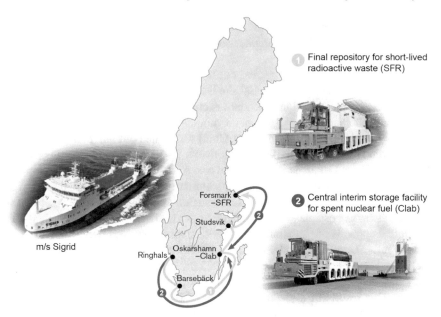

Fig. 7.14 Transport of conditioned radioactive waste between conditioning plants and repositories. *Source* SKB AB, public domain

Fig. 7.15 KBS-3 Concept Forsmark repository, Sweden Cross section. *Source* SKB AB, public domain

is, therefore, based on a repository with 6000 canisters, corresponding to around 12,000 tonnes of fuel.

The long-term safety assessment is underlying a reference scenario. According to the SKB spent fuel reference scenario, over 8300 tonnes of BWR fuel and about 2800 tonnes of PWR fuel need to be deposited by year 2045 (year 2045, time for shut down of the last nuclear power plant). This assumes a volume of 19,600 m^3 spent fuel elements is presented to be encapsulated and the total number of (average) canisters (in copper casks) is 6103 [49].

Table 7.19 lists thirteen radionuclides from the spent fuel assemblies that are important for decay behaviour and radiotoxicity. Their total activity Ai is 5.3E+19 Bq, of which 1.4E+19 are alpha emitters. This shows the long-term risk in the long-term safety assessment. The probability that one out of the 6,000 canisters has failed at the end of the initial 1000 year period is estimated at 2.4×10^{-5}, i.e. hypothetically 40,000 repositories, each with 6000 canisters would have to be constructed in order for there to be an expectation of one failure during the initial 1000 years. Despite this extremely low probability, a risk contribution was calculated for the first 1000 years [29].

For the Swedish reference final storage concept KBS-3 V (vertical storage of individual final storage containers in vertical boreholes from the bottom of a storage section), the variant with a copper cask with a wall thickness of 50 mm was selected. Either 4 PWR-FE or 12 BWR-FE can be stored in a Fuel Element (FE)-final storage container. The temperature at the surface of the containers should not exceed 100 °C in the repository, see [50], in order not to impair the geotechnical barrier of bentonite, see Figs. 7.16 and 7.18.

Crystalline rocks offer a high mechanical stability that protects the waste packages during the post-operation phase. At the same time, however, the rocks often contain

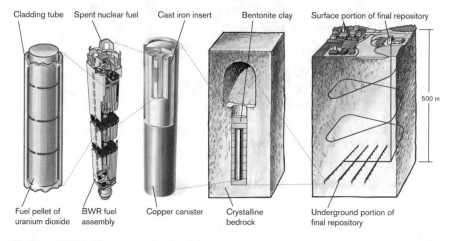

Fig. 7.16 KBS-3 Concept of safety barriers[(safety) barriers]. *Source* SKB AB, public domain

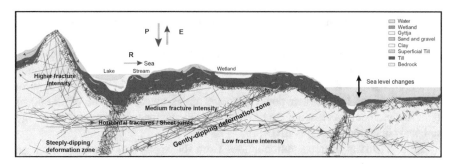

Fig. 7.17 Geological Barrier-principle. *Source* Follin et al. [51]; SKB AB, public domain P = precipitation, E = evapotranspiration, R = runoff

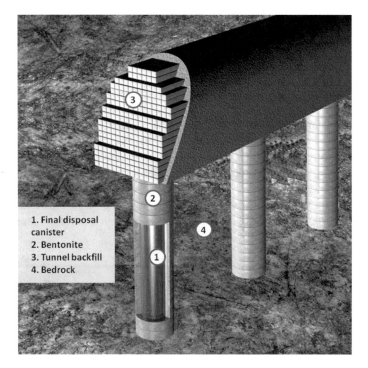

Fig. 7.18 KBS-3 Concept of emplacement scheme/technical/geotechnical barriers Forsmark repository/Sweden. *Source* SKB [29], public domain

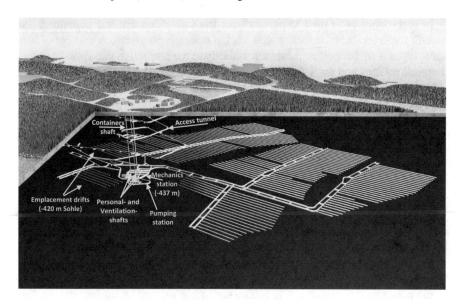

Fig. 7.19 Schematic diagram of repository for HAW Olkiluoto. *Source* Posiva Oy processed

water-carrying crevices, so that the retention functions for the radioactive inventories by technical and geotechnical barriers must be ensured, see Figs. 7.15, 7.16 and 7.17.

Essential elements of the safety concept in crystalline rock are, therefore, corrosion-resistant repository containers with a service life of at least 100,000 years and a bentonite buffer around the repository container to prevent or delay solution access to the container, see Fig. 7.16. Further details are provided in Chap. 8.

In addition to these two measures, the backfilling and closure concepts provide for backfilling of the storage sections with bentonite or a bentonite/rock fragment mixture as well as the closure of the storage sections with a closure construction in the connecting drifts.

To date, only rough concepts have been developed for the backfilling of the remaining mine workings and shafts. In principle, the backfilling and sealing measures developed for the storage drifts are also to be used here. A verification concept for the barrier function was developed for the backfilling and sealing of the emplacement boreholes and -drifts. This concept is based on safety functions, criteria for compliance with the safety functions and methods for proof, see Fig. 7.18.

In the previously submitted long-term safety assessment for Forsmark, it was necessary for the approval to provide evidence on the emplacement method and site selection as well as covering all relevant safety factors. A detailed presentation of all relevant aspects and influencing factors is required for a period of up to 1000 years and a reduced presentation for up to 100,000 years; in addition, the period of up to one million years is considered. As a normal development, it is assumed that all containers and geotechnical barriers function according to their design. Container and barrier defects are regarded as alternative scenarios.

In accordance with their importance for the safety concept, the containers and backfill material form the focal points for the research programs. A number of bentonite types were investigated for their stability in the respective site-specific hydrochemical milieus and the safety concept was optimised. In addition, the production of bentonite form elements and the transport and introduction of these elements were (are) being tested in in situ experiments; furthermore, cements with the lowest possible pH values were tested and developed, which are required for the abutment of barriers, for the sealing of faults and cracks by injections and as a buffer in the super container concept.

In Sweden, too, the principle is followed that the safety of the repository must be subjected to a continuous optimisation process with periodic safety checks from the planning stage to the closure of the repository.

As a radiological criterion beyond 10,000 years it has been determined that the annual dose is less than an order of magnitude above the risk limit of 0.014 mSv. This corresponds to the ICRP reference limit dose of 0.1 mSv/a.

SKB estimates the total cost of the concept at SEK[6] **136 billion.** Of these, 39 billion have already been invested, 56 billion are in a state-managed funds set up to finance final disposal and 41 billion have been secured by nuclear power plant operators to be lodged with the funds. The final storage of spent fuel will cost around SEK 37 billion.

(C)   Selection of approved European repository sites for HAW—Olkiluoto/Finland

As in Sweden, in Finland the responsibility for site selection and final disposal lies exclusively in the hands of liable private companies. Here, the state only acts in a supervisory role, which it exercises through the Radiation Protection Authority and the Ministry of Labour and Economic Affairs. The private company "Posiva Oy" was founded for the operational realisation of a central repository for spent fuel elements, of which the nuclear power plant operators—Fortum Oyj [40%] and Teollisuuden Voima Oyj (TVO) [60%]—together hold 100% of the shares.

The Finnish Nuclear Energy Act provides for a phased procedure for the establishment of a repository for high-level radioactive waste. The first decision is the political decision of the Council of State to establish a repository for radioactive waste in Finland. The Finnish Nuclear Energy Act stipulates the involvement of the municipalities, regional and supra-regional administrations and organisations concerned in the subsequent siting process. A public hearing must then be organised once the respective opinions have been submitted. The final location decision of the State Council must be ratified by Parliament. The final building permit and the operating permit are again issued by the State Council.

By government decision, from 1986 to 1992 Posiva Oy has already investigated the first sites for a potential repository. The investigations concerned the geological properties of the crystalline host rock of the potential sites and their environmental factors. From 1993 to 2000, four of these potential sites were explored in detail both above ground and with various wells, including the two nuclear power plant sites of

---

[6]Currently: 1 SEK corresponds to 0.10 Euro, total costs correspond to approx. 14 billion Euro.

Loviisa and Olkiluoto, where the existing interim storage facilities are also located. After all four sites had proved to be suitable in principle, Posiva Oy selected Olkiluoto to minimize the required transports. Transport logistics play an important role in waste management because transport is an essential part of the public perception of final disposal.

The government approved the choice of site in December 2000 and set a maximum capacity of 4000 tonnes of spent nuclear fuel. Parliament ratified this government decision almost unanimously in May 2001 with 159:3 votes. The construction of a further reactor block at the Olkiluoto site meant that an increase in the storage volume to a total of 5400 containers was included in the application documents. The decision in favour of Olkiluoto was supported by the local council with a large majority. The local council of the receiving municipality has a legal right to file a veto against a location decision.

On 28 December 2012, Posiva Oy submitted the application for a building permit for the repository to the Finnish Government. The Finnish Nuclear Safety Authority STUK (Säteilyturvakeskus) confirmed the safety of the submitted repository concept to the Finnish Ministry of Economy and Labour in February 2015. Within the framework of the statement, the submitted documents were checked for conformity with the Finnish Nuclear Energy Act. In November 2015, building permission was granted for the Olkiluoto repository. The construction of the necessary infrastructure (shafts and ramps) for an underground laboratory at the site (ONKALO) was successfully completed in 2014.

Following an opinion of the Finnish Ministry of Economy and Labour and the consultation of other stakeholders, the Finnish Government will decide whether to grant the operating licence application in 2020. The final storage operation is scheduled to commence in 2022. This would make Olkiluoto the world's first repository for HAW from civil use of nuclear energy.

According to current proposals, the repository is to be operated until 2112 and will be closed securely in the year 2120. The Finnish repository concept in granite was developed in close cooperation with the Swedish nuclear fuel disposal company Svensk Kärnbränslehantering AB (SKB). Not only are the containers- and emplacement concepts of the two countries very similar, but the Olkiluoto repository will also be constructed, operated and sealed according to the SKB concept, see Fig. 7.18.

The underground research laboratory ONKALO is directly integrated into the repository as a pilot plant within the project. During the development of the repository, geological, geophysical, rock-mechanical and geochemical investigations will be carried out to determine the suitability of the rock and the emplacement techniques will thus be tested under real conditions before emplacement begins.

Posiva Oy also submitted the long-term safety certificate with the building permit application to the approval authority and this was confirmed with the building permit, but with an updated obligation. The IAEA and the OECD-NEA were also involved in its review. It is incumbent upon Posiva Oy to constantly incorporate the knowledge gained from both construction and operation into the long-term safety record. In Finland, too, the principle is followed that the safety of the repository must be subjected to a continuous optimisation process with periodic safety checks from the planning phase to the closure of the repository.

Some facts about the Finnish repository concept: Only spent fuel elements (BE) will be taken into account. The maximum quantity of HAW to be stored is given as 9000 tU, which is to be conditioned in 4500 packages, see [30]. In the approved repository layout, there is space for 5400 containers. In Table 7.20, the total activities of the thirteen radionuclides, which are of significance with regard to decay performance and radiotoxicity, are determined from the reference inventory in accordance with [30], based on a decay time of 30 years. The total activity Ai is 7.15E+19 Bq, of which 2.03E+19 are alpha emitters. This represents the long-term risk in the long-term safety assessment [52]. The probability that one of the 4500 canisters will have failed at the end of the initial 1000 year period is estimated at 1.6×10^{-5}. Despite this extremely low probability, a risk contribution was calculated for the first 1000 years. The long-term safety assessment follows in the same way as for Forsmark [30].

Also, in the Finnish repository concept, the temperature at the surface of the containers in the repository should not exceed 100 °C, see [50], in order not to impair the geotechnical barrier of bentonite. The entire emplacement technology is designed for this purpose, see Figs. 7.14 and 7.20.

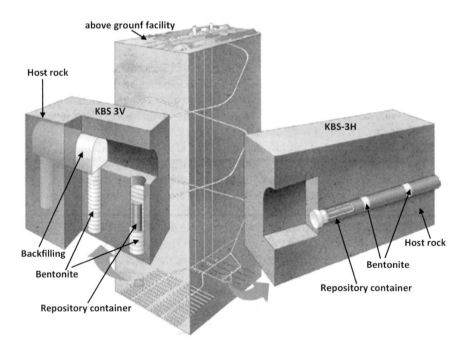

Fig. 7.20 Vertical and horizontal container emplacement and the envisaged geotechnical barriers in Olkiluoto's long-term safety concept. *Source* © SKB AB; public domain

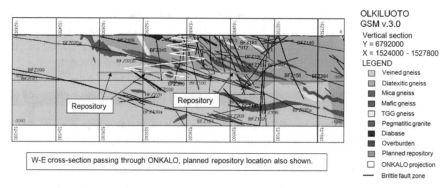

W-E cross-section passing through ONKALO, planned repository location also shown.

Fig. 7.21 Geological W-O section with location of the repository area. According to POSIVA [53, 52]; public domain

The Olkiluoto repository is built within the granite formation "Äspödiorit". It has remained unchanged in site for 1.8 billion years and is thus one of the oldest geological formations in Europe. However, the granite formation "Äspödiorit" is also crossed by water-bearing fissures, so that the retention functions for the radioactive inventory must be ensured by technical and geotechnical barriers, see Fig. 7.15. According to Posiva Oy, only about 40 l of water per minute are measured in Olkiluoto in already existing repositories (LAW/MAW), which corresponds to 2.4 cubic metres per hour or 58 cubic metres per day. In this respect, a relatively dense granite formation is found in the Olkiluoto repository area, see Fig. 7.21.

The entire repository consists of a system of transport and emplacement sections accessible from the surface via a ramp. In addition, three shafts are planned—a shaft for container transport, a shaft for passenger transport and a ventilation shaft [30]. The drifts driven for storage are run at the final storage level at a distance of 25 m will be oriented, in direction NW-SE and in E-W parallel to the main horizontal stresses in the rock. The drifts will run 400–450 m below ground level and should have a maximum length of 350 m. The boreholes in which the canisters will be sunk will have a diameter of 1750 mm. The total length of the emplacement drifts should be approx. 41 km, see Fig. 7.19.

The essential elements of Olkiluoto's safety concept are corrosion-resistant repository containers with a service life of at least 100,000 years and a bentonite buffer around each repository container to prevent or delay solution access to the container, see Fig. 7.15 Further details on the containers are provided in Chap. 8.

The long-term safety concept consists of two parts, which are summarised in the safety case:

– The radionuclide release and -dispersion up to the exit into the biosphere. A reference dose of 0.1 mSv/a is prescribed. Finland thus follows the ICRP proposal,
– solution inflow scenario—prevention of the mobilisation of pollutants.

In principle, retrievability is only guaranteed during the emplacement phase, whereby this would require the removal of the bentonite and the development of suitable removal techniques. After completion of the emplacement phase, which is expected to last about 100 years, the repository is to be closed in such a way that unauthorised retrieval is made effectively impossible. An authorised retrieval of the stored waste after closure is no longer provided for in the current concept.

In the so-called ice age-scenario, the penetration of permafrost to a depth of 800 m is assumed, which would endanger the safety of the repository. Posiva proposes to displace the containers in this case.

Based on the reactor blocks currently approved in Finland, the **cost of final disposal** is estimated at around EUR 6 billion, of which around EUR 3.5 billion is accounted for by the Olkiluoto repository for high-level radioactive waste. The remaining EUR 2.5 billion will be spent on the disposal of low and medium level radioactive waste and on the decommissioning of nuclear power stations. These costs form the basis for calculating the levy, which is also levied in Finland as a surcharge on nuclear power and contributes 67 million euros annually to the Finnish Disposal Fund. The law requires that there must always be enough funds available in the fund at the end of the year to cover the total disposal costs from that date forward. The fund currently has around EUR 2 billion in it. The operating expenses of the operating company are borne directly by its shareholders and not by the fund.

7.5   Temporal Change in Radioactivity in Waste

7.5.1   Basic Principles

Long-lived radionuclides with high radiotoxicity play the decisive role for the period under consideration in determining long-term safety. Therefore, the decay behaviour of radioactive waste and its influenceability are of great importance. In the following, some remarks on this topic will be presented.

In case of final disposal of radioactive waste and residues the material is a waste mixture, also known as nuclide mixture. The composition of the waste naturally depends to a great extent on the composition present according to Fig. 7.21, or as prepared (conditioned) for final disposal. Nuclide mixtures are listed in Tables 7.1 and 7.3, which are typical for the respective species.

The radioactive nuclei (radionuclides) decay by emission of radiation (α-, β-, γ- and neutron beams). They are transformed into other nuclei, some of which are also radioactive. Sooner or later, however, stable isotopes are formed that do not continue to radiate. The half-life of an isotope is the time after which half of the material has decayed. The isotopes found in radioactive waste have extremely different half-lives—from less than one hour to millions of years.

Waste mixtures exhibit a different (natural) decrease in radioactivity than individual radionuclides (decay times of nuclide mixtures). For example, the mixture of U-235 and U-238 contained in uranium oxide pastilles of fuel elements reacts as follows after bombardment: U-235 can fehhibit fission. U-238 cannot exhibit fission but may transform into fissile elements. In this way, further mixtures of different radiating nuclides are formed in the irradiated fuel elements in the reactor. The smaller and often highly radioactive nuclides are called fission products. The radioactive nuclides converted from U-238 are referred to as transuranic elements (minor actinoids: neptunium, americium and curium—the isotopes Np-237, Am-241, Am-243 and Cm-245), see Tables 7.3 and 7.14. This is the basis for considering the radioactive inventory in the wastes and thus the physical properties of the nuclide mixtures, in this case highly radioactive waste (HAW).

Fission products are usually much more short-lived than transuranics. The fragments, which are for example. formed during the fission process of U-235, are mostly short-lived or very short-lived and are often transformed into stable non-radioactive nuclides via further radioactive short-lived fission products. This generates a great deal of decay heat, which in turn is used in the reactor to heat water to generate steam which is used in turbines to generate electricity; this decay leads to the spent fuel elements having to be cooled for years and perhaps decades in wet storage, whereby the water both dissipates the heat and shields a significant part of the radiation, see Fig. 7.1. The fuel elements are only transportable when the heat generation has decreased significantly. They can then be taken to an interim storage facility in special air-cooled casks (Castors). However, significant heat generation only occurs in the case of highly radioactive waste, see Fig. 7.22.

Atomic fission also produces fission products with "medium" and "long" half-lives. In the "middle" group two elements are decisive: Cs-137 and Sr-90 (and the daughter product Yt-90). Both elements have half-lives of about 30 years: after 300 years one thousandth after 600 years one millionth and after about 100 years less than one billionth of the original starting nuclide are still present. In addition, however, there are 7 long-lived fission products, such as technetium or Ma-99 (half-life 211,000 years) and I-129 (half-life 15.7 million years). Tc-99, derived from a decay chain around Y-99, see Tables 7.1 and 7.2.

Figure 7.23 illustrates the (natural) decay process of the fission products by means of the so-called toxicity index of highly active solidified waste from the reprocessing of fuel elements from pressurized water reactors with a burn-up of 33,000 MWd/tHM (Megawatt-day per Metric Ton).[7]

The toxicity index is defined as the ratio between the quantity of water and the volume of waste required to dilute radioactive substances in order to meet radiation protection standards. It is therefore an empirical quantity that is adjusted according to the radiotoxicology findings.

In the first 1000 years or so, the toxicity index of the fission products decreases by a factor of about 1 million, see Fig. 7.22. After that, the long-lived fission products dominate over the further millions of years and more. This observable decay behaviour has also been taken into account in the long-term safety assessment for the Forsmark and Olkiluoto HG-HAW repositories, in which the observation-period is divided into 3 time periods: up to 1000 years, up to 100,000 years, up to 1 million years.

---

[7] 1 MWd = 24,000 kWh; Metric Ton; tHM i.e. MTU = metric ton of uranium metal.

Fig. 7.22 Fission products and decay properties measured by toxicity index: Sr = Strontium; Cs = Caesium; Y = Yttrium; Eu = Europium; I = Iod; H-3 = Tritium; Tc = Technetium; Zr = Zirkon; Nb = Niobium

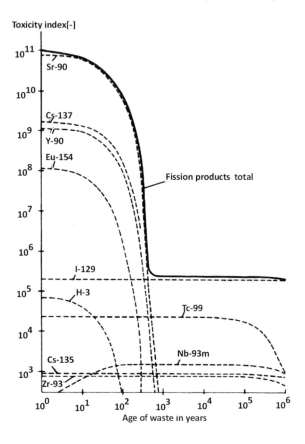

7.5.2 Targeted Change in Radioactivity Through Transmutation

Meanwhile available research results and implementations must be taken seriously and incorporated into a repository concept, including the nuclear treatment of radioactive waste [nuclear transmutation consisting of partitioning and transmutation (P&T)]. Here are a few remarks. P&T refer to technical processes for the treatment of spent fuel elements from nuclear power plants which aim at the separation of groups of substances (partitioning) and the subsequent conversion of the actinide elements plutonium (Pu), neptunium (Np), americium (Am) and curium (Cm) by irradiation (transmutation). Radioactive isotopes with a long half-life are converted by irradiation with neutrons into radionuclides which are either stable or have significantly shorter half-lives. Figure 7.24 shows the temporal development of a radiotoxicity indicator (measured in the dose unit Sievert, Sv). By P&T, the decrease of the radiotoxicity of actinides (here: plutonium, uranium, blue curve in Fig. 7.24) can significantly run much faster (yellow curve in Fig. 7.24). This shortens the time in which the radiotoxicity of HAW drops to a level corresponding to that of

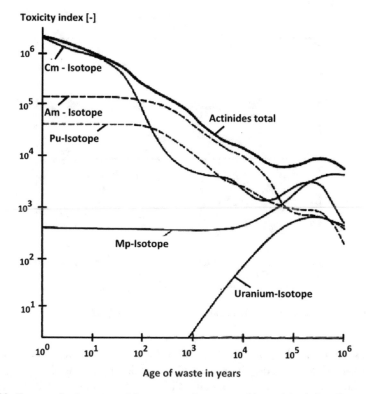

Fig. 7.23 Transuranic elements and decay properties measured by toxicity index: Cm = Curium; Am = Americium; Pu = Plutonium; Np = Neptunium; U = Uran

natural uranium (ore) from approx. 16,000 years to only 330 years. This also reduces the heat release of HAW much faster to a level at which the thermal release due to decay heat in the repository can be largely avoided.

This means that extensive measures to reduce the heat-induced load on the rock can be dispensed with without lowering the safety standard. This would make it possible to resort to a long-term safety proof, how it was applied to and approved for radioactive waste with low heat generation. With the modular repository concept presented in the following, the safety standard of the repository will be considerably increased.

Technical analyses show, that the reduction of volume of heat generating waste, by reduction of thermal power and total activity as a result of a subsequent implementation of P&T can be significant for a German repository scenario; the minimum reduction of these is between one and two orders of magnitude, and in case of the radiotoxicity even more significant, see Acatech STUDIE [18].

Of course, the radioactive residues are not eliminated by transmutation and they must also be disposed of, but with a significantly altered radionuclide vector and reduced radiotoxicity of HAW. Thus remain, C-14, Cl-36, Cs-135 and I-129, among

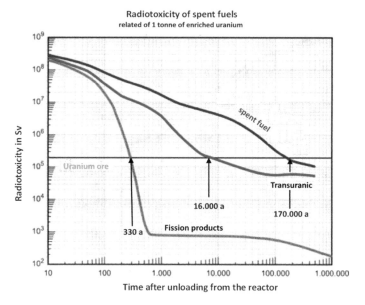

Fig. 7.24 Transmutation, course of radiotoxicity; according to KIT and [40]

others, as mobile and long-lived radionuclides for which a long-term proof of safety must be provided. However, the increased safety due to the faster degradation of radiotoxicologically relevant radionuclides of actinides is an important aspect with regard to long-term safety. Research programs are currently underway on a large scale, e.g. in France, which suggest the application of transmutation in a repository concept. First assessments of the advantages and disadvantages of (P&T) in different reactor configurations depending on the given specific targets are compiled in Acatech STUDIE (2014). This report states: "A total of four technical scenarios were developed in order to estimate and evaluate the various options for action for the Federal Government in the question of the future handling of P&T. These scenarios were examined for their social, ecological, legal and economic implications on the basis of a broad mix of methods".

Although the study deals with numerous questions still to be solved, the application of P&T in a German repository concept cannot be expected from today's perspective. There are, however, three facts that argue against it:

– From today's perspective, P&T uses light water reactors with oxide fuel specially developed for this purpose. Current scenarios assume that the existing light water reactors will gradually be replaced by third generation reactors of the EPR[8] type, until from 2050 successively fast fourth generation reactors will be introduced, see [20];

---

[8]EPR—European Pressurized Water Reactor.

- There is currently no ready-to-use technology available. The ADS technology[9] and the production of transmutation fuel from radioactive waste from light water reactors are planned for 2050, see [54].
- Since 1 July 2005, the reprocessing of fuel rods from Germany has been completely prohibited by law. According to the current legal situation therefore, a P&T ban should also apply;
- The P&T process will take a very long time, several decades, for the HAWs that have occurred in Germany, so that only integration into a nuclear power generation process would be worthwhile from today's point of view, financially as well as technically.

P&T technology should be integrated into a modular concept of final disposal until the final decision on the final disposal of HAW has been taken and considering the expected further development of the technology. P&T are thus classified as "can be omitted" if the procedure is not suitable for the repository concept.

7.6 Conclusions

The construction of repositories for high-level radioactive waste (heat-generating waste) is not an insurmountable problem, as various examples have shown. Some countries have already made great progress. Three of them, namely Finland, Sweden and France, are very close to achieving the target. For political reasons, Germany has started searching for a location for a repository from scratch. After almost 50 years of repository research for HAW (heat-generating), Germany is almost starting from "zero". Does the so-called new start give cause for hope? We will have to wait and see. A concept of how to achieve long-term safe, stable final storage cannot be discerned. Here, the elements of international disposal solutions are to be put together, without which a repository concept has hardly any chance of being realised. For all the repository projects presented, these are optimised both during the construction phase and until they are decommissioned. This is an accepted basic principle of repository design.

The objective which both the location selection act (StandAG) and the End-lagerkommission founded with the StandAG have formulated is: "…to find the site for a final storage facility in the Federal Republic of Germany in accordance with the Atomic Energy Act, Section 9a Paragraph 3 Sentence 1, which guarantees the best possible safety for a period of one million years". However, there is no selection procedure worldwide that would have aimed at "the best possible safety". This also applies to the successful solutions presented here. There is also no generally accepted definition of this term in the entire scientific literature. The term "best possible safety" is irrelevant for the objective to be achieved, namely that "a repository concept for radio-toxic waste and residues must permanently prevent radio-toxic contaminants

---

[9]Accelerator-driven Systems (ADS) and Fast Reactors [46].

from entering the biosphere and, if so, only within socially and legally accepted limits". Rather, it becomes apparent that the organisational and technical aspects must first be sorted out, without which a repository concept for a permanently safe storage solution cannot be found. The permanent criterion is the approval capability of the repository system.

Following the activities of the Repository Commission pursuant to StandAG [55] and with the "Act on the Reorganisation of the Organisational Structure in the Field of Repository Waste Management", in 2016 [37] the Federal Government reorganised the responsibilities in the field of repository waste disposal. The Act stipulates that the state tasks of supervision and licensing in the field of nuclear technology, interim storage, site selection and repository surveillance are to be handled by one authority— the Federal Office for Nuclear Safety of Disposal (BfE). Thus, the BfE became the central licensing and supervisory authority independent of the repository operator. A Bundesgesellschaft für Endlagerung (BGE) mbH [37] was founded as the project executing agency for the operative tasks of site search, construction and operation of the repositories and the Asse II mine as well as the operation of the Gorleben exploratory mine. This company assumes the tasks of Asse-GmbH, DBE mbH and the BfS. The BfS then concentrates on the state tasks of radiation protection or the monitoring networks for radioactivity in the environment.

This re-structuring is now completed.

The selection of the international repository projects made here is not accidental. Either the repository projects have been completed (WIPP) or they are under construction or the repository site has been selected, so that administrative procedures have started. Yucca Mountain in the USA is also back in this position. The successful European projects have one thing in common: there has been a site selection procedure that was open and transparent and in which the population of the receiving community not only has the final decision on the site, but also the prospect of an appropriate compensation for any disadvantages. Anyone who undertakes to ensure the long-term safe disposal of HAW (heat-generating radioactive waste) on the community territory for the citizenship of a country should also be able to count on a high financial reward from the financing fund. This was organised in the same way in Finland and Sweden. Only in the USA (Yucca Mountain) has the situation so far been the same as in Germany, that is non-transparent and without public or regional participation. In the end, courts had to decide. It can be consistently proven that a positive mood towards the peaceful use of nuclear energy is not decisive for the acceptance of repository projects, but rather that the statement that possible negative effects of the final disposal of HAW are regional and over a time period of 1 Million years the reliability of behavioral predictions is often questioned and thus the final decision on the approval of a repository in the territory of a municipality should lie with the municipality itself.

Most countries that are making intensive efforts to find a final disposal site for HAW operate nuclear power plants. Before deciding on a suitable repository site, all these countries asked themselves whether sites at which nuclear power plants are operated or at which nuclear power plants have been operated are not particularly suitable. All these countries answered the question in the affirmative. They had two strong arguments in their favour: The existing infrastructure, including interim storage facilities, could be integrated into the repository concept and that one repository site combines several tasks, namely the conditioning of the radioactive waste (encapsulation) and the transport logistics (waste management) between the conditioning facilities and the repository site. These are to be optimised and become an essential part of acceptance by the public. The solution offered in Sweden for sea transport between the conditioning plant in Oskarshamn and the final repository in Forsmark, see Fig. 7.13, is original; the solution to transport the containers directly from the conditioning plant underground to the final repository is also original.

In most international repository projects, spent nuclear fuel is not transferred directly from the interim storage facility to the conditioning plant. Transitional storage facilities are being built. In Germany, interim storage facilities do not meet the requirements of a transitional storage facility.[10] Due to the length of the procedure, the operating licences for the interim storage facilities will expire, so that the highly relevant questions are raised: where and how many interim storage facilities will be involved and should the former power plant building be able to be used etc.?

What temporal relevance do these questions have? The Federal Government has declared that the site search should be completed by 2031 at the latest and the Repository Commission states in the final report on page 104: "In the opinion of the Commission, the disposal of highly radioactive waste at the site sought could begin with the best possible safety in the year 2050,..." and then insert the general insurance in the second half sentence: "... if there are no unforeseen delays", see [55]. For emplacement operations, 20-30 years are forecast. In view of the fact that errors in the formulation of the selection criteria for a repository site can lead to considerable time delays because the citizenship refuse to consent, the authors do not consider the dates given to be realistic, based on plausible time requirements for licensing procedures, for public participation, for voting and deliberation processes, for legal protection procedures, for post-clearance collection of data and for the exploration of possible siting regions. If this is taken as a basis, the exploratory approach will lead to significantly different periods of time. The commissioning (beginning of the storage of the waste) would be expected to last perhaps for the next century, a closure time therefore is far into the next century. By way of comparison, the planned construction time for the Konrad mine repository will more than triple in the end, and the costs will have quadrupled.

---

[10]The German Federal Government's National Disposal Plan provides for a receiving storage facility into which the HAW from the interim storage facilities are to be transferred, in particular due to the expiration of operating licences. The interrelationships and derived questions raised here do not play a role therein. The final report of the Repository Commission on page 248 states: "In terms of time, the removal from the interim storage facilities must be coordinated with the conditioning required at the repository site in accordance with the repository concept. It is uncertain whether and to what extent the receiving storage facility provided for in the National Disposal Plan will exist. If this storage facility is built before the final repository has a legally binding permit, the impression of a preliminary decision is created that may cast doubt on the legality of the procedure. If a large receiving storage facility is built, this could also be perceived as the greater burden compared to the final storage facility in the discussion on site...". Of course, this statement also raises the question, if one had not may expect more detailed work from the Repository.

All the international repository projects presented here have the characteristic that they are being built in countries that have integrated energy generation from nuclear fuels into their energy concept on a long-term basis and where the production of spent nuclear fuel will continue for an unforeseeable period of time. In addition to the now approved HAW repositories, more are likely to follow. In Finland, Oklituoto will remain in operation for about 100 years. The Oklituoto repository is therefore large in terms of the annual volume of HAW in Finland. France has decided to store as far as possible only radioactive material originating from reprocessing in the Cigéo (Bure) repository; direct storage is the exception, as the heat generation and the relative radiotoxicity of the inventory are correspondingly low. France is also planning to use the P&T process in the future. The common factor is that these countries need space for intermediate storage of the newly accumulating HAW. However, Germany will not be in this position after 2050 because no new HAW will be produced from the nuclear power plants. According to information published by the BMU, the interim storage capacity is sufficient. Furthermore, paragraph 6 (5) of the Atomic Energy Act stipulates: "The storage of nuclear fuels in nuclear installations in accordance with paragraph 3 in conjunction with paragraph 1 shall not exceed 40 years from the beginning of the first storage of a container. Licences in accordance with sentence 1 may only be extended for irrefutable reasons and after the prior referral to the German Bundestag". The German Bundestag must decide at the earliest in 2042, when the operating licence for the interim storage facility at the Emsland site expires, and at the latest in 2048, when the operating licence at the Unterweser site expires, whether the non-existent final storage capacities are to be regarded as irrefutable reasons. However, the operating licences of the decentralised interim storage facilities expire n 2032 (Ahaus) and in 2039 (Rubenow). The next generation will therefore have to deal with this issue. But in the meantime, however, the federal government has taken over all interim storage facilities from the power plant operators in accordance with the law. The fact is that the capacity for a German repository is fairly accurately calculable and it is very likely at this stage that Germany will have no further need for the final disposal of HAW. Is there therefore a reason to hurry when it comes to finding and constructing HAW disposal capacities? Not from the author's point of view. What is there a need for in Germany, namely a considerable need for a balanced repository concept which raises all existing potentials, which offers the required degree of safety, in order to achieve the goal of permanent, complete isolation of the stored radioactive inventory from the biosphere with a probability bordering on safety, which can be expected to be accepted by the citizenship and which also keeps an eye on the costs incurred. The potential concerns also include those of nuclear power plant operators. They and their qualified personnel can contribute considerable know-how to the process and should also be given the opportunity to earn money from the public law fund for services rendered, as in other countries.

So why is it that part of the community is in such a hurry? Well, because, in their opinion, a commissioning date after 2050 misses the goal of "**not shifting the disposal issue to future generations**". At first it sounds like a precaution, **but it is an empty phrase.** The disposal of the radioactive waste generated in this gener-ation will affect many, many subsequent generations per se. A subject has already been formulated above, the operating permits for interim storage facilities, possibly for transition storage facilities, conditioning facilities, P&T, etc. We must therefore rather ensure that subsequent generations do not suffer any damage as a result of the disposal of radioactive waste of this generation, not even financial damage. Safety must have top priority in all phases of the final disposal of HAW (heat-generating), interim storage, conditioning, emplacement, decommissioning and sealing phase. This applies to the environment of the repository, for which a long-term safety proof is required, but also and especially to the operating personnel in all areas (interim storage, conditioning, transport, reception, emplacement and sealing). According to the current understanding, a rapid decommissioning of the repositories is not advisable. An immediate closure of the plant according to the principle "Close your eyes and hope for the best!" implies the message that the intention is to get rid of the radioactive waste as quickly as possible and to transfer the problem to future generations. If subsequent generations succeed in developing better methods for the final disposal of HAW, they must be able to use them, e.g. nuclear transmutation as an regular operation. This is most likely to succeed if free access, permanent monitoring and information retrieval from the isolated HAW are made possible over a period of at least 500 years. It is interesting to note here that the French repository concept contains this consideration, but from the point of view of retrievability. In this concept, retrievability is also possible after closure of the repository for future generations, who have to calculate the resultant risk. This will occur because these generations may also operate nuclear power plants and therefore also have the problem of final disposal of HAW. This means that future generations can benefit from the repositories that are currently developed by combining the existing repository concepts with their own approaches and presenting new and better repository solutions.

In the case of international repository projects, it is noticeable that underground laboratories and/or pilot plants are integrated into these projects, without exception. Underground laboratories and/or pilot plants are indispensable in the approval procedures of the respective countries. Only Germany has so far almost completely done without them, apart from a few experiments carried out in the Asse "research mine" at the time. There, between 1983 and 1985, the effects of highly active waste on the surrounding salt were simulated at four test sites using electric heaters and cobalt-60 sources. The cobalt-60 sources are now part of the Asse inventory, see Table 7.7. Otherwise, little is left to be used; instead, the Asse "research mine" was set up to drastically reduce the acceptance of final disposal. With the determination of the repository location, the establishment of an underground laboratory and/or a pilot plant must therefore be applied for. This application must become part of the site permit. Geological, geophysical, rock-mechanical and geochemical investigations into the suitability of the rock can then be carried out there and the storage techniques can be tested under real conditions before storage begins. In the case of a pilot plant, this can later be integrated into the final repository. With the establishment of transition storage facilities, this approach is proposed and combined with a mandatory long-term monitoring system to gather information on the behaviour of isolated HAWs, the stability of containers and their interaction with the environment, as well as their inclusion effectiveness, etc. For this reason, the barrier-free retrievability of containers over a period of 500 years should not be abandoned. In countries such as France, Sweden and Finland, it would be easier to maintain such a long "transition" period, because the peaceful use of nuclear energy will continue over a very long period and with it the accumulation of HAW. This would certainly be a contribution for future generations to optimise the final disposal of HAW. This also includes the effectiveness of technical and geotechnical barriers and their integration into a repository concept.

The international repository concepts presented cover a range of possible host rocks. The authors do not see the choice of site and thus the determination of a host rock as a central question in a scientifically based repository concept. For them, this question is merely the definition of boundary conditions under which a repository concept is developed. The solution of the acceptance problem is of utmost importance when selecting a site. Therefore, it seems reasonable to ask at the beginning whether municipalities would not be prepared, as in Sweden and Finland, to offer themselves as storage sites! It goes without saying that the terms and conditions of support must first be determined in such a way that a municipality can also benefit from above-average long-term advantages from the fact that a repository is to be built in its own district.

Geological formations that appear suitable for a repository site should be characterised by their high insulating capacity and by the fact that they largely prevent the formation of a reactive mass transport environment that enables the transport of radionuclides from the ECZ into the biosphere. These are water or aqueous solutions. The consistency and nature of the aqueous solutions also have an influence on the transport mechanism. There are several assessment matrices that compare the repository-relevant properties of the potential host rocks, see Table 7.21. The result can only be that there is no perfect host rock that guarantees sufficient, permanent safety under the various scenarios to be considered. A repository concept must always integrate a multibarrier system in such a way that the weaknesses of the host rock do not come to bear. Host rocks are not the result of geological processes that they become host rocks for repositories. They are there, just as they are, because our earth was not created for final disposal.

National repository concepts can only take into account the host rocks that are present inside the national borders. In Scandinavia these are the granite formation "Äspödiorit" and in Germany primarily the enormous salt domes in Northern Germany and granite formations in South-East Germany. In both geological formations there are large areas from which raw materials can be extracted. This to prevent dispersal for over 1 million years is a challenging task.

The age of the geological formation is often highlighted as an advantage of host rocks. The granites of the Baltic Shield (Finland, Sweden), for example, are 1900–1125 million years old. The mighty salt domes in the North German Basin are about 250 million years old. Technical materials or geotechnogenic, geotechnical constructions are of course much younger. What is much more important, however, is that there has not been the time to prove their long-term properties with experiments. However, this alone is not a deficiency. It is therefore a matter of starting correspondingly long test series which reliably prove these long-term properties and which permit extrapolation to periods of >100,000 years. For this reason in particular, the establishment of an underground research laboratory and/or a pilot plant is indispensable. In a multi-barrier system of a repository concept, the individual barriers are coordinated so that the isolation of the radioactive waste is permanently guaranteed. The technical and geotechnical barriers compensate for the imperfection of the geological barrier. Only the properties of the technical and geotechnical barriers can be influenced and designed and developed by man. In particular, there is a considerable need for research in Germany with regard to container development and the selection of long-acting and effective immobilising agents. SKB's double-manifold copper tanks equipped with bentonite buffers would certainly be transferable to German conditions; at least the Swedish and Finnish research is pointing in this direction.

A comparison of the radionuclide vectors of the HAWs intended for final disposal in Germany, Sweden and Finland shows that the radionuclide vectors as such are not decisive for the selection of the host rock for the repository to be accommodated, see Table 7.22. For this purpose, the various reference times in Tables 7.4, 7.12 and 7.13 had to be normalised. This was done by removing the radionuclides with a half-life of <1000 years.

The selection of host rocks is primarily determined by the isolation capacity of the geological formations (permeability, sorption behaviour) in which the ECZ is created and by the permanent prevention of the development of a reactive mass transport environment, of transport mechanisms, so that radionuclides can escape from the ewG into the biosphere to a damaging extent (accident scenario: solution inflow into the ECZ). However, the fact must be mentioned here that a simultaneous, spontaneous discharge of all radionuclides from the container is not to be expected following corrosion of the container matrix. Although radionuclides may escape from the waste product, as a result of corrosion in the containers, they can be incorporated into the new solid phases and thus immobilised. The extent of release and retention is determined by the geochemical environment, as radionuclides can be subject to a variety of processes. How the source term will develop is one of the great secrets that we try to decipher with a number of assumptions. Which path the aqueous solutions may ultimately take is also subject to a number of assumptions until a mass transport model can provide information about the possible spread of radionuclides in the ECZ and finally about possible release into the overburden. Any assumption between 100% immobilisation and 100% mobilisation is possible, only a part of the radionuclides can be mobilised and a part of it is immobilised by the processes described above.

The influence of high temperatures on the host rock, especially in the ECZ, played a major role in the discussion on the final disposal of heat-generating radioactive waste. The questions were: temperature-induced changes in rock properties; thermally induced stresses that can lead to the formation of water pathways; the formation of secondary permeability under the influence of temperature; migration of fluid inclusions in the salt under temperature increase, etc. In addition to the host rock's properties of temperature resistance, thermal conductivity and the influence of temperature on the technical and geotechnical barriers that have led to the determination of design temperatures for the individual host rocks, the effect of temperature on the

Table 7.22 Comparison of radionuclide vectors of HAW intended for final disposal in Germany, Sweden and Finland

Reference values			Germany (Location not defined)		Sweden (Forsmark)		Finland (Olkiluoto)	
Isotope	Half-life $T_{1/2}$ [a]	Exemption limit FGi (Bq)	SUM [Bq]	Sum Ai/FGi	SUM [Bq]	Sum Ai/FGi	SUM [Bq]	Sum Ai/FGi
Am-241[a]	4.33E+02	1.00E+04	3.20E+18	3.20E+14	1.7E+18	1.7E+14	1.74E+18	1.74E+14
C-14	5.73E+03	1.00E+07	4.38E+14	4.37E+07	5.2E+14	5.2E+07	1.45E+15	1.75E+08
Cl-36	3.00E+05	1.00E+06	1.30E+13	1.30E+07	2.3E+12	2.3E+06	2.37E+13	2.37E+07
Sr-90[a]	2.91E+01	1.00E+04	1.03E+19	1.03E+15	2.3E+19	2.3E+11	3.11E+19	3.11E+15
Cs-137[a]	3.00E+01	1.00E+04	1.73E+19	1.73E+15	1.4E+13	1.4E+09	1.72E+19	1.72E+09
I-129	1.57E+07	1.00E+05	2.14E+13	2.15E+08	1.3E+18	1.3E+13	2.38E+13	2.38E+13
Pu-238[a]	8.78E+01	1.00E+04	2.12E+18	2.12E+14	1.4E+17	1.4E+13	1.28E+17	1.28E+13
Pu-239	2.41E+04	1.00E+04	1.66E+17	1.65E+13	2.5E+17	2.5E+13	2.81E+17	2.81E+13
Pu-240	6.54E+03	1.00E+03	3.73E+17	3.73E+14	1.1E+19	1.1E+16	1.58E+19	1.58E+16
Pu-241[a]	1.44E+01	1.00E+05	4.70E+18	4.70E+13	1.6E+19	1.6E+14	2.01E+19	2.01E+14
U-234	2.45E+05	1.00E+04	8.05E+14	8.05E+10	6.0E+14	6.0E+10	4.98E+14	4.98E+10
U-238	4.47E+09	1.00E+04	1.12E+14	1.12E+10	1.3E+14	1.3E+10	1.05E+14	1.05E+10
Sum			3.82E+19	3.73E+15	5.34E+19	1.14E+16	7.15E+19	1.93E+16
Sum α-emitter			5.86E+18	9.22E+14	1.31E+19	1.12E+16	1.79E+19	1.60E+16

(continued)

Table 7.22 (continued)

Reference values			Germany (Location not defined)		Sweden (Forsmark)		Finland (Olkiluoto)	
Isotope	Half-life $T_{1/2}$ [a]	Exemption limit FGi (Bq)	SUM [Bq]	Sum Ai/FGi	SUM [Bq]	Sum Ai/FGi	SUM [Bq]	Sum Ai/FGi
Radionuclide with Half-life >1000 years								
C-14	5.73E+03	1.00E+07	4.38E+14	4.37E+07	5.2E+14	5.2E+07	1.45E+15	1.75E+08
Cl-36	3.00E+05	1.00E+06	1.30E+13	1.30E+07	2.3E+12	2.3E+06	2.37E+13	2.37E+07
I-129	1.57E+07	1.00E+05	2.14E+13	2.15E+08	1.3E+18	1.3E+13	2.38E+18	2.38E+13
Pu-239	2.41E+04	1.00E+04	1.66E+17	1.65E+13	2.5E+17	2.5E+13	2.81E+17	2.81E+13
Pu-240	6.54E+03	1.00E+03	3.73E+17	3.73E+14	1.1E+19	1.1E+16	1.58E+19	1.58E+16
U-234	2.45E+05	1.00E+04	8.05E+14	8.05E+10	6.0E+14	6.0E+10	4.98E+14	4.98E+10
U-238	4.47E+09	1.00E+04	1.12E+14	1.12E+10	1.3E+14	1.3E+10	1.05E+14	1.05E+10
Sum			5.40E+17	3.90E+14	1.26E+19	1.10E+16	1.85E+19	1.59E+16
Sum α-emitter			5.40E+17	3.90E+14	1.13E+19	1.10E+16	1.61E+19	1.58E+16

Radiotoxicity determined on the basis of the exemption limits of the German StrlSchV, Annex 4 Table 1

[a]The various emplacement or reference time be normalised, by eliminate of radionuclides with a Half-life <1000 years

In Germany, reprocessing of fuel elements was permitted until 2005. This reduces activity and radiotoxicity of the radionuclide vector

The final storage design in Finland is based on an emplacement period of approx. 100 years, measured from 2022. During this period, the amount of waste to be stored increases considerably, so that the radiotoxicities which have to be stored in a long term safe and stable manner at Forsmark and Olkiluoto are almost identical host rock can be counteracted by the emplacement conditions, insulation and backfill material and the integration of P&T into a repository design, so that the development of limit temperatures or even their exceedance will be completely ruled out. In the Finnish, Swedish and French concepts this is implemented following the realization that man cannot change the thermal properties of the host rocks.

Two scenarios have acquired special significance in the repository concepts under consideration:

- Human Intrusion Scenario and
- Scenario of possible future climate changes (humid climate instead of desert climate; ice age etc.)

The human intrusion scenario, the deliberate or unintentional intrusion of future generations into the repository area (ECZ), cannot be separated from the recognition that mining activities have always taken place in Europe. While this problem is not discussed with the necessary rigor in Germany, this scenario occupies an important place in the repository concepts of the USA (Canada). Thus, after decommissioning, the WIPP in the USA is to be protected with a large-scale facility on the surface above the repository, with about 50 monoliths, 3 km long earth walls and a comprehensive marking, further see Sect. 7.4.1. This also shows the different social developments (traditions) in the USA and in Germany. While it is relatively easy in the USA to obtain a mining license (General Mining Act—rights of mining claims), the Federal Mining Act poses higher hurdles. In addition, the Germans gold-digger mentality is not quite so pronounced. The mining registers in Germany are excellent for this purpose. This should also lead to the final documentation being kept in the register of the respective mining office. In Finland and Sweden, the sites have been selected in such a way that no significant mineralisation has been proven there. The human intrusion scenario requires long-term monitoring of all activities in the repository area. A permanent security guard concept cannot therefore be ruled out.

In the long-term safety assessment for the Yucca Mountain repository, only little attention was paid to the climate change scenario. The low annual precipitation, the almost complete above-ground runoff combined with the high evaporation rates, the very low groundwater level and the closed basin structure, which does not affect groundwater catchment areas in neighboring regions, were decisive factors for the selection of the site in the mainly Miocene volcanic tuffs of the Yucca Mountains of Nevada. The formation of the Miocene volcanic tuffs are extremely dry, solution influx into the ECZ under the formulated conditions is practically impossible under the prevailing conditions. For this purpose, a sophisticated container and transport system was integrated into the building application for Yucca Mountain, see Fig. 7.7. When reviewing the Yucca Mountain repository concept, the failure to take into account various scenarios, in particular possible future climate changes (humid climate instead of desert climate), earthquakes and volcanic eruptions as possible hazards for the repository led to the NRC (Nuclear Regulatory Commission) rejecting the building application for Yucca Mountain.

Update 2019: The construction application for Yucca Mountain submitted by the DOE to the competent licensing authority NRC in June 2008 has been suspended since 2011 due to a lack of budget funds. Meanwhile, the NRC has submitted a proof of eligibility for Yucca Mountain: "Safety Evaluation Report Related to High Level Radioactive Wastes in a Geologic Repository at Yucca Mountain, Nevada", [56] showing the site's overall suitability.

In 2017, $120 million was approved for the 2018 financial year for the progress of the licensing process and further work on the Yucca Mountain site. A bill dated May 10th 2018 directs the DOE to continue its licensing process for Yucca Mountain while moving forward with a separate plan for a temporary storage site in New Mexico or Texas. The possibilities of including the warm-cold-time scenario has discussed in Sect. 7.4.1.

In the Swedish and Finnish repository concepts, too, the climate change scenario, here the formation of an ice age, were treated as part of the long-term safety assessment and was the subject of the approval procedure. The observation period is divided into up to 100,000 years and beyond. In the period up to 100,000 years the occurrence of an ice age is unlikely, see Sect. 8.5.1 and Fig. 8.10, in the period >100,000 years such an event is probable Based on an freezing depth of 800 m, the scenario "relocation of waste" was developed as part of the safety case and for the approval of the project. A further conclusion from this could be to provide deeper emplacement levels for the HAW from the beginning or to investigate the freezing of the aqueous solutions and their influence on the transport mechanism in more detail. In the last cold period, the Weichsel cold period, over approx. 50,000 years there was a deviation of −4 K and less, 40,000 years of −6 K and less and 10,000 years of −8 K and less from the 500,000 year average temperature. Answers to questions such as: at what temperatures does the repository horizon freeze and when does it thaw again and when is it likely that a transport mechanism will stop and start again, are still open. The composition of the aqueous solutions plays a decisive role here. For example, a saturated saline solution has a freezing point of −21 °C. Bentonite buffers can be designed to lower the freezing point of the entire system so that freezing can possibly be averted. Bentonite container shields are used in the Swedish and Finnish repositories, see Figs. 7.16 and 7.17. A number of bentonite types were investigated for their stability in the respective site-specific hydrochemical environments. In addition, the production of bentonite-form elements as well as the transport and inserting of these elements are tested using in situ experiments. Cement, which is required for the abutment of barriers, for the sealing of faults and cracks by injections and as a buffer in the super container concept, should have as low a pH value as possible. Since the ONKALO underground laboratory is directly integrated into the repository, the tests can also be continued during the construction phase.

The period under consideration for the long-term safety proof plays a major role with regard to the predictability of developments in the long periods under consideration. On the one hand there is the limited predictability of geological developments, consider climate change; on the other hand there is the service life of isolation systems and possibly initiated transport mechanisms and their course. These are often associated with a probability of occurrence that is also dependent on location and time. In order to increase the reliability and credibility of the statements, numerical solutions are used and supplemented by natural observations or analogues. Nevertheless, the risk assessment, the proven long-term safety for a period of observation of 10 million years, is not quantifiable and involves considerable uncertainty. It is at least suggested to assume a shorter observation period of, say, 10,000 years, for which the probability of occurrence of processes and events can be quantified. From the long-term safety thus determined, it is then possible to extrapolate to longer observation periods. The long-term safety factors determined in this way are then assigned uncertainties which are equivalent to the reliability with which the long-term safety factor was verified.

There are considerably diverging international views on long-term monitoring and its duration. In this context, the incorporation of a monitoring concept into the licensing documents should be pointed out as a matter of urgency, also as a trust-building element at the repository concept. The French Cigéo repository concept has provided some suggestions. The following structure is proposed as the basis for the life phases of a repository:

– **Construction phase**—Monitoring of compliance with the requirements of the planning approval decision, quality requirements for the structures, execution, radiation protection standards,
– **Operating phase**—the monitoring programs should monitor all safety-relevant parameters of the repository during the entire operating period (working speed of the machines, hydrogen concentration, performance of the highly effective filters, air of the ventilation system, etc.). This must include the monitoring of the external (possibly also the internal) radiation exposure of the personnel.
   In addition to monitoring in the narrower sense, parameters are monitored that may influence the development of the repository in the medium and long term: Ambient air temperature, extension of the plant (meeting of the galleries), durability of the concrete, corrosion of the steel, etc. During the operating phase, the entire ECZ, including the barriers, should be extensively instrumented, programmed and energetically equipped for internal long-term monitoring.
   If a **pilot project** is operated during the operating phase, this is equipped with extensive measurement programs, the results of which serve both to optimise the repository concept and to develop the repository in the medium and long term. The pilot project can be integrated into the repository.
   If an **underground research laboratory** is operated, this can be equipped with a measuring program similar to that of a pilot project. It can also fulfil the same functions as the pilot project, provided that it is also operated during the operating phase. The integration into the repository is exempted.
– Decommissioning phase—in this phase the function, observance with the objective is monitored by internal and external long-term monitoring. For external monitoring, an observatory will be set up to record and evaluate the air, soil and water over large areas via extensive measuring networks and to archive the relevant data. This also includes monitoring flora and fauna, agricultural production in the observation area and monitoring the physical-chemical and biological quality of the soil. The

observatory should work according to best practice. During the decommissioning phase, health monitoring of the general population in the region is also carried out, including a network for monitoring local dose rates and background radiation. Monitoring of the general population should begin during the operational phase.

The author considers a balanced, repository and site specific monitoring to be developable and feasible. With the development possibilities currently available and those described, there should be hardly any limits with regard to instrumentation (including energy supply) and data management, further reference is made to Chap. 9.

7.7 Responsibility and Costs

The decommissioning of nuclear installations is the responsibility of the plant operators. The energy supply companies operating nuclear power plants as well as the public authorities and private operators of other nuclear facilities who are obliged to deliver nuclear waste are obliged as waste producers to bear all costs of decommissioning (including dismantling) of their nuclear plants and facilities as well as the disposal of radioactive waste.

According to Section 6 of the Ordinance on Advance Payments for Repositories, see [57], the necessary expenditure is currently distributed as follows (as of 31.12.2012):

For a repository for (highly radioactive) in particular heat-generating waste

(a) 96.5% of nuclear power plants;
(b) 0.7% from the Karlsruhe reprocessing plant;
(c) 2.8% of other nuclear installations (industry, medicine and research).

In its cabinet decision of 14 October 2015, the Federal Government appointed a commission to review the financing of the phase-out of nuclear energy (KFK). On behalf of the Federal Government, the Commission should examine and develop recommendations for action on how the financing of decommissioning and dismantling of nuclear power plants, as well as the disposal of radioactive waste, can be arranged in such a way that the responsible companies are economically able to fulfil their nuclear obligations in the long term. They must be granted legal and planning security.

Initially, the commission to review the financing of the phase-out of nuclear energy estimated the cost of disposal at 2014 prices to be 47.5 billion euro.

As a result, KFK then submitted the following recommendation and handed it over to its client on 27 April 2016, see [35]: The state should take over the interim and final storage and the operators of nuclear power plants should transfer the funds to a public law fund for security. The Act on the Reorganisation of Responsibility in Nuclear Disposal has in the meantime acquired legal force [37], according to which the nuclear power plant operators transfer 17.4 billion Euro to the public fund. Against payment of a risk surcharge of about 35 percent, the companies can also terminate their obligation for any additional payments for cost and interest risks to the fund. The nuclear power plant operators are to pay this risk premium to the fund by the end of 2022. This does not include the costs of decommissioning and dismantling the nuclear power plants and packaging the radioactive waste. These remain the responsibility of the companies. Two laws, the Post-Liability Act and the Transparency Act, are intended to provide more security for the financial reserves of the operators for these tasks.

All described tasks and financial risks have meanwhile been transferred to the Federal Government against payment of a flat rate of 24.1 billion euros. This concerns further intermediate storage, conditioning and transport between transition storage and disposal. All that remains are the decommissioning and dismantling of the nuclear power plants and facilities by the four operators, who have to organize these for their own account.

Initially, this fund endowment—24.1 billion euros—is to be placed in the international comparison. Chosen as a reference figure is the amount of radioactivity deposited, see Table 7.23.

From this it can be seen that the financial resources to be spent in Germany for the final disposal of HAW exceed the financial resources used for this purpose in comparable countries in Europe, without it being possible to conclude, that the repository, which will be constructed in Germany will also be much safer compared to the countries of comparison. This statement is also independent of whether cost adjustments are still made in the countries of the comparison.

Another obvious difference is that these countries will have established a repository for HAW long before the construction of the repository will begin in Germany. Thus, the cost of development in the individual countries will have different impacts (Fig. 7.25).

It should be noted that the comparison is still acceptable, considering:

– the considerable uncertainties in the assessment of costs, because:

Table 7.23 Planned final disposal costs including research, construction and operation

	?/Germany (Table 7.22)	Cigéo/France (Table 7.18)	Forsmark/Sweden (Table 7.22)	Olkiluoto/Finland (Table 7.22)
Activity (TBq)	5.4E + 05	4.8E + 06	12.6E + 06	18.5E + 06
Costs (Billion Euro)	24.1[a]	36.0 (144[b])	14.0[c]	6.5 (3.5[d])
Underground laboratory	**???**	**Bure**	**Äspö**	**ONKALO**
Thousand-Euro/TBq	44.63	7.50	1.11	0.35

[a]In funds paid in, real costs not known
[b]Estimate, share to be paid by waste producers, with LL-MAW [58]
[c]Operating costs not included
[d]Operating costs not included
Research and URL—Olkiluoto pro rata 3 billion euros

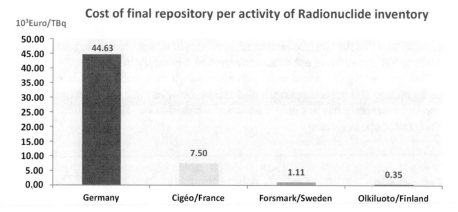

Fig. 7.25 Costs of final disposal of HG-HAW—comparison (How the costs will develop over a period extending into the next century is fraught with great uncertainty)

- a significant price increase may be expected over such long periods of time;
- general price changes to be expected as a result of changes in social conditions and
- unexpected costs likely to arise in the planning and construction of a repository for HAW.

On the one hand, in Germany experience is available with the ERA Morsleben sites, the Konrad mine facility and the Asse II mine facility, where there have already been considerable cost increases; on the other hand, reference can be made to a general average price increase of approx. 500% in the Federal Republic of Germany between 1990 and 2018, see Fig. 7.26. However, these cost increases could be partially offset by the interest on the fund. According to our own calculations, with an annual fund interest rate of 6.5%, this would be complete balanced.

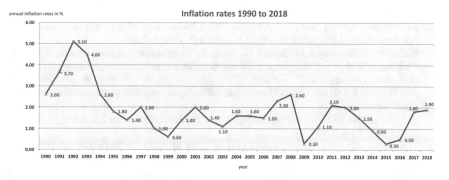

Fig. 7.26 Inflation rates 1990–2018. *Source* According to information from Statista GmbH, Hamburg

Not included in this list are the 1.7 billion euros already spent on the Gorleben exploration mine. From the author's point of view, it seems completely out of the question now that the Gorleben exploration mine will be selected as the best possible location. Yet it could find use as an underground laboratory in future.

This cost comparison can only provide a rough comparison and is of course not complete. The cost comparison also shows, however, that the construction of a repository structure has not only a technical dimension but also a commercial one which cannot be neglected.

References

1. Classification of radioactive waste - General Safety guide - No. GSG-1, IAEA, Vienna 2009.
2. Micah D. Lowenthal; Radioactive-Waste Classification in the United States: History and Current Predicaments; Center for Nuclear and Toxic Waste Management University of California Berkeley, California 94720-1730.
3. World Nuclear Association: http://www.world-nuclear.org/information-library/nuclear-fuel-cycle/nuclear-wastes/radioactive-waste-management.aspx.
4. U.S. Government Accountability Office (GAO): https://www.gao.gov/key_issues/disposal_of_highlevel_nuclear_waste/issue_summary.
5. Yucca Mountain Repository SAR; DOE/RW-0573; „WASTE FORM AND WASTE PACKAGE".
6. U.S. Government: Code of Federal Regulations No. 10 CFR 61 § 55 und 56: Waste classification. U.S. Government, Washington 2001.
7. ANNUAL TRANSURANIC WASTE INVENTORY REPORT – 2018 (Data Cutoff Date 12/31/2017) DOE/TRU-18-3425; November 2018.
8. QUANTITIES, SOURCES, AND CHARACTERISTICS OF SPENT NUCLEAR. FUEL AND HIGH-LEVEL WASTE IN THE UNITED STATES; Chapter 5 http://lobby.la.psu.edu/066_Nuclear_Repository/Agency_Activities/EPA/EPA_Yucca_Ch5.pdf.
9. Gmal, B., Hesse, U., Hummelsheim, K., Kilger, R., Krykacz-Hausmann, B., Moser, E. F.: Untersuchung zur Kritikalitätssicherheit eines Endlagers für ausgediente Kernbrennstoffe in unterschiedlichen Wirtsformationen. Im Auftrag des Bundesamtes für Strahlenschutz (BfS) im Vorhaben 1005/8488-2, GRS-A-3240. Gesellschaft für Anlagen- und Reaktorsicherheit (GRS) mbH. Köln, 2004.
10. International Atomic Energy Agency (IAEA): Classification of Radioactive Waste. Safety Standards Series GSG-1, ISBN 978-92-0-109209-0: Vienna, 2009.
11. Peiffer, F., McStocker, B., Gründler, D., Ewig, F., Thomauske, B.,: Havenith, A., Kettler, J.: Abfallspezifikation und Mengengerüst. Basis Ausstieg aus der Kernenergienutzung (Juli 2011). Bericht zum Arbeitspaket 3, Vorläufige Sicherheitsanalyse für den Standort Gorleben, GRS-278, ISBN 978-3-939355-54-0, Gesellschaft für Anlagen- und Reaktorsicherheit (GRS) mbH: Köln, September 2011.
12. „Gesetz zur Neuordnung der Verantwortung in der kerntechnischen Entsorgung"; vom 27. Januar 2017 (BGBl. I S. 114, 1222).
13. TIMODAZ – Thermal Impact on the Damaged Zone Around a Radioactive Waste Disposal in Clay Host Rocks (Contract Number: FI6 W-CT-2006-036449), Paris, June 2007.
14. Neles J.: Endlagerung wärmeentwickelnder radioaktiver Abfälle in Deutschland Anhang Abfälle - Entstehung, Mengen und Eigenschaften von wärmeentwickelnden radioaktiven Abfällen, Anhang zu GRS -247, ISBN 978-3-939355-22-9, September 2008.
15. Reaktorsicherheitskommission, RSK-Stellungnahme: „Sicherheitstechnische Aspekte konzeptioneller Fragestellungen zur Endlagerung von bestrahlten Brennstäben mittels Kokillen in

Bohrlöchern anhand eines Vergleiches mit dem Konzept , Streckenlagerung von dickwandigen Behältern'", 380. RSK Sitzung, Berlin, 31.03.2005.

16. Freiersleben, H.: „Endlagerung von radioaktiven Abfall in Deutschland", TU Dresden, Dez. 2011, verfügbar im Internet unter: https://iktp.tu-Dresden.de/IKTP/Seminare/IS2011/Endlagerung-IKTP-Sem-01-12-2011.pdf.

17. Mönig, J., Buhmann, D., Rübel, A., Wolf, J., Baltes, B., Fischer-Appelt, K.: Vorläufige Sicherheitsanalyse für den Standort Gorleben – Sicherheits- und Nachweiskonzept. Bericht zum Arbeitspaket 4, GRS-277. Gesellschaft für Anlagen und Reaktorsicherheit (GRS) mbH, Köln, 2012.

18. Partitionierung und Transmutation, acatech STUDIE, Forschung – Entwicklung –Gesellschaftliche Implikationen; Zentrum für Interdisziplinäre Risiko- und Innovationsforschung (ZIRIUS) der Universität Stuttgart, Herausgeber: Prof. Dr. Dr. hc Ortwin Renn, acatech – DEUTSCHE AKADEMIE DER TECHNIKWISSENSCHAFTEN, 2014.

19. Peter Brennecke; Stefan Steyer: Endlager Konrad - Bilanzierungsvorschrift für Radionuklide/Radionuklidgruppen und nichtradioaktive schädliche Stoffe; SE-IB-33/09-REV-1; Bundesamt für Strahlenschutz; Stand: 07. Dezember 2010.

20. Lersow, M.: Energy Source Uranium – Resources, Production and Adequacy, Glueckauf Mining Reporter 153(3):286-302 · June 2017.

21. Bericht der Bundesrepublik Deutschland für die fünfte Überprüfungskonferenz, Kommission Lagerung hoch radioaktiver Abfälle, K-MAT 14, Mai 2015.

22. Dr. U. Gerstmann (2002), H. Meyer, M. Tholen; Abschlussbericht: Bestimmung des nuklidspezifischen Aktivitätsinventars der Schachtanlage Asse, August 2002, GSF – Forschungszentrum für Umwelt und Gesundheit, see also https://www.greenpeace.de/sites/www.greenpeace.de/files/GSF_Bericht_nuklidspezifisches_Aktivitaetsinventar_0.pdf.

23. Lersow, M.; Gellermann, R (2015).; Langzeitstabile, langzeitsichere Verwahrung von Rückständen und radioaktiven Abfällen – Sachstand und Beitrag zur Diskussion um Lagerung (Endlagerung); Ernst & Sohn Verlag für Architektur und technische Wissenschaften GmbH & Co. KG, Berlin · geotechnik 38 (2015), Heft 3, S. 173–192.

24. BMUB Prognose (einschließlich Bestand) der Mengen radioaktiver Abfälle aus der Wiederaufarbeitung, die in der Bundesrepublik Deutschland endgelagert werden müssen (Stand: 31. Dezember 2014); Nationales Entsorgungsprogramm, 2015.

25. The Waste Isolation Pilot Plant: A Potential Solution for the Disposal of Transuranic Waste (1996); Chapter: 1 INTRODUCTION; https://www.nap.edu/read/5269/chapter/3#9.

26. Roddy, J. W., H. C. Claiborne, R. C. Ashline, P. J. Johnson, and B. T. Rhyne. 1986. Physical and decay characteristics of commercial LWR spent fuel. ORNL/TM-9591/V16R1. Oak Ridge, Tenn.: Oak Ridge National Laboratory.

27. Joonhong AHN; Deterministic Assessment for Environmental Impact of Yucca Mountain Repository Measured by Radiotoxicity; ISSN: 0022-3131 (Print) 1881-1248 (Online) Journal homepage: https://www.tandfonline.com/loi/tnst20.

28. Accelerator-driven Systems (ADS) and Fast Reactors (FR) in Advanced Nuclear Fuel Cycles. A Comparative Study. Nuclear Energy Agency (NEA), 2002.

29. Follin S, Johansson P-O, Hartley L, Jackson P, Roberts D, Marsic N, 2007. Hydrogeological conceptual model development and numerical modelling using CONNECTFLOW, Forsmark modellingstage 2.2. SKB R-07-49, Svensk Kärnbränslehantering AB.

30. Aaltonen, I., Engström, J., Front, K., Gehör, S., Kärki, A., Kosunen, P., Mattila, J., Paananen, M., Paulamäki, S., 2016. Geology of Olkiluoto. Posiva Report 2016–16. Posiva Oy, Eurajoki. 398 p.

31. Wilhelm Bollingerfehr et. al., Entwicklung und Umsetzung von technischen Konzepten für Endlager in tiefen geologischen Formationen in unterschiedlichen Wirtsgesteinen (EUGENIA) – Synthesebericht; DBE TECHNOLOGY GmbH, 2011 http://www.andra.fr.

32. Beuth, T. et. al. „Untersuchungen zum menschlichen Eindringen in ein Endlager" - Bericht zum Arbeitspaket 11 Vorläufige Sicherheitsanalyse für den Standort Gorleben, GRS 280; ISBN – 978-3939355-56-4, Köln, Juni 2012.

33. Gesetz zur Suche und Auswahl eines Standortes für ein Endlager für hochradioaktive Abfälle (Standortauswahlgesetz - StandAG); vom 23. Juli 2013, zuletzt geändert am 20. Juli 2017.

34. Anselm Tiggemann; Gorleben als Entsorgungs- und Endlagerstandort - Der niedersächsische Auswahl- und Entscheidungsprozess Expertise zur Standortvorauswahl für das „Entsorgungszentrum" 1976/77; erstellt im Auftrag des Niedersächsischen Ministeriums für Umwelt und Klimaschutz, 2010.

35. Abschlussbericht der KFK: Verantwortung und Sicherheit - Ein neuer Entsorgungskonsens Abschlussbericht der Kommission zur Überprüfung der Finanzierung des Kernenergieausstiegs (KFK); Berlin, 27.04.2016.

36. GRS „Vorläufige Sicherheitsanalyse Gorleben (VSG)", Köln 2013.

37. Gesetz zur Neuordnung der Verantwortung in der kerntechnischen Entsorgung, vom 27. Januar 2017 (BGBl. I S. 114, 1222)).

38. Witherspoon, P.A.; Bodvarsson, G.S.: Geological Challenges in Radioactive Waste Isolation - Fourth Worldwide Review. - Lawrence Berkeley National Laboratory, LBNL-59808, 283 S., Berkeley, USA, 2006.

39. Updating the NEA International FEP List An Integration Group for the Safety Case (IGSC); Technical Note 1: Identification and Review of Recent Project-specific FEP Lists; Radioactive Waste Management NEA/RWM/R(2013)7; OECD-NEA, January 2014.

40. David Snow, Unsafe Radwaste Disposal at WIPP, Engineering Science University of California-Berkeley, update 2014.

41. Bericht der Blue Ribbon Commission on America's Nuclear Future, Januar 2012 http://energy.gov/sites/prod/files/2013/04/f0/brc_finalreport_jan2012.pdf.

42. Strategy for the management and disposal of used nuclear fuel and high-level radioactive waste. U.S. Department of Energy, Januar 2013 http://energy.gov/sites/prod/files/Strategy%20for%20the%20Management%20and%20Disposal%20of%20Used%20Nuclear%20Fuel%20and%20High%20Level%20Radioactive%20Waste.pdf.

43. CURRENT INFORMATION CONCERNING A POTENTIAL WASTE REPOSITORY AT YUCCA MOUNTAIN; Chapter 7 http://lobby.la.psu.edu/066_Nuclear_Repository/Agency_Activities/EPA/EPA_Yucca_Ch7_part_1of4.pdf.

44. Thomas C. Hanks, Isaac J. Winograd, R. Ernest Anderson, Thomas E. Reilly, and Edwin P. Weeks: Yucca Mountain as a Radioactive Waste Repository; Circular 1184; U.S. Department of the Interior; U.S. Geological Survey; https://pubs.usgs.gov/circ/1184/pdf/c1184.pdf.

45. Agence nationale pour la gestion des déchets radioactifs; Inventaire national des déchets radioactifs et des matières valorisables; 2006; http://www.andra.fr/download/site-principal/document/rapport-synthese.pdf.

46. Takagi, J.; Schneider, M.; Barnaby, F.; Hokimoto, I.; Hosokawa, K.; Kamisawa, C. et al. (1997): Comprehensive Social Impact Assessment of MOX Use in Light Water Reactors. Final Report of the International MOX Assessment. IMA Project. Citizens' Nuclear Information Center (CNIC).

47. Long-term safety for the final repository for spent nuclear fuel at Forsmark Main report of the SR-Site project Volume I; Svensk Kärnbränslehantering AB; Technical Report TR-11-01; March 2011.

48. Chun-Liang Zhang et. Al.: Thermal Effects on the Opalinus Clay; A Joint Heating Experiment of ANDRA and GRS at the Mont Terri URL (HE-D Project); Final Report; GRS – 224; ISBN 978-3-931995-98-0; March 2007.

49. Data report for the safety assessment SR-Site; Svensk Kärnbränslehantering AB; Technical Report TR-10-52; December 2010.

50. POSIVA 2012-09; Safety Case for the Disposal of Spent Nuclear Fuel at Olkiluoto -Assessment of Radionuclide Release Scenarios for the Repository System; December 2012; Posiva Oy.

51. Integrated account of method, site selection and programme prior to the site investigation phase. SKB, Technical Report, TR-01-03, December 2000.

52. ABSCHLUSSBERICHT der Kommission Lagerung hoch radioaktiver Abfallstoffe, Kommission Lagerung hoch radioaktiver Abfallstoffe gemäß § 3 Standortauswahlgesetz; K-Drs. 268.

53. POSIVA 2014-03; Safety Case for the Disposal of Spent Nuclear Fuel at Olkiluoto FEP Screening and Processing; January 2014; Posiva Oy.
54. OECD - NUCLEAR ENERGY AGENCY (NEA): Actinide and fission product partitioning and transmutation - Thirteenth Information Exchange Meeting Seoul, Republic of Korea 23-26 September 2014; OECD-NEA NEA/NSC/R(2015)2, Paris, 2015.
55. Projektkonzepte für die Lagerkammern und Versiegelungsstrecken und deren Bewertung Arbeitsbericht NAB 16-45, Nationale Genossenschaft für die Lagerung radioaktiver Abfälle (nagra), Juli 2016.
56. US NRC safety evaluation report Yucca Mountain: „Safety Evaluation Report Related to High Level Radioactive Wastes in a Geologic Repository at Yucca Mountain, Nevada"; https://www.nrc.gov/reading-rm/doc-collections/nuregs/staff/sr1949/.
57. BMUB, Bericht über Kosten und Finanzierung der Entsorgung bestrahlter Brennelemente und radioaktiver Abfälle, Berlin, August 2015.
58. Die Kosten der Kernenergie; Öffentlicher thematischer Bericht; Januar 2012; 13 rue Cambon 75100 PARIS CEDEX 01 - tel: 01 42 98 95 00 - www.ccomptes.fr.

Chapter 8
Long-Term Safety of Geotechnical Environmental Structures for the Final Disposal of Radioactive Waste and Residues in Germany

In some countries the time for undertaking a long-term safety assessment for a repository for high-level radioactive waste is set down in law, while other countries set out only the framework upon which a long-term safety assessment must be based. In many countries, the reference solution is preferred for the long-term safe disposal of radioactive waste in technical deep geological repositories. The decision-making and social acceptance of site-specific deep storage depends on the level of trust in the safety assessment of such repositories. In this context, it is important that the effects of repository failure are regionally limited. The consent of the receiving municipality that will host the site is an absolute must. Reference scenarios[1] must be analysed and evaluated taking into account the location conditions, today's social conditions, local and national, and the current state of science and technology. In this assessment, the number of persons affected, the spatial and temporal extent of any possible contamination and the possibility of reducing the impacts through planning countermeasures must be included. From one point of view this regional impact can then also be used to explain the regional commitment of citizens related to the site selection of a repository.

The term reference method can have several meanings in science, e.g. as a method whose result quality is so high that it can be used to assess the result quality of other methods, or by which a measurement or test quantity is defined, or it may be prescribed for a safety analysis to meet legal requirements. In the long-term safety assessment for geotechnical environmental structures for the final disposal of radioactive waste and residues, the entire range of the term "reference method" is used: Reference persons, geological reference profiles, see [1] BGR 2012, reference scenarios (reference cases), reference local dose rates, etc. A sensitivity analysis is therefore used to investigate which differences in the results for the reference case result from the variation of an individual model parameter whose value cannot be specified precisely (e.g. the time after which a repository container becomes leaky). A reference solution generally represents, if possible, the best estimation of the parameter θ; however, the true value θw remains unknown generally.

---

[1]Compare also Glossary: Reference Scenario.

© Springer Nature Switzerland AG 2020
M. Lersow and P. Waggitt, *Disposal of All Forms of Radioactive Waste and Residues*, https://doi.org/10.1007/978-3-030-32910-5_8

Errors (deviations) are classified into gross errors caused by human behavior, as well as random (statistical) and systematic errors (bias). Bias is the difference between the expected value of a test result and the recognized reference value.

A reference solution may therefore be subject to bias. This indicates the need, e.g. in the validation of a calculation system, to evaluate a larger number n of reference measurements or benchmarks, provided the benchmark and use case are not identical.

In order to meet the demand for avoidance of unacceptable radiation hazards to the population, in Germany the planning, construction, operation, decommissioning, safe enclosure and dismantling of installations or facilities are required to keep the effective dose for individuals of the population in a calendar year due to discharges of radioactive substances with air or water from these installations or facilities below 0.3 mSv/a.

This can also be considered as a reference value, see § 47 (1) German radiation protection ordinance. The determination of the long-term safety refers to this reference value and the reference solution.

The IAEA has set itself the task of formulating the framework requirements. More precise proposals have been submitted by the Nuclear Energy Agency (NEA) of the Organization for Economic Cooperation and Development (OECD Nuclear Energy Agency), [2]. The NEA, [2] formulates this as follows:

> Safety assessment is an interdisciplinary approach that focuses on the scientific understanding and performance assessment of safety functions as well as the hazards associated with a geological disposal facility. It provides crucial technical and scientific information to guide site investigation, research and development at various stages of repository development. Safety assessment is an essential component of the disposal safety case, providing inter alia the technical evidence to achieve confidence in the decision-making process.

The assessment of the long-term safety of a repository, in particular for highly radioactive—heat-generating waste, requires a comprehensive understanding of the system and a qualified and powerful program package in order to carry out this assessment [3]. That needs to be constantly adapted to the site conditions during the process. The models of this integrated instrument must take into account the relevant processes involved in the mobilization and release of radionuclides from the Effective Containment Zone (ECZ) of a repository and the role they play for their transport through the host rock and adjacent formations as well as for exposure in the biosphere. It is not the aim of these statements to present a long-term safety record for a permit application of a repository, but to highlight the essential structures and requirements for such activity, see also [4] Bonano et al.

8.1 Task/Aim

In the preceding chapters, various geotechnical environmental structures were presented, which are designed to permanently contain or store defined radioactive inventories as well as what evidence exists to demonstrate this credibly and on the basis of which the Geotechnical Environmental Structures are authorised and approved. For tailings ponds, the general procedure for a long-term safety assessment is presented in Chap. 5, but this will be done here, especially for the final disposal of high-level radioactive waste. A binding procedure has not yet been presented in Germany and is perhaps also undesirable because the long-term safety record must be carried out site-dependently so that the repository structure for high-level radioactive, heat-generating waste can also be approved. The framework for long-term proof of safety should, however, be formulated in a binding manner. In the following, this framework will be presented on the basis of the currently available findings.

The goal of the final disposal is to safeguard radioactive material subject to supervision in a Geotechnical Environmental Structure[2] and to permanently prevent the transfer and spread of radionuclides from the inventory of the repository in the biosphere. This means that the geotechnical environmental structure and the storage of the radioactive material must be carried out so that this goal is achieved with high certainty. Geotechnical environmental structures that are to take radioactive waste for final disposal undergo various phases, which are all helping to determine the long-term safety of the geotechnical environmental structure:

- The phase of planning and constructing a repository, which must meet the requirements of state-of-the-art science and technology for such waste, which is required under the Atomic Energy Act.
- The operating phase (with a specified operational safety and an operating permit), in which the radioactive waste is accepted and stored permanently in the storage areas.
- The post-operational phase (decommissioning phase), during which the geotechnical environmental structure is shut down, closed and monitored.
- The post-closure phase,[3] for which the geotechnical environmental structure is designed for long-term safety.

This process must allow for optimization until the final closure of the repository so that the target is reached with the highest possible level of safety, including any necessary Recoverability (recovery) of the inventory.

The Geotechnical Environmental Structure is being built as a multi-barrier system. It consists of the geological, technical and geotechnical barriers, which together form the effective containment zone (ECZ) and which, in their interaction, should to ensure the isolation of the waste for the required period. Such a multi-barrier system consists of various system components, each of which has its specific function in

---

[2]A Geotechnical Environmental Structure, also repository construction, is characterized in that it was built with geotechnical methods and procedures in the environment. The environment as such is both a boundary condition and part of the repository construction. A Geotechnical Environmental Structure is built with the aim of both serving the environment and sustainably protecting it from existing harmful effects. Since a geotechnical environmental structure can not be built without gentle interventions in the environment, it must be proven that the goal: serve and protect, justifies a gentle intervention. A repository must live up to this description. The radioactive waste and residues as such are not part of the Geotechnical Environmental Structure, but initially part of the environment.

[3]Compare also Glossary: Closure and repository sealing.

the isolation of the waste. No element of the multi-barrier system is absolute and totally impermeable. It is therefore important to combine the barriers, if necessary, in such a way that an overall isolation capacity of the effective containment zone (ECZ) is achieved for as long a period as possible. For high-level radioactive waste (HAW), according to current regulations in Germany the barriers must be able to guarantee an isolation period of the order of one million years. The fact that such a long period of observation (proof period) does not lead to more certainty seems to be undisputed from the discussion. Thus, it seems more appropriate to divide the entire period of observation into periods of different probability of failure, since the probability of failure does not remain constant over a period of 1 million years. This approach should be put up for discussion here.

The achieving of the objective, the warranty compliance with the protection criteria, requires both permanent monitoring of the environment of the repository site and crisis management in the event of limit violations (accident scenarios).

This chapter deals almost exclusively with the long-term safety of radioactive waste repositories. Long-term safety for the disposal of residues from uranium ore beneficiation was discussed in Chap. 5. It was shown that it is not the inventories, the hazard potential, and therefore the safety criteria that are decisive for the evaluation of the different Geotechnical Environmental Structures. Only the underlying legal regulations and guidelines are decisive, which are the same for all types of radioactive material but are interpreted differently. This approach does not do justice to the resulting hazard situations for the biosphere, not least because these tasks are perceived as public tasks in Germany. This applies both to the final disposal of radioactive waste and to the long-term safe and stable storage of residues from uranium ore processing, especially in Saxony and Thuringia.

In Chap. 5, using the example of the uranium tailing ponds of the SDAG Wismut, it was shown that a long-term safety assessment for these residues cannot be achieved because there is no base sealing and radionuclides are continuously transferred from the inventory into the biosphere, via the water path and the diffuse exiting contaminated water amount has to then be collected and sent to the water treatment plants. Both the collection of the contaminated waters and their treatment is described in detail in Chap. 5, so that in Sect. 8.3.1 only the concept for a long-term safety assessment of the Geotechnical Environmental Structures for the Uranium Tailings Ponds is summarized here.

The claim of this presentation here is not to present an algorithm for the provision of a long-term safety record for a license application for the disposal of radioactive waste and residues. The purpose of this presentation is to show the essential requirements and relationships for a long-term safety assessment, so that the dimension of the task which is to be presented for approving a radiotoxic waste and residue final disposal system becomes clear. The basic requirement for a permit is that it can be demonstrated that the repository system in all likelihood will permanently prevent the spread of radionuclides into the biosphere from the repository's inventory; based on different probable and less probable scenarios of the protection criteria it is may be demonstrated that compliance with the protection criteria is ensured by permanent monitoring of the surroundings of the repository site and that crisis management is installed and maintained permanently functional in the event of limit values being exceeded (incidents/accidents).

The statements can also provide some suggestions for the design of a long-term safety assessment for a future repository for heat-generating HAW in Germany and beyond.

8.2 Long-Term Safe and Stable Storage of the Uranium Tailing Ponds of the SDAG Wismut

The depictions here for the long-term security of uranium tailings ponds are only a summary of the representations in Chap. 5. For this reason explanations and descriptions of contexts are not included here but are summarized in Chap. 5.

SDAG Wismut's industrial settling ponds (IAA—Uranium mill Tailings ponds) were originally set up at locations without consideration of the demands of a later long-term safe, long-term stable storage.[4,5]

Site selection and the construction of uranium mill tailings final storage facilities are based on site specific existing conditions. The evidence is site-specific in each case. The main functional elements of the tailings ponds for tailings of the uranium ore processing, see Chap. 5, are: tailings dams, cover (encapsulation), base sealing—as geological barrier and/ or as geotechnical barrier, water gathering and water treatment as well as the discharge of the treated waters (by observance of pre-set discharge values) in the receiving water, revegetation and greening, monitoring system, characterized as a (multi-barrier system with long-term monitoring).

Both the dam structures and the multi-functional covers of the uranium mill tailing ponds of the SDAG Wismut are constructed as technogenic loose rock bodies. The individual layers of the entire loose rock body are constructed differently but each with classified material. Nevertheless, the material parameters of the functional elements (mineral, geosynthetic, etc.) of the technogenic loose rock bodies are stated variables and thus time-dependent. Thus, an aging of the mineral layers during the period under consideration cannot be ruled out. This must be included in the proof of long-term safety for the Geotechnical Environmental Structure, e.g. on the proof of

---

[4]This contradiction could be resolved by clearing out the uranium tailings pond and relocated of the uranium tailings to another site. However, this is out of the question due to the magnitude of the task and the associated risks to the environment (personnel) here. The in situ deposition of the uranium tailings is here a "good" solution at of all challenges.

[5]The Moab uranium mill tailings pile is probably the best known relocation project and a part of the UMTRA program. Moab is located in the state of Utah/USA. The reason for the relocation is its proximity to the Colorado River. Otherwise, a contamination escape into the Colorado could not be prevented because an aquifer was connected to the Colorado. The relocation takes place using special containers on a train over a distance of 30 miles to a prepared location, Crescent Junction, Utah. The storage concept has been prepared and arranged by the US Department of Energy (DOE). The supervision is the US Environmental Protection Agency (EPA).

the design against internal and external erosion, which also have special significance for radon exhalation, but not only for this purpose.

The long-term safety certification of the Geotechnical Environmental Structure for the storage of radioactive residues comprises the following individual certificates, which should be summarized as an assessment of the probability of failure, see also Chap. 5.

• **Proof of the permanent functionality of the multibarrier concept**

Note 1: Evidence of the long-term stability of the multi-functional coverage of a tailing pond should be provided with an observation period[6] of 1000 years. As an assessment value (limit value) for the radiation exposure, Wismut GmbH has established a local dose rate ODL $\leq 0.15\ \mu Sv/h$, see [5] and Chap. 5.

– Designation and monitoring of the long-term stability of dam structures
– Designation and monitoring of the long-term stability of the surface sealing (multi-functional coverage)
– Permanent encapsulation of the uranium tailings to minimise escapes of radio-toxic contaminants and radon exhalation into the biosphere[7]: Assessment of the shielding effect protecting against contained radioactive minerals, and limitation of the external radiation exposure (ground radiation) to $\leq 0.15\ \mu Sv/h$.

Note 2: The layers of the multi-functional cover can move against each other over time, or penetration can take place at the layer boundaries, so that the barrier system can be damaged, which then results in e.g. development of radon pathways. Aftercare will be necessary. For this, a proof of the stability against internal erosion is to be provided.

• **Proof of long-term stability of the base sealing**

Note 3: All uranium mill tailings ponds of the SDAG Wismut lack a base sealing. Base sealings cannot be retro-fitted. Possible contamination discharge into groundwater is therefore expected and needs to be detected early and uncontrolled leaks, especially into groundwater, prevented or limited. In addition to an appropriate groundwater monitoring system, the creation of well- and surface-, leachate- and groundwater measure points and galleries, with protection- and production wells and deep drainage for collection of groundwater and transfer to the water treatment plants; in addition mass transfer models are also included as a prognosis tools for the long-term safety concept. In order to prevent or limit contamination escapes from the tailings body, the groundwater level, based on the site specific hydrological conditions in the tailings pond area, can be lowered by means of wells to produce a defined safety distance between tailings body and the highest groundwater level and contamination passages into the groundwater body. In the case of diffuse escapes and ingresses of the leachate, the creation of delivery wells is problematic. Mass transfer models must be constantly readjusted to the changing site conditions. The results from long-term monitoring

---

[6]Reason see Chap. 5.
[7]US EPA standard: Radon emission rate $\leq 20\ pCi\ m^{-2}s^{-1} = 0.74\ Bq\ m^{-2}s^{-1}$.

serve as grid points for verification. From these data, so-called location models are developed with which a high forecasting certainty can be achieved.

– permanent collection of the contaminated water, in particular the groundwater and leachate, and transfer to a water treatment plant,
– proof of the permanent protection of the groundwater and the receiving water (run off protection).
– Permanent treatment and discharge of water into the receiving water within the exemption limits.
– Long-term monitoring consisting of exposure monitoring, measurement of the composition of the collected and treated waters, monitoring of radon exhalations, the functionality of the multibarrier system.
– Aftercare and replacement investments, especially the water gathering and treatment.
– The documentation (final documentation) shall include specifications which guarantee the reliability of the repository structure, guarantee the protective functions over its service life and indicate in good time any aftercare services which may become necessary. The time of dismissal of the site from the mining supervision is part of the long-term safety for the Geotechnical environmental structure. The closure and release of the geotechnical environmental structure must be applied for by the operator with the responsible mining authority. Due to the lack of base sealing, an "perpetual task" must be assumed.
– The duration of the monitoring period is determined by the relevant mining office, in which the operator's application for release from mining supervision is approved or rejected. Due to the lack of base waterproofing, a permanent supervision of the mining authority must be assumed.
– The file (final documentation) is kept in the register of the mining authority and remains there permanently.

The permanent collection of seepage water, water treatment, long-term monitoring including the prognosis tools for the possible spread of contamination and its limitation, as well as the securing of the area are components of the long-term safety system and represent a de facto replacement for the missing base sealing.

Unplanned, extraordinary events—worst cases, which greatly reduce the functionality of the respective geotechnical environmental structure, so that radio-toxic can reach the biosphere pollutants in considerable amounts:

– Earthquake of a previously unknown scale
– aeroplane crashes
– Bombing (explosions), terrorist attacks
– tectonic events
– Climate change, extreme weather events
– External interventions; terrestrial interventions—human intrusion scenarios: (drilling, exploration, etc.), except terrestrial interventions—impact events: (meteorite impacts), see [6]
– etc.

A site rehabilitation concept for uranium mill tailings ponds also contains the restrictions promulgated by the responsible mining authority, with which any possible reuse of remediated areas have to comply. It may be possible to prohibit reuse, see "Human Intrusion Scenarios[8]".

The concept pursued in the long-term safe and long-term stable storage of the uranium mill tailings ponds of the SDAG Wismut is referred to as a guardian-concept by the Commission for the Storage of Highly Active Waste, [7] EndKomm (2016) and rejected. This is justified by the fact that the monitoring, here the ceded uranium tailings ponds, must be passed on to future generations. This must be done on the basis that interventions from the outside (drill holes, holes etc.), affecting the cover and/or the dam structure, so-called "human intrusion scenarios", must be prevented. The guardian-concept is seen here as uncritical, also because the registers of the mining authority are among the best kept data collections in the German administrative system and because "guarding—long-term environmental monitoring" should be part of every repository concept. The long-term environmental monitoring is part of the long-term safe and stable custody of the uranium mill tailings ponds of the SDAG Wismut and therefore follows a guardian-concept. Here, however, this concept is referred to as a "concept of long term, tracked permanently to the data situation, proof of the functionality". This also includes the worst case scenario in which the geotechnical environmental structure loses its function (wholly or partially) and radio-toxic contaminants enter the environment (air, groundwater, soil), unhindered to the extent that the limitation of the external radiation exposure no longer exists (accident scenario). The period during which the limits for external radiation exposure are exceeded must be avoided in the long term. For further information see Chap. 5.

8.3   Long-Term Safe and Stable Storage of Radioactive Waste in Germany

8.3.1   Compilation of the Currently Valid Requirements for Final Disposal

The national legal and subordinate regulations determine the essential frameworks for the final disposal of radioactive waste. These include the Atomic Energy Act (AtG), the Radiation Protection Ordinance (StrlSchV) and the Federal Mining Act (BBergG) with the associated Federal Mining Ordinance (ABBergV). To this end,

---

[8]Especially in sparsely populated regions of the USA, attempts were made to unmistakable to install the corresponding information about the hazard potential of the repository by means of appropriate signage elements (linked to the respective design elements surrounding of the final disposal construction) also for periods, for which administrative measures of the repository are no longer effective. This is intended in particular to prevent unintentional human intrusion into the tailings ("human intrusion scenario").

the relevant international recommendations of the ICRP, the IAEA and the OECD NEA must be taken into account, especially if they contain additions or modifications to the national regulations. The presentation here can only be based on the existing legal standards and other applicable documents in Germany for the existing and/or erected repository, and should reflect the final disposal concept to be developed for heat-generating HAW, including not only the current legal requirements but also the scientific findings about it, which decisively determine a safety philosophy, [8–12].

In the following chapters it will be shown that in the respective areas of assessment are individual proofs for the various geotechnical environmental structures which are highly location-dependent. Similarities in the basic requirements for the long-term safety of the geotechnical environmental structures are highlighted and the differences in the requirement structure are shown. The fact that these cannot be justified with the respective radio-toxic inventory shows, however, that a uniform requirements concept for all types of radioactive waste and residues was not considered in Germany with the consequence, that a very diffuse picture has resulted by reason of different safety standards and interpretations of geotechnical environmental structures for radioactive waste and residues, see also Chaps. 5, 6 and 7.

8.3.2 Central Safety Requirements for a Repository System to Be Developed in Germany

According to BMU [8], the central safety requirements to be met currently by a repository system to be developed are:

- For a million years, it must be shown that only very small quantities of pollutants will be released from the repository. For this purpose, the integrity of the effective containment zone must be proven and the risk out of escapes from the repository evaluated and presented.
- The safety of the repository, from planning to closure of the repository, must be subjected to a continuous optimization process with periodic safety checks.
- A multi-barrier system must be established that follows the principles of redundancy (dual safety systems) and diversity (independent mechanisms of action) that are common in the nuclear field.
- A control and evidence protection program must be carried out even after the repository has been decommissioned.
- The retrievability of radioactive waste must be possible during the operation of the repository. In addition, the final disposal canisters must be able can be recovered safely for up to 500 years after closure of the repository.

Except for the facts, that

- the observation period (detection period) is currently being discussed internationally in a manner diverging from the above requirement, without questioning the security situation as such, and

- a control and evidence preservation program must include long-term environmental monitoring, which is close to a guardian-approach, but which, according to the current safety philosophy, is rejected, these requirements can be fully endorsed here. For both the period under consideration and the "guardian"- concept, an assessment is presented below.

8.4 Long-Term Safe and Stable Storage of Radioactive Waste in Germany

8.4.1 Final Disposal of Radioactive Waste with Low (Negligible) Heat Generation

In Germany, there are two licensed repository sites for radioactive waste with low (negligible) heat generation, the Morsleben repository for radioactive waste (ERAM) and the Konrad shaft in Salzgitter/Bleckenstedt.

The Repository for radioactive waste Morsleben (ERAM) was selected by the predecessor of the State Office for Atomic Safety and Radiation Protection (SAAS) of the former GDR, from ten considered salt mines and also approved as a nuclear facility. The SAAS commissioning approval was granted in 1978/79, although it had already been used before for emplacements of low to intermediate level radioactive waste during a trial operation. At present it is the only repository for low- and intermediate-level radioactive waste in the Federal Republic of Germany, which was operated on the basis of an atomic license. The operator on behalf of the federal government is the BGE. The BGE is currently preparing the decommissioning of the ERAM. However, the Federal Office for Radiation Protection (BfS) has drawn up the plan approval documents, and the application already in September 2005 submitted to the Ministry of the Environment of Saxony-Anhalt (MLU) acc. § 9 b of the Atomic Energy Act to implement a plan approval procedure for decommissioning. This includes proof of the long-term safety of the Geotechnical Environmental Structure. The BGE now continues the procedure.

The federally-owned company "Bundesgesellschaft für Endlagerung" (BGE) mbH was founded as the project executing agency for the operational tasks of site search, construction and operation of the repositories and the Asse II mine. This company assumes the tasks of Asse-GmbH, Deutsche Gesellschaft zum Bau und Betrieb von Endlagern für Abfallstoffe mbH (DBE) as well as the operator tasks of the Federal Office for Radiation Protection (BfS), i.e. also the operation of the Gorleben exploration mine. The BGE has taken over the procedure from BfS, see Chap. 6 for further details. The Länder of Lower Saxony and Saxony-Anhalt remain the nuclear licensing authorities and thus responsible for the supervision of the Asse II mine, Konrad repository and ERAM projects. The responsibility of the Länder ends for the

Konrad repository with the commissioning and for the Morsleben repository with the conclusion of the ongoing planning approval procedure for decommissioning.

In the case of repositories for radioactive waste with low heat generation, the long-term safety assessment also has the objective of permanently preventing the transfer of radionuclides from the repository inventory into the biosphere. However, due to the situation described above, it is not expedient to present a long-term safety assessment for the disposal of radioactive waste in general. Instead, reference is made to the available evidence and these are evaluated. The justification is as follows:

- The selection of the location and thus also the type of final disposal has been made, it is in each case a matter of storage in deep geological formations,
- Either no information was given in the respective planning approval documents regarding the duration of accessibility to the storage areas or it is unnecessary. Concepts of Reversibility, Retrievability and/or Recoverability play no role in the repository concepts of the ERAM or the Konrad mine,
- A probability of failure has not been determined for these repository sites. Only deterministic methods were used,
- The consideration of the containers as a technical barrier in the multibarrier concept of a repository was not applied. The contribution of the waste matrix and the containers to the isolation of the radionuclides and thus to the limitation of the release of the radionuclides from the waste matrix and from the container into the effective containment zone was not implemented in the repository concepts, see Brennecke [13].

Within the framework of the licensing procedure for the Konrad repository, it was determined, among other things, that the Konrad mine is suitable for conversion into a repository for radioactive waste with low heat generation, especially because:

- the rock-mechanical conditions in the area of the emplacement chambers and the mine works do not anticipate any faults will be expected and thus do not allow any gas pressure build-up,
- the properties of the geological formations serve as barriers against the spread of radionuclides (permeability and sorption behaviour),
- a long-term seismic stability of the site can be predicted,
- because the mine is exceptionally dry, etc.

The calculations for the long-term safety analysis for the shaft Konrad repository showed that the water pathways emanating from the mine workings can reach the biosphere at the earliest after approx. 300,000 a and Thus, long-lived radionuclides from the emplaced inventory on this water path can cause a potential radiation exposure of the population that will remain far below the fluctuation range of the natural radiation exposure of ≤ 0.3 mSv/a in the long term. The system of natural and geotechnical barriers thus ensures that an impermissible load on the biosphere from the shaft Konrad repository can be excluded in the long term, see Physikalisch-Technische Bundesanstalt (PTB) (1982), application for initiation of a planning approval procedure.

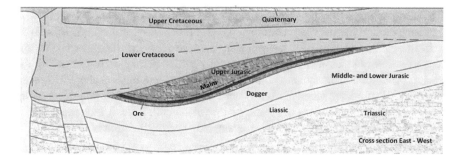

Fig. 8.1 Geological section in the area of the Konrad mine; iron ore deposit thickness of (4–18) m with a clay barrier approx. 400 m thickness. *Source* BfS

Only radioactive waste with negligible heat-generating capacity may be stored in the Konrad repository. Negligible heat generation means that the so-called 3 K criterion is complied with. 3 K-criterion refers to the limitation of the thermal influence on the host rock to 3 K at butt (side wall of the mine works). From this it follows that the radionuclide vector of the future inventory—as far as heat development is concerned—is described largely.

A consistent concept for low- and intermediate-level radioactive waste has not yet been defined separately. One reason may be that Germany did not follow the principle of a single repository site. For this reason, reference is made here to the presented concept of determining long-term safety for HAW, which to a large extent can also be transferred to low- and intermediate-level radioactive waste or will find application in the final disposal of waste which is being stored in the Asse mine.

An advantage for the process shown can be seen, however, in that the repository sites for low and intermediate level radioactive waste are located in different host rocks: in rock salt, a ductile host rock, the ERAM and in an iron ore horizon (coral-lenoolite—sedimentary oolithic iron ore of the minette type), geological section of the Konrad mine, see Fig. 8.1, in order to gain experience with different host rocks.

Chapter 6 described some site-specific individual evidence, how it was presented at the different sites which is also of general interest.

8.4.2   Repository for Heat-Generating Radioactive Waste

In order to develop a repository for the BR Germany, a site search procedure was initiated in 1973, see Chap. 7. The then Federal Government carried out a site search for this purpose. In view of existing knowledge and experience from salt mining and research work at the Asse mine, the focus of the investigations was on evaporite rocks (rock salt). The advantages for salt include the high thermal conductivity and ductility of the rock salt. Since the rock salt deposits of the Federal Republic of Germany were mainly located on the territory of Lower Saxony at that time, the government of Lower Saxony carried out its own selection procedure parallel to the process of the Federal Government. This took into account the selection criteria adopted by the Federal Government. In February 1977, the then government of Lower Saxony finally named the Gorleben salt dome as the only site for a repository and a disposal centre. There is ample evidence that safety arguments, geoscientific criteria and ethical objections were not the only decisive factors. The Site Selection Act [9], completed the exploration of the Gorleben salt dome in July 2013 and stipulated that the salt dome, like any other site in Germany, would be included in a new site selection procedure. The previous investigations on the Gorleben site are summarised in "Preliminary Safety Analysis Gorleben (VSG)" [14] for further use. The site selection process is now to be carried out in an open-ended manner, including the Gorleben site.

8.4.3 Requirements for Long-Term Safe and Stable Storage and Possible Areas for HAW Repository Sites in Germany

The long-term safety proof must always be carried out on a site-specific basis, this is regardless of the individual federal state. Nevertheless, the conditions found in Germany will form the basis for further discussion. This enables the long-term safety proof to be compared with the proofs in other countries.

8.4.3.1 Site Selection Procedure

Location Search In Germany

Pursuant to Sect. 8.3 of the StandAG, a "Commission for the Storage of Highly Radioactive Waste" was set up to prepare the new site selection procedure, in which had to prepare a report that should comprehensively address all issues relevant to the decision. The Commission had to subject the StandAG to an examination and to submit corresponding recommendations for action to the German Bundestag and Federal Council, § 4 StandAG:

1. It is to assess and decide whether, instead of an immediate disposal of high-level radioactive waste in deep geological formations, other possibilities for an orderly disposal of this waste should be scientifically investigated and if the wastes in above-ground interim storage facilities are to be stored until the investigations have been completed,
2. for the basis for decision-making (general safety requirements for storage, geoscientific, water management and spatial planning exclusion criteria and minimum requirements with regard to the suitability of geological formations for final disposal as well as host rock-specific exclusion and selection criteria for the possible

host rocks: salt, clay and crystalline as well as host rock-independent consideration criteria and the methodology for the preliminary safety investigations to be
carried out), see also Baltes et al., [15].

3. for criteria for possible error correction (requirements for the design of
   the storage facility, in particular with regard to the issues of Reversibil-
   ity/Retrievability/Recoverability of radioactive waste as well as the issue of possible back tracking for some stages in the site selection procedure (Reversibility),
   see Chap. 9, Fig. 9.1),
4. for requirements for the organisation and procedure of the selection process and
   for the consideration of options,
5. for requirements for the participation and information of the public as well as to
   ensure transparency as well as to discuss socio-political and technical-scientific
   questions and make recommendations on how to deal with decisions taken and
   specifications made in the repository question and international experiences and
   conclusions to be drawn from this for a storage concept.

The final report, [7] EndKomm (2016), of the Commission on the storage of high
level radioactive waste has been published in the meantime. It is striking, that the
Commission's basic requirements for the repository site, which are laid down from
BMU [8], have not been subjected to a complex critical evaluation, see Chap. 7.
Although these represent the basic matrix for long-term safety and thus for the best
possible repository location:

– one or more repository sites?
– 500 years retrievability after closure of the repository structure,
– Observation period (detection period) 1 million years?
– Positioning of geotechnical barriers, containers, packages in the long-term safety
  concept,
– Deep geological formations or near-surface storage?
– Creation of transition storage?
– Implementation of the long-term safety assessment.

There may be doubts as to whether an open-ended selection procedure for the disposal of high-level radioactive waste can be derived from the Commission's instructions for action. Too many fields have already been excluded for the site search and
for the repository concept without a sound scientific analysis having been carried
out. Thus, essential and coordinated criteria, see BMU [8] and AkEnd [16, 17], were
ignored and watered down. However, it cannot be excluded that the new organisational structure in the field of radiation protection and final disposal will face up
to this task. This hope is nurtured by the fact that the Commission, in accordance
with Sect. 8.3 of the Standing Ordinance on Nuclear Safety and Waste Management
(StandAG), is focusing on reversibility and retrievability on a process path leading
to a repository.

Reorganisation of the organisational structure in the field of radiation protection and final disposal

According to StandAG § 7, a Federal Office for the Safety of Nuclear waste management (BfE) had be established. It provides for the state tasks of supervision and licensing in the field of nuclear technology, interim storages, site selection and repository surveillance to be concentrated mainly in the BfE. The service, legal and technical supervision of the BfE is exercised by the Federal Ministry of Environment, Nature Conservation and Nuclear Safety (BMU).

According to Section 9a (3) sentence 2 of the Atomic Energy Act, the project sponsor is the third party responsible for implementing the site selection procedure. In 2016, the "Gesetz zur Neuordnung der Organisationsstruktur im Bereich des Strahlenschutzes und der Endlagerunges" [18] [Act on the Reorganisation of the Organisational Structure in the Field of Radiation Protection and Disposal] came into force. A Bundesgesellschaft für Endlagerung (BGE) mbH serves as the project executing agency for the operational tasks of site search, construction and operation of the repositories and the Asse II mine. The fundamental contradiction remains that application, licensing, construction execution and supervision, release, operation and decommissioning with aftercare in Germany are bundled in the public sector.

International site selection

The recommendations of the International Atomic Energy Agency (IAEA) and the Nuclear Energy Agency (NEA) of the Organization for Economic Co-operation and Development (OECD) are of fundamental importance for the site search, planning, operation and decommissioning of radioactive waste repositories. From this work, central elements for the realization of a repository with a high level of safety also result in Germany.

Finland, for example has opted for direct disposal from the very beginning. At the end of 2012, the Finnish Nuclear Safety Authority STUK (Säteilyturvakeskus) at the nuclear licensing authority issued a positive safety assessment in January 2015 for the Finnish repository for HAW at Onkalo. On 12 November 2015, the Finnish Government approved the construction of the repository. The operating license application is due to be submitted in 2020. The storage at a depth of 400–700 m will take place from the 2020s in a granite formation (crystalline), see also Chap. 7 (Fig. 8.2).

In Finland, the responsibility for site selection and final disposal lies exclusively in the hands of the liable private companies. The state only acts in an authorising and supervising role, which it performs through the Radiation Protection Authority and the Ministry of Labour and Economic Affairs. The private company "Posiva Oy", in which the nuclear power plant operators—Fortum Oyj [40%] and Teollisuuden Voima Oyj (TVO) [60%]—together hold 100% of the shares, was created for the operational realisation of a central repository for spent fuel elements, see [19].

The following presentations on long-term safety follow the safety requirements for the final disposal of heat-generating radioactive waste from [8, 17, 16]; they can serve as support for solving of the task set out above.

Fig. 8.2 Test gallery (tunnel) in the host rock granite, underground research laboratory (URL) Onkalo/Finland. *Source* Posiva Oy

8.4.3.2   Basic Structure of the Suitability Proofs for a Repository System

The proof of suitability for a repository system in Germany is based on the primary objective of completely isolating the stored inventory by effectively sealing the emplacement area against water ingress.

The aim is to develop and optimise a geotechnical environmental structure designed as a multi-barrier system. Such a multi-barrier system consists of various system components, each of which has its own, specific part in the insulation of the waste. It can be assumed that no element of the multibarrier system is absolute and impermeable forever. It is therefore important the barriers so to put together, coordinate them with each other and equip them with immobilization in such a way that the total insulation capacity of the effective containment zone (ECZ) remains effective for as long as possible. For high-level radioactive waste (HAW), the barriers must be able to guarantee an isolation period in the order of one million years according to current specifications.

Implementation of the primary objective: It is therefore necessary to develop a repository system consisting of the repository mining plant, the ECZ and the surrounding or above layered geological layers up to the earth's surface and the biosphere with a multi-barrier system, in which the individual barriers are so out together, coordinated and equipped with immobilization in such a way that an overall insulation capacity of the ECZ is achieved for the entire period and as long as possible, in Germany currently for 1 million years. It is therefore necessary to demonstrate the long-term safety for the repository system with the repository construction and the individual barriers, natural (geological), geotechnical and technical, see Fig. 8.6. The demonstration of the long-term safety for the repository system is therefore based on the individual proofs.

If one follows the currently favoured proposal that heat-generating HAW in particular are to be deposited in deep geological formations, a repository mine will be excavated in the horizon of a suitable host rock formation. The first step is to find a suitable location, which is characterized by a known and suitable host rock formation. Internationally, three host rock formations are being considered for the disposal of radioactive waste, in particular heat-generating HAW: salt (salinar), clay (claystein) and crystalline rocks (granites). These rock formations are all found in Germany, while Scandinavia, for example, can only rely on crystalline rock formations. In arid zones such as in the USA (Nevada), where solution inflows to the storage areas are not expected to any considerable extent, it is also possible to access different formations, see Yucca Mountains—see Chap. 7, which, however, led to the consideration of the scenario of climate change.

Each of these rock types has its specific advantages and disadvantages with regard to the primary objective and thus on its suitability as host rock for the storage of heat-generating radioactive waste. In principle, safe final disposal is possible in all these host rocks, provided that the repository concept is adapted to the geological formations in question. From the author's point of view, at least two repository concepts in two different host rock formations have to be compared to find out whether one repository concept is better than the other. This makes it clear that an isolated site selection will always have the flaw that arguments will be found to reject it, without the general suitability of this site being refuted, because the assumption will be expressed that a repository concept can be found in a corresponding different host rock formations, which could be better suited.

The following individual proofs in the long-term safety demonstration are described in more detail below:

Selection of suitable repository sites (ENW 1),[9] at least 2 in different host rocks, which are then fed into the final selection procedure, for which the consent of the respective host communities should also be obtained. For the pre-selection of suitable repository sites in at least two different host rocks, a multi-stage selection procedure with various basic and comparison-specific criteria should be applied. The various basic and comparison-specific criteria have not yet been formulated in Germany.

At final disposal in deep geological formations, "rock preserving" cavities will be excavated, similar to a mine; as underground laboratory on the one hand in order to be able to carry out the corresponding rock investigations (geomechanical, geohydraulic, geothermal, geochemical, etc.) and on the other hand in order to be able to carry out the necessary geological investigations. Parameters to be investigated include: water-bearing, insulating layers, thickness of host rock formations, overburden and dirt bed, position of aquifers, fissures, etc. as well as for the construction of emplacement areas, drifts, chambers, conveyor shafts; chamber seal- and dam structures, etc. Thus the mine will be equipped, with various barriers provided to meet the primary objective. A mine equipped in this way is called a repository mine. The repository structure must be characterised by robustness and a high degree of integrity, see also [20].

---

[9]ENW 1-Single proof 1.

The stability and serviceability of the repository structure as well as the integrity of the geological barrier are prerequisites for the guarantee of the specified normal operation as well as for proving the permanent isolation of the waste for the post-closure phase.

Proof of robustness and integrity of the repository structure (ENW2): A repository system is designed as a multi-barrier system consisting of various system components, each with its own time-dependent specific contribution to the isolation and spreading prevention of the radioactive inventory. For the individual barriers, natural (geological), geotechnical and technical barriers, which must also be adapted to the selected host rock, the proof of integrity and robustness must be provided. Thus, repository waste containers, which are used in rock salt, have a greater robustness against acting rock pressure than that in the crystalline rock, because rock salt converges and thus the waste containers are completely enclosed in the post-closure phase of rock salt. Storage containers developed for rock salt are not generally suitable for crystalline host rocks and vice versa.

Proof of the integrity and robustness of the individual barriers of the multi-barrier system of a repository (ENW3)—effective containment zone (ECZ).

Examples of Single Proofs

- **Selection of suitable repository sites (ENW 1)**

The selection of suitable repository sites should take place in a multi-stage selection procedure, in which different fundamental and comparison-specific criteria are applied. These different fundamental and comparison-specific criteria have not yet been formulated in Germany.

For the final decision at least 2 locations in different host rocks should be evaluated. In the final decision, from the author's point of view, the consent of the host community should be sought if the process of construction is not to be significantly disturbed, as shown by the procedure for the Gorleben exploratory mine. But the development of a repository for civilian HG-HAW in the US—Yucca Mountain nuclear waste repository—is similarly politically influenced, see Chap. 7. For the development of a repository system in Germany, both comparable site-independent and site-specific criteria must be formulated. The development of a repository system must, in addition to the current safety philosophy, also allow for the reversibility of decisions by reversing decisions that have already been made in order to be able, if necessary, to switch to other disposal paths. In the BfS[10] safety philosophy presented in BfS [21], see Table 8.1, measures taken in the event of extraordinary developments, has introduced the reversibility of decisions and the reversibility of emplacement: recovery and retrievability, other than dangerous-situations-depending decisions.

---

[10]The tasks for final disposal have been transferred from BfS to BfE. This means that the BfE has also adopted the safety philosophy for the final storage of HG-HAW. I.e. of course not that the BfE will not develop this further.

Table 8.1 Development categories for the post-closure phase of a repository according to BfS [21], adjusted

Development Categories	Events and Processes	Measures
1	Expected developments	Selection of the site in a multi-stage procedure Safety distance to the loading limits of the geological barriers Sealing of the repository according to the state of the art
2	Exceptional developments	Evaluation of the safety reserves of the site Arrangement of complementary barriers Reversibility of decisions *)
3	Human Intrusion Scenario --------- Special cases --------- Climate change (Ice age, Warm period)	Transmission of information Location selection Design of the plant ------------------------------------- Location selection Recovery; Retrieval *)

* are not included in [21]

By **Reversibility** of decisions one understands once made decisions being undone and if necessary to be able to change to other disposal paths.

The term **Retrievability** refers to the act of retrieving the waste containers from the repository, even if it is partially backfilled and closed.

Under **Recoverability** we understand the possibility of recoverability of containers with HAW, if the repository mine is already completely closed.

Recoverability and Retrievability argue for a comprehensive investigation of near-surface disposal paths for the heat generating HAW. The exclusion of these disposal paths contradicts the listed principles from the final report of the commission according to § 3 StandAG.

Internationally, 3 host rock formations have proved to be fundamentally suitable: rock salt (Salinar), clay/mudstone, crystalline (granite, gneiss). In Germany it is

Table 8.2 Influence of host rock on system components of a repository according to BGR [22]

Host rock/ System components	Salt rock	Clay/Claystone	Crystalline rock (e.g. granite)
Emplacement floor	(600 – 1.000) m	(ca. 500) m	(400 – 1.300 m)
Type of emplacement	Drifts and Deep boreholes	Drifts or short boreholes	Boreholes or Drifts
Max. Design temperature *)	Max. 200° C	Max. 100° C	Max. 100° C (if Bentonite used)
Backfill- (Fill-)material*)	Salt grit	Bentonite	Bentonite
Necessary interim storage time (acc. max. design temperature)	Min. 15 years	Min. (30 – 40) years	Min. (30 – 40) years
Support systems	Not required	Required, if necessary very elaborate	Required, but not continuous
Concept of repository containers	Available (Gorleben)	Available, but not in German	Available, but not in German
Experience in the excavation of a repository mine	Available	Scarcely available	Available

favorable characteristic	unfavorable characteristic	neutral characteristic

* is adapted to the respective host rock

known where these can be found. Surprises are not expected due to the density and depth of exploration.

The investigation of the suitability of the repository formation is based on the investigation of the host rock typical parameters with regard to the expected effects and insolation properties. It is known that with the development of waste containers in connection with site-specific conditions, significant improvements in the insulation properties of the multi-barrier system can be achieved. Table 8.2 lists influences on the configuration of the repository system derived from the site selection and host rock formation, BGR [22].

The assessment carried out of the influence of the host rock on the system components of a repository by the BGR in 2007 seems outdated. The compilation of the BGR also appears to tend to make the rock salt (salinar) appear advantageous over clay/claystone and crystalline rock. Points of interest:

– Emplacement method: is completely unproblematic and can be optimized for the respective final storage configuration,
– Backfilling (filling) material*): The use of bentonite in the barriers: drifts, chambers, final storage containers appears to promote integrity, see Chap. 7,
– Max. Design temperature*): With the final repository configuration, the temperature influence of the HG-HAW on the host rock can be considerably reduced, see also [14],
– Necessary interim storage time: this is of no interest for the German repository concept because the commissioning of a repository in Germany is delayed to such an extent that the required interim storage time is fulfilled,
– Final storage container concept: e.g. in Sweden and Finland there are container developments (based on to the host rock granite) with very long isolation times, see also Chap. 7.

• Proof of robustness and integrity of the repository structure (ENW 2)

In a proof and safety concept, the first step is to provide here quantitative safety proofs (e.g. verification of serviceability, cavity stability, barrier integrity), for a repository structure. An essential requirement for long-term safety is that a repository system and its subsystems, such as repository structure and barriers, must be characterised by robustness. Robustness refers to the reliability and quality and thus the strength of the safety functions of the repository system and its barriers to internal and external influences and disturbances as well as the insensitivity of the results of the safety analysis to deviations from the underlying assumptions.

For the repository system and its subsystems, integrity must be guaranteed over the entire observation period (detection period). Integrity describes the preservation of the properties of the containment capacity of the effective containment zone of a repository, i.e. the repository system consisting of natural and technical and geotechnical barriers in their interactions, their redundancy and diversity. The first of the central safety requirements according to BMU [8] describes the task comprehensively: "For one million years it must be shown that at most very small quantities of pollutants can be released from the repository. For this purpose, the integrity of the effective containment zone must be verified and the risk emanating from the repository evaluated and presented". The observation period (detection period) is currently 1 million years.

The proof of robustness and integrity of a repository structure is a special comprehensive geotechnical safety proof, which not only represents a great challenge, but is also of great importance and requires a great deal of special experience. This always involves the holistic consideration and evaluation of theoretical, experimental and visual investigations (e.g. geological and engineering geological investigations, laboratory tests on the material behaviour of the rock strata, determination of the thermal-hydraulic-mechanical effects, rock-mechanical modelling, numerical model calculations, inclusion of in situ measurements and local observations if necessary). The model calculations are of particular importance, since the long-term behaviour of repositories in the post-closure phase can no longer be described and assessed by measurements and observations, but only by theoretical prognosis models including thermal, hydraulic, mechanical and chemical effects (coupled THMC processes). The correlations are not only highly non-linear but also are stochastical in nature. At time t, the material properties characterise a state of the geotechnical body which is subject to changes in the course of time. The material properties are distributed statically over, or in, the rock area under consideration. The effects are associated with probabilities of occurrence, which often do not follow any laws. The proof of the robustness of a repository structure can be investigated according to the method described in Fig. 8.12. According to the current understanding, the repository waste containers deposited in the storage fields are initially completely filled with salt grit. The emplacement fields are then closed. The convergence of the surrounding rock mass exerts considerable pressure on the waste containers. The backfill assumes the same strength as the surrounding rock. The investigation of the convergence behaviour of the rock is part of the methodology according to Fig. 8.3.

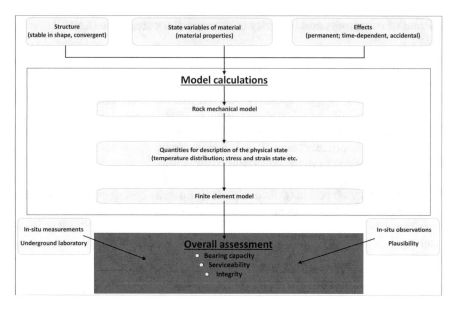

Fig. 8.3 Methodological procedure for geotechnical safety proofs, according to a Federal Institute for Geosciences and Natural Resources (BGR) proposal

The methodical procedure shown in Fig. 8.3 can also be applied to the integrity verification. However, it must be adapted accordingly. Model calculations are then used for hydrological and mass transport models.

A hydrological model represents a simplified description, that realistically describes the physical, chemical and biological processes (hydrological process) or sub-processes taking place in the hydrological system. The stated variables put into the hydrological model include the permeability of the rock element; the dynamic viscosity and density of the fluid; the thermal conductivity and the specific heat capacity of both as well as the rock element and the fluid (advection and diffusion), the milieu conditions (temperature, pH value, etc.).

These stated variables are derived from the area-differentiated investigations of the rock and fluid properties (distribution of parameters). For the consideration of random changes in the hydrological system, a distinction has to be made between stochastic and deterministic models and between linear and non-linear models according to the dependence, as well as after consideration of the dependence of the model parameters from temporal system behaviour, in time-invariant and time-variant models.

Table 8.3 lists some repository-relevant properties of potential host rocks according to BGR [22]. These are applied in the integrity investigation of the repository structure.

Table 8.3 also shows that this compilation tends to make the host rock salt (salinar) appear advantageous over clay/claystone and crystalline rock. If fissured and strongly fissured granite formations, which are not included under "best possible" in the pre-selection, are removed from the evaluation, this formation obtains further, favourable,

Table 8.3 Repository-relevant properties of potential host rocks compiled according to BGR [22]

Host rock/ Property	Salt rock	Clay/Claystone	Crystalline rock (e.g. granite)
Thermal diffusivity	high	low	middle
Permeability	practically impermeable	very low up to low	very low (unfractured) up to permeable (fractured)*)
Strength	middle	low up to middle	high
Deformation behaviour	viscous (creeping)	plastic up to brittle	brittle
Shape stability of cavern	Inherent stability	Supporting system necessary	high (unfractured) up to low (strong fractured)*)
In-situ stresses	lithostatic isotropic	anisotropic	anisotropic
Dissolution behaviour	high	very low	very low
Sorption behaviour	very low	very high	middle up to high
Temperature resistance	high	low	high

	favorable characteristic		unfavorable characteristic			characteristic

*) a fractured or strongly jointed granite base would fail in the selection of locations, because this does not fall under the category "best possible"

properties. From area-differentiated geotechnical investigations of the rock and fluid properties it becomes clear that with a considerable bandwidth in the distribution of parameters, both location- and time-related, anomalies are to be expected. It has happened here, before during detail investigations that there is a need to further investigate the area by appropriate geophysical field studies.

When developing the security and proof concept, suitable methods for the systematic handling of inherent uncertainties must also be applied. These uncertainties affect inter alia not only the present location data, see Table 8.3, but also the future development of the repository system over the long periods to be considered, and the knowledge of the processes in progress. Some of the uncertainties that arise when describing the processes in progress are the result of investigations obtaining the necessary data required for modeling, especially in the parameterization of the models used. For their treatment, these data and parameter uncertainties are identified in connection with the description of the corresponding models or treated during the evaluation of model calculation results. To take account of data and modeling uncertainties, probabilistic analyzes need to be carried out. For example, probabilistic methods are used for taking into account parameter uncertainties in the assessment of the inclusion of radionuclides in the ECZ and possibly in the radiological consequences analysis and partial safety factors in the consideration of uncertainties of the load cases for the geotechnical structures, GRS [23].

In addition, a repository concept should take into account the uncertainties of the occurrence, timing and characteristics of future events at the site, for example in connection with future cold periods. The occurrence of such events is to be expected with a probability, so that with the systematic handling of inherent uncertainties a failure probability of the repository system can be identified.

The determination of a failure probability of repository systems is currently not part of the standard design procedure. However, it is not unlikely, that the probabilistic model calculations might also lead to opening up this area for consideration.

- **Probability of failure of the repository system or its subsystems**

A failure probability for both the repository system or its subsystems points to remaining risks (residual risk) in a probabilistic investigation. The probability of failure results from the probability of the occurrence of the unplannable, extraordinary events and the probability that two (several) of these events can occur simultaneously. Some principles for their determination are given below.

In order to be able to evaluate the safety of a repository site with a probability measure, two conditions are introduced which are mutually exclusive, i.e. complementary to each other. These are, on the one hand, the failure event, which is designated V, and, on the other hand, its complementary event, the non-failure, which is designated $\tilde{V}$. There is then the difference quantity V to $\tilde{V}$.

Both events are assigned corresponding probabilities: The probability of failure (P_f) and the probability of surviving (reliability) (P_r).

The following applies here:

$$p_r + p_f = 1 \tag{8.1}$$

The prerequisite for this is the safety, in this case of a repository structure, and to be able to evaluate with a probability measure, which enables the formation of a mathematical relation between its input (action process) and its output (action process) and for which a sufficiently large amount of statistical information on the basic variables is available. The condition of the repository structure depends on a number of variables that are considered to be statistically detectable. These are also called base variables (or random variables). For cavity stability, these include the mechanical properties of the host rock, the effects, e.g. the static equivalent loads, the geometry of the cavity and the geometric imperfections, the effects from the environment, overburden, etc., as well as the effects of the host rock. The state variables can be summarized in a vector vekX.

$$\text{vekX} = \begin{pmatrix} X_1 \\ X_2 \\ X_3 \\ \dots \\ X_n \end{pmatrix} \tag{8.2}$$

Using suitable probability theoretical approaches, it is not formally difficult, to determine the probability measure for the occurrence of a certain, clearly defined limit state g(vekX) for a mechanical calculation mode,

$$P_f = P(g(X_1, X_2, \dots, X_n) \leq 0) = P(Z \leq 0), \tag{8.3}$$

$$P_f \leq 0 - failure; \; \underset{\leftarrow}{\rightarrow} \quad P_f > 0 - \text{no failure}$$

that is, the probability that the failure condition is satisfied or the security zone is negative.

Z is referred to as the safety zone, X_i as random variables whose probability distribution F_i and -densities f_i are known.

$$Z = g(X_1, X_2, \ldots, X_n) \leq 0 \qquad (8.4)$$

In n-dimensional space, the boundary state function $g(vekX) \equiv 0$ represents an $(n - 1)$-hyperface which divides the whole space into an unsafe, i.e. failure area $\{vekX \, | g(vekX) < 0\}$ and a safe area $\{vekX \, | \, g(vekX) \geq 0\}$. Geometrically, is a limit state the surface of the safe area.

At many boundary states it is not difficult to consider the resulting resistance R and the resulting action S as stochastically independent partial states in the state function, so that the two-dimensional example calculation shown in Fig. 8.13 results. If the base variables are random variables or random processes, this also applies to R and S. It applies:

$$g(vekX) = R - S = X_1 - X_2 \qquad (8.5)$$

The difference $Z = R - S$ is called the safety distance. If $Z \equiv 0$, then the limit state is reached. If $Z < 0$, failure is present; if $Z \geq 0$, this is not the case. According to the above considerations, the safety distance can also be understood in its most general interpretation as a random variable or random process $Z \equiv g(vekX)$, see Fig. 8.4.

If one set e.g. earthquakes and tectonic events as static equivalence loads, one can consider the calculation method as a calculation of probability of failure static systems with random, stochastically dependent properties and under random effects defines, see Spaethe [24].

The correlations presented above can be regarded as the basis of probabilistic safety analysis (PSA) or probabilistic risk analysis (PRA), which investigates the risks of repository structures using methods of probability calculation and systems analysis, see further Sect. 8.4.2. A number of probabilistic models have become known with which risk analyses of repository structures can be performed, GRS [25].

It is important in probabilistic calculations that a very high sensitivity of the theoretical failure probability to changes in the selected stochastic models can be demonstrated, see also Sect. 8.4—Sensitivity Analysis.

- **Proof of the integrity and robustness of individual barriers of the multibarrier system of a repository system (ENW 3)**

According to BMU [8], the search space for a repository is currently specified as follows: "The safety requirements specify, which safety level a repository for heat-generating radioactive waste in deep geological formations has to comply to fulfill the

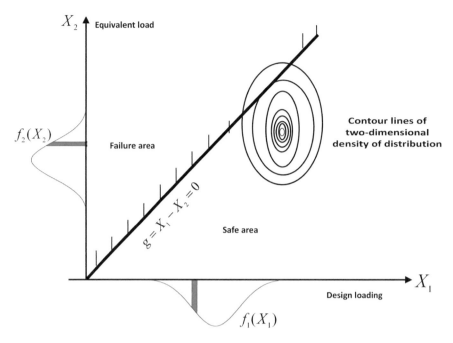

Fig. 8.4 Calculation of the failure probability P_f for a two-dimensional example calculation according to Spaethe [24]

nuclear requirements. This safety level is not determined solely by general protection objectives and protection criteria, …. These safety requirements are based on a repository concept in which the radioactive waste is permanently stored of in a deep geological formation with high isolation capacity". According to this concept, the effective containment zone is defined as follows: "… is the part of the repository system which, in conjunction with the technical closures (shaft closures, chamber closure structures, dam constructions, backfilling, …), ensures the containment of waste".

A long-term proof of safety shall be based on the central safety requirements according to BMU [8]. From today's point of view, it is not possible to assess with certainty whether these requirements will ultimately be implemented, so it should be pointed out here what can also be assessed differently without reducing the safety level of the repository for heat-generating HAW. Near-surface repositories are excluded according to the currently valid philosophy. Whether this can be maintained in the long term is currently uncertain.

The long-term safety proof for a repository in deep geological formations must therefore be carried out for a multibarrier system and must meet the requirements of redundancy and diversity. Depending on the host rock and the type of waste, adapted multibarrier systems can be developed for suitable sites.

Barriers are natural or (geo)technical components of the repository system. Barriers are, for example, the waste matrices, the waste containers, the chamber and shaft closure constructions, the effective containment zone (ECZ) and the geological layers surrounding or overlying this ECZ.

A barrier can perform various safety functions. The safety function of a barrier can be a physical or chemical property or a physical or chemical process. For example, the obstruction or hindrance of access of liquids to the waste or the protection of the effective containment zone from erosion can all be considered safety functions.

For a repository in salt, for example, the barrier system would look different compared to a site in granite or clay and, in particular, for a repository near the surface. Salt as a host rock has the advantage that the geological conditions, in particular the dryness, provide a much greater barrier effect than with a repository in granite, through which water flows. The geotechnical and technical barriers therefore play a much more important role in granite than in salt. Dissolved nuclides are retained relatively well in the claystone. Due to the moisture present in the clay, however, the requirements for the final storage container are also significantly higher than for salt.

- **Natural Barriers**

The mode of action of the natural barriers is determined by the isolation potential of the host rock. The behaviour of rocks and deposits in nature, even over long geological periods, has been well researched. We therefore know that the retention capacity of different types of rock (salt, clay, granite) is different. This is just as important for the selection and planning of a repository as the comprehensive investigation of geology at a specific site, see Fig. 8.6.

Barrier assessment host rock:

- Reducing conditions (no free oxygen), i.e. no or only minor corrosion of repository containers
- Concentrated and heavier salt solutions in relation to water in deep geological subsoil above the host rock
- Different barrier effect of formations with different water permeabilities located above each other (e.g. in overburden).

- **Technical and geotechnical barriers**

- Conditioning, with the aim of reducing the volume and immobilising the pollutants
- Packages, containers, final disposal containers
- Mineral mixtures such as bentonite, which keep incoming water away from the containers or those which bind nuclides dissolved in the water/immobilisates
- Tight backfill material for filling boreholes and chambers/volcanic ash (industrial bentonite)/immobilizing materials
- Borehole or chamber seals
- Drift closure structures, dam constructions, see Fig. 8.5
- Shaft filling
- Shaft seals.

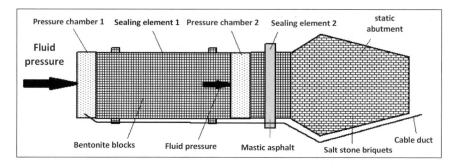

Fig. 8.5 Experimental shaft seal construction (dam) according to SITZ [26]

The interaction of natural and geotechnical barriers is determined on the one hand by the host rock and on the other hand by the safety requirements resulting from the safety analysis. The search for a suitable site, reduced to a suitable host rock, is therefore also dependent on the state of development of the technical and geotechnical barriers. Here man has the greatest potential for influence. The argument that natural barriers have proven their barrier effect over a long period of time must be countered by the fact that technical and geotechnical barriers have not yet been given the time to do so. It would be unfounded to deduce from this that these barriers cannot develop a long-term barrier effect. Rather, the development shows that they have an important function in isolating the radioactive inventory.

There are counterarguments that the barrier effect (barrier parameter) is lowered by the heat-generating waste, such as:

- Permeability increases can occur when existing pathways in a barrier rock or in the material of a technical barrier are widened due to thermal volume changes or if, in the event of a hindered change in volume, tensile stresses occur which form new networked paths, when local strength overshoots occur.
  The advective transport of substances is made easier by a lower viscosity.
- Sorption capacity: The sorption capacity of the bentonite introduced plays a role in the host rock types clay and crystalline. It must therefore be ensured in these concepts that the necessary sorption capacity is not impaired by thermally induced mineral transformations.
- Stability of the glass matrix, the transition temperature of glass coquilles should not be exceeded, etc.

It is hereby reiterated that heat generation of the waste can be reduced to such an extent by transmutation, decay storage or lower emplacement density of the heat-generating waste such that negative effects on the barrier are limited or even excluded and the barrier effect is maintained throughout the observation period. As evidenced by the building permit for the Finnish repository Onkalo for HAW, in which copper containers are used with an enveloping layer of prefabricated bentonite rings, the second intent has already been taken into account (Fig. 8.6).

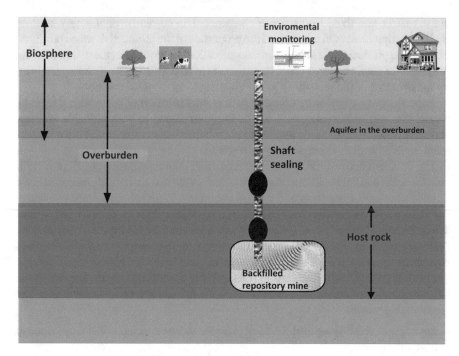

Fig. 8.6 Sealed repository mine in the post-closure phase

- **Geotechnical barrier—final disposal container**

The multi-barrier system of a repository consists of natural (geological) and geotechnical barriers. The geological barriers are essentially determined by site selection. The geotechnical barriers are adjusted in an iterative process. For example, [8] specifies as a protection criterion that for the post-closure phase it can be shown that for probable developments, as well as for less probable developments, release of radionuclides originating from the stored radioactive waste, may result in additional effective dose in the range of only 10 microsieverts per year for individuals of the population.

In the case of uncertainties, such as input data for the consequence analysis, the type and location of the impairment as well as their temporal classification must be determined. In the case of partial failure of the geological barrier and the related likely developments, such as the inflows of surface-, overburden- and formation solutions, it should also be examined whether and to what extent the sealing system is so impaired, so that a release of dissolved radionuclides in the biosphere cannot be excluded. A special task is assigned to the repository waste containers, see also [27, 28].

For the final disposal of the heat-generating HAW, these are conditioned and waste packages consisting of radioactive waste product and waste container are produced, see Chap. 2. After conditioning, the radioactive waste product is enclosed by the waste matrix and fixed in the waste container. The currently valid safety concept BMU [8] stipulates: "On the basis of the properties of the radioactive waste produced or still to be produced and the appropriate conditioning procedure, the safety-relevant properties of the final stored waste packages are to be derived from the safety analyses by the operator of the repository and implemented as disposal conditions".

A particularly noteworthy criterion here is the exclusion of criticality: it must be shown that self-preserving chain reactions are excluded in both probable and less probable developments. This applies to each individual container and the emplacement areas as well as the repository as a whole.

For the Gorleben repository concept, repository container developments are available for drifts and borehole emplacement as well as for a so-called difference analysis, in which the use of transport and storage containers (TLB) as repository containers for Pressurized Water Reactor (PWR), Boiling Water Reactor (BWR) and Water-Water Energy Reactor[11]-fuel elements was considered. These transport and storage casks are approved for the transport and interim storage of spent fuel elements and reprocessing waste. The proof of suitability of such casks for final disposal has not yet been provided. Self-shielding POLLUX® and CASTOR® casks, for which the following requirements apply, are provided for the storage of fuel assemblies and reprocessing waste:

– mechanical stability against the rock pressure to be expected in the emplacement horizon
– gas-tight containment of the radioactive substances packed in the final storage container
– sufficient shielding of the radioactive substances packed in the final disposal container during transport, intermediate storage and handling
– Long-term stability to corrosion until other barriers ensure safe confinement, see GRS [29].

The POLLUX casks, with the dimensions listed in Table 8.4, were developed by the enterprise GNS mbH for the direct final disposal of spent fuel elements in salt. The POLLUX is equipped with a forged steel inner cask, which is designed to receive compacted waste and ensure the containment of the radionuclides. A second outer container made of spheroidal graphite iron GGG 40 serves to shield the radioactive radiation, but has no sealing function and therefore only has a comparatively simple screw cap. It seems extremely unlikely that Pollux containers will ever be used in this development. For example the long-term stability against corrosion is insufficient, to meet the criteria of retrieval and recovery. These final storage containers effect only a low long-term isolation of the radionuclides.

It should be noted that the previous German repository container concept does not meet the requirements for the crystalline host rock formation (Fig. 8.7).

The contribution of the waste matrix and the container to the isolation of the radionuclides consists in limiting the release of the radionuclides from the waste

---

[11] Russian reactor type.

Table 8.4 Dimensions and masses of final storage casks for drift emplacement according to GRS [29]

Dimensions/Final disposal containers	Length/Height (mm)	Diameter (mm)	Package-volume (m³)	Max. package-mass (Mg)	Transport package-volume (m³)
POLLUX®-10 (POLLUX®-9*)	5517	1.560[a]	10.55	65	10.55
CASTOR® THTR/AVR	2784	1.380[a]	4.16	26	4.16
CASTOR® KNK	2784	1.380[a]	4.16	26	4.16
CASTOR® MTR 2	1631	1.430[a]	2.62	16	2.62

[a]lifting lugs; *) assumed equal dimensions

Fig. 8.7 General construction of a double-jacketed copper waste container. *Source* Raiko et al. [30]

1,050 mm

4,835 mm

matrix and the container into the effective containment zone. A basic requirement is that the waste matrix currently meets the criteria of long-term stability. A release of radionuclides from the waste matrix can only begin when the waste container has already failed. A release of radionuclides from the waste matrix, however, requires water inflow at the waste matrix. Water ingress to highly radioactive waste leads to corrosion of the waste products in the long term.

However, it is now known that the resulting corrosion products can also bind radionuclides and thus effectively prevent their mobilisation. It will therefore be necessary to investigate the possibilities of adding immobilizing agents to the container that largely prevent or considerably delay the mobilization of the radionuclides. A number of immobilizing agents are known, many of which do not fulfil the requirement of corrosion inhibition, see also vitrified waste.

Vitrified waste from the reprocessing of fuel elements from German nuclear power plants is also to be transferred to a repository for heat-generating HAW. If aqueous solutions come into contact with the glass matrix, corrosion will also occur. The question of which container solution can be found for use with the vitrified waste is still open. In principle, the same procedure could be used as for the fuel rods.

A platinum-catalyzed, room temperature-curing silicone foam DOE [31] was used for the conditioning of wastes containing a high proportion of nitrates and chlorides. For this purpose, the contaminated waste is mixed with the silicone foam and poured into a container in the liquid state. The investigations carried out so far show that the mixture produced is highly resistant to leaching, which is necessary for long-term stability. In addition, good compatibility of the salt-containing waste with the polysiloxane material was determined, see [32]. Internationally, container development, as a technical barrier in connection with storage technology, is given outstanding importance because it can provide a high proportion of insulation for the radiotoxic material. This is related to the fact that even nature does not offer an immaculate host rock, even if the impression arises from some expert reports that this the case.

An interesting solution has become known from Sweden and Finland, which in principle can also be transferred to Germany, Bollingerfehr et al. [33]. The spent fuel elements are placed in a double container with a welded cap. The inner cask is made of corrosion-resistant steel to ensure mechanical stability, while the outer cask is made of copper to ensure chemical stability. The stability of steel containers is assessed at 1000 years [34, CROP], see also Table 8.6; that of copper containers at least 10^5 years [35, NAGRAb].

The copper container described above is surrounded by an enveloping layer of prefabricated bentonite rings. Water entering the container first reaches the swellable bentonite barrier and cannot therefore reach the double container. The retrievability of this container type was investigated by Pettersson [36]. The final selection of the material should be made according to the site-specific boundary conditions, see also Chap. 7.

The containers are accepted in accordance with the acceptance conditions of the repository operator and placed in prepared emplacement chambers. The protection of the personnel in the repository plays a special role here.

- **Interim storage**

The fuel rods removed from the cooling pond are packed in a prepared Castor container and taken to the interim storage facility. There are different types of Castors.[12]

---

[12]Castors—cask for storage and transport of radioactive material, [37].

Fig. 8.8 CASTOR® V/19
containers. *Source* GNS [39]

The CASTOR® V/19 cask is designed for the transport and storage of irradiated fuel assemblies from Pressurized Water Reactors (PWR). The fuel assemblies are at 400°–500° when loaded into the castors. The casks are therefore equipped with cooling fins to ensure that the surface temperature does not rise above 85°. Moderator rods are also installed in the container walls to slow down the gamma and neutron radiation.

The loading capacities is max. 19 PWR fuel elements. Total heat output is 39 kW, total activity: 1900 PBq. The Castor container is approx. six metres long, has an outer diameter of approx. 2.5 m and a wall thickness of 42 cm. This Castor type consists of stainless steel and weighs approx. 126 tons when loaded, see Fig. 8.8. The CASTOR® V/19 has all necessary permits for transport, permits for long-term interim storage as well as for Loading capacity in nuclear facilities, see [38] www. gns.de Product info. The CASTOR® V/19 container is not suitable for final disposal.

Irrespective of whether or not interim storage facilities are assigned to a final storage concept, in Germany the duration of the approval of interim storage facilities is 40 years according to Atomic Energy law (AtG), measured from the first storage of a container which will not be sufficient until a repository for heat generating HAW is available in Germany, 12 interim on-site storage facilities are available at the nuclear power plants and two central interim storage facilities at Ahaus and Gorleben. The permit for Ahaus expires in 2032 and that for Gorleben in 2035. The permits may only be extended for irrefutable reasons and after prior referral to the German Bundestag. From the author's point of view, the fact that there is no public participation here is in stark contradiction to the remarks of König [40], who ascribes the interim storage facilities to the German waste management concept. In the meantime, all interim storage facilities have been transferred to the Federal Republic of Germany.

It is proposed here to construct 3–5 shallow interim storage facilities in caverns or bunkers, whereby at least one at the same time could be a pilot project for a shallow final disposal. For further details, see Chap. 9.

The Castor containers placed in interim storage have to ensure a prescribed shielding, both for transport and for storage. In [39], the IAEA has laid down the framework for national regulations for this purpose. For this purpose, the surface local dose rate of the containers is limited. For the on-site interim storage facilities, containers only receive a permit for a maximum permitted surface local dose rate of 0.35 mSv/h.[13,14] The Castor containers, however, shield the neutron radiation relatively poorly. A second limit value applies to shielding in Germany, namely that the radiation exposure at the fence of the interim storage facility should not exceed 0.3 millisieverts per calendar year.

It is to be assumed that other stricter limitations will be prescribed for the final storage containers. The decisive factors for this are probably the long operating period of the repository and the 500 years retrievability currently specified, which would result in unacceptably high loads for the personnel at unmodified surface dose rates, see Chaps. 2 and 3. However, there is currently no proof method for long-term safety, which that also guarantees compliance as far as possible.

The storage chambers are located in the effective containment zone (ECZ) of the repository. They are equipped as transition locations (intermediate disposal). It can be assumed that in the case of an open-ended site search, even near-surface repository solutions must be subjected to a detailed scientific investigation before a final decision can be made, especially if the transmutation with the aim of deliberately changing the radioactivity of the inventory is integrated into a repository concept, see Chap. 7. However, the [7] EndKomm (2016) proposes not to pursue transmutation further in a repository concept with the justifications that this is an independent treatment procedure for HAW and it is too long term and too expensive, see also Chap. 7. It must be pointed out that transmutation can only be a partial aspect of disposal. First of all, it must be assumed that no quick solution for HAW can be found in Germany and that it is not necessary to find one. But, it is to be expected that the safety requirements for a repository site may lead to equipment requirements that far exceed the costs for transmutation. Thus [7] EndKomm (2016) has determined a limitation of the outside temperature of the containers at 100 °C until other proofs are available, without knowing the effects on the host rock and its physically maximum

---

[13] An one-hour stay immediately adjacent to the container results in a dose of approximately 0.35 mSv. This corresponds to around one-seventh of the normal annual radiation exposure of the population of 2.5 mSv.

[14] Of these, a maximum of 0.25 mSv/h may be caused by neutron radiation.

tolerable temperature as well as the resulting requirements. Due to the longer interim storage periods and the lower storage density, the temperature in Germany falls below 100 °C anyway. The maximum limit temperature in the respective host rocks and geotechnical barriers is thus becoming less important, of course because transmuted waste does not cause a heat-induced process, see BGR [41]. The extent to which international P & T capacities for the transmutation of the German HAW can be used has not yet been exhaustively investigated, see Chap. 7. For the sake of completeness, it must be noted here, that near-surface repositories, supported by the longer interim storage periods and a lower emplacement density, could also be considered more closely.

Comprehensive Safety Proof for All Operating States

Safety proofs for geotechnical structures (tunnels, dams, caverns, etc.) depend on the risks and threats they pose to "innocent bystanders" and future generations. Accidents or failures in risk technologies (structures) produce irreversibilities (deaths, contamination of land, water and air, genetic damage to humans, animals and plants) that invalidate the expected handling of technical accidents, namely that they are repairable and reversible. In addition, even the normal operation of risk systems requires safety precautions that far exceed the organisational requirements of the environment to these technical systems. According to the author's assessment, the interim storage, the transport of heat developing HAWs to be finally stored, the conditioning plants, the operating phase and the decommissioning phase can be absolutely regarded as risk systems. In the post-closure phase, a repository is assessed as a system with development uncertainties, but its probability of occurrence is so low that irreversibility as a result of failure can practically be ruled out. The following safety proofs can also be derived from this, Halfmann and Japp [42]. According to the current deterministic repository philosophy, the probability of occurrence of failure is not considered.

According to BMU [8] "a comprehensive safety proof shall be performed for all operating conditions of the repository including the above-ground facilities. In particular, plant-specific safety analyses shall be performed for emplacement operation and decommissioning, taking into account defined design basis accidents which prove the protection of operating personnel, population and environment required by the Radiation Protection Ordinance. This includes analysing and presenting the robustness of the repository system. Furthermore, for the safety-related systems, sub-systems or individual components, the respective probabilities of impacts, failures or deviations from the expected case (reference case) shall be calculated or estimated as far as possible and the effects on the respective associated safety function shall be analysed".

Proof of Long-Term Safety for the Repository System

Long-term safety is the condition of the repository structure for which the relevant safety requirements are met after decommissioning. Proof of long-term safety is essential before any substantive determination is taken, and a comprehensive, site-specific safety analysis and safety assessment covering a period of one million years should be performed.

Numerical analyses of the long-term behaviour of the repository shall be performed for:

– Integrity of the effective containment zone (ECZ)
– Radiological consequences of the escape of radionuclides from the inventory of the ECZ
– Mobilisation of natural radionuclides from the geological barrier, the overburden and dirt bed
– The properties of container and backfill as part of the geotechnical barriers
– The properties of the sealing constructions.

According to the current repository philosophy, deterministic calculations are to be carried out on the basis of modelling that is as realistic as possible (e.g. median values as input parameters).

As results of these numerical calculations the following outcomes are expected:

– Demonstration of the expected system behavior
– Derivation of time-dependent requirements, if any, for the components of the repository system
– Optimization of the repository system.

In addition, uncertainty and sensitivity analyses must be carried out in order to show the possible solution space and to be able to assess the influence of the uncertainties. The model uncertainties have to be considered. Compliance with numerical criteria, taking into account the uncertainties resulting from or derived from these safety analyses, determines the reliability of the numerical results. These and any numerical failures of these criteria resulting from the analyses shall be evaluated for their relevance.

For the period of the optimisation process in which the uncertainty in the input data and calculation models is too great, reference models (e.g. reference biosphere) should be applied if necessary. For this period and in the final evaluation, qualitative arguments should also be consulted.

8.5 Elements of a Long-Term Safety Assessment

8.5.1 Analysis of the Repository System (System Analysis)

The core of the repository system analysis is the long-term safety analysis. This is the analysis of the long-term behaviour of the repository after decommissioning. It comprises all the information, analyses and arguments which prove the long-term safety of the repository and must explain why confidence in this evaluation is justified. The system analysis shall include the following key areas of analysis:

- **Long-term statement on the integrity of the effective containment zone (ECZ)**

The integrity of the effective containment zone is proved for probable developments on the basis of a long-term geoscientific forecast. This must be ensured over a period of one million years. For this purpose, the effective containment zone must be clearly defined both spatially and temporally.

- **Radiological long-term statement**

For probable and less probable developments it has to be proven that according to the current German safety concept BMU [8] the following criteria are met:

(a) "For the post-closure phase, it shall be demonstrated that for probable develop-
    ments due to the release of radionuclides originating from the stored radioactive
    waste, only an additional effective dose in the range of 10 microsieverts[15] per
    year can occur for individuals of the population. Individuals with today's life
    expectancy who are exposed during the entire lifetime should be considered";
    ICRP [43].

(b) For less likely developments in the post-closure phase, it shall be demonstrated
    that the additional effective dose caused by the release of radionuclides from
    the stored radioactive waste does not exceed 0.1 millisieverts[16] per year for
    the people affected thereby. Individuals with a present-day life expectancy who
    are exposed during the entire lifetime must also be considered, ICRP [44]. To
    the extent that sufficiently reliable statements can be made for a period of one
    million years about the effectiveness of safety functions of the overburden and
    dirt bed of the effective containment zone, these can be included in the long-term
    radiological statement.

- **Long-term robustness of technical components of the repository system**

The proof of the robustness of technical components of the repository system must be predicted and presented on the basis of theoretical considerations. If technical barriers take on important safety functions with regard to long-term safety and are subject to special requirements and there are no recognised rules of technology for

---

[15]See Chap. 2.

[16]The effective dose is higher here because of the low probability of occurrence, thus lower risk for individuals (risk $< 10^{-5}$/a), see also Chap. 2.

this, their manufacture, construction and function must be fundamentally tested. These investigations are currently not conducted in Germany, in particular not for different host rocks.

- **Exclusion from criticality**

It should be shown that self-sustaining chain reactions are excluded in both probable and less probable developments.

- **Control- and evidence collection program**

A control- and evidence collection program during emplacement operation, decommissioning and a limited period after decommissioning shall be used to demonstrate that the input data, assumptions and statements of the safety analyses and safety demonstrations performed for this phase are complied with. In particular, this measurement programme shall record the effects of the thermomechanical reactions of the rock on the heat-generating wastes and the technical measures as well as the rock mechanical processes. An extrapolation to an observation period of 1 million years cannot be scientifically proven from the current point of view and is therefore subject to considerable uncertainties. A long-term monitoring programme should therefore also be included. With the maintenance of an underground laboratory (transition storage) for 500 years, a sufficiently long period would be defined, which goes hand in hand with the period of retrievability to be guaranteed.

A central aspect is the analysis of the isolation capacity and reliability of the repository system. It includes, for example, the development of conceptual models, scenario development, consequence analysis, uncertainty analysis and comparison of the results with specified safety principles, protection criteria and other proof requirements.

The requirements for long-term safety analyses are

- the collection of data (site surveys, experiments, research and development),
- ensuring that the requirements for technical and geotechnical barriers are met,
- the determination of safety-relevant processes and events—scenarios development (natural and technical analogues, site- and plant-specific research),
- the scientific description of these processes and events (basic research, natural analogues, experiments, in situ experiments),
- the implementation of adequate models,
- the qualification of the models (natural analogues, experiments, in situ experiments, reviews) and the associated computer programs,
- the qualified treatment of the problem as well as an evaluation by arguments and analyses.

Uncertainties occur in each procedural step of the long-term safety analysis, which must be presented and taken into account in the decision-making processes, see Fig. 8.9.

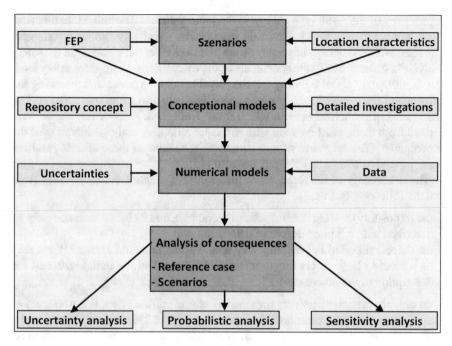

Fig. 8.9 Main steps of a long-term safety analysis according to BMU [8]

These uncertainties result from, Baltes et al. [16], as a consequence:

- the limited knowledge of geological development, the given geological conditions in situ and the characteristics of sub-areas, siting regions and sites,
- a full of fissures knowledge of the safety relevance of geological conditions,
- the limited predictability of geological developments in view of the duration of the periods under consideration,
- the different quality of the respective data (input data) for the comparative assessment of sub-areas, siting regions and sites.
- From the current point of view, the geological conditions represent boundary and initial conditions, so that the numerical calculation methods are methods with variable boundary and initial conditions. The prediction of geological developments at the site is also a prediction of boundary and initial conditions, which ultimately cannot be verified.

The methodology of the long-term safety analysis in its main steps is summarized in Fig. 8.9. Besides analyses for the evaluation of uncertainties in the data and models (uncertainty analysis), for the evaluation of the robustness of the repository system against changes of data or boundary conditions, determinations regarding the sensitivity to individual input parameters (sensitivity analysis and probabilistic analysis) must also be carried out. A sensitivity analysis is a calculation method, which delivers a certain result with fixed input values. In a sensitivity analysis, the dependency of the result on a change in the input values is determined. In this way, those input values to which the result reacts most sensitively can be determined. The sensitivity and uncertainty analysis are therefore essential elements in the safety analysis in order to verify the statements of the calculations. Long-term safety analysis is ultimately about pointing out the possible solution space and assessing the influence of the uncertainties. Model uncertainties are also taken into account. The observance of numerical criteria which result from these safety analyses, or were derived from them, must be done with sufficient reliability and consideration of the uncertainties. The relevance of any numerical variations of these criteria resulting from the analyses must be assessed, see also [45].

The uncertainty of the forecasts of future developments inevitably results from that the following factors:

- the (repository) system is extraordinarily complex and its initial state can only be described with a limited degree of detail, and
- the detection period is generally very long, so that temporal extrapolations may be subject to large errors due to uncertainties regarding the initial state and the description of the processes.

In order to carry out analyses for a period of one million years, it is necessary to identify possible developments of the repository system. This is based, for example, on forecasts of developments at the site (geological, anthropological) as well as on changes in the natural barriers associated with the construction and operation of a repository. From this, scenarios are derived which are incorporated as assumptions in the subsequent analyses. This must also include unusual events that cannot be planned such as extraordinary events and worst cases.

Many extraordinary events can be excluded on the basis of comprehensive evidence that these impair the integrity of a repository, see also Hertzsch [6].

Two of these extraordinary events which cannot be planned for and which also point to the remaining risks of a repository should not be absent from a long-term safety demonstration:

- Terrestrial interventions from outside (drilling, test pittings, etc.)—Human Intrusion Scenarios, see ESK [46]
- Time and any history of climate change (glacial and warm periods scenario), Petit et al. [47]; Cyclic occurrence of glacial and warm periods, interglacial periods, Fig. 8.10.

In the investigation of tectonic scenarios AkEnd [16] came to the following conclusion: "Areas in which the barrier system of a repository located at a depth of about 1000 m is significantly influenced over millions of years or whose development cannot be predicted over this period should be identified as particularly unfavourable".

The demarcation of areas with obviously unfavourable geological conditions, which are excluded from the search areas for repository sites, was defined as follows:

- Large-scale vertical movements: The final repository region must not exhibit large-scale elevations of more than one millimetre per year on average in the foreseeable period.

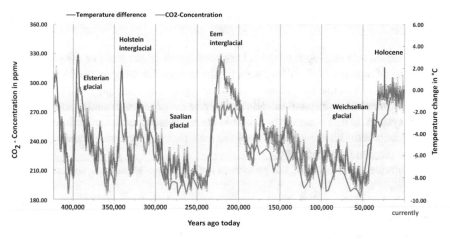

Fig. 8.10 Change in temperature (deviation from the mean temperature of the Holocene) and CO_2 concentration of the Antarctic atmosphere in the last 420,000 years after ice core measurements at the Antarctic station Vostok according to Petit et al. [47] and Barnola et al. [48]

- Active fault zones: There must be no active fault zones in the repository area.
- Seismic activity: The seismic activity to be expected in the repository area must not exceed earthquake zone 1 according to DIN 4149.
- Volcanic activity: In the repository region, no quaternary or future expected vulcanism may be present.
- Groundwater age: No young groundwater may be present in the effective containment zone. The groundwater must therefore not contain tritium and/or C-14.

System analyses must take into account all probable and less probable scenarios such as partial failure of sealing structures, see Fig. 8.5, or release of radionuclides from the waste matrix into the effective containment zone. For the scenarios, the influence on the effectiveness of the barriers is determined by means of special geomechanical and hydraulic simulation programs. In particular, analysis will be used to determine whether the protective function of the effective containment zone is maintained or not.

The hydraulic calculations and radionuclide transport analyses include the following aspects:

- Verification of the barrier effect of the sealing system (shafts and drifts closures, compacted salt breeze backfilling)
- Clarification if it is possible that solution inflows may occur in the emplacement areas
- Analysis of the release dynamics and transport of dissolved and gaseous radionuclides
- Analysis of the effect of gas pressures due to corrosion processes
- Assessment of radiological consequences using scientifically recognised exposure models.

A long-term safety analysis should include the following main steps, GRS [49], see also Fig. 8.9:

– Preparation of a catalogue of the relevant scenarios on the basis of the properties, events and processes (FEP)[17] relevant for the temporal development of a repository system as well as the site properties and the repository concept,
– Development of conceptual models to describe the processes running in the repository system. In this context, the possible chemical/geochemical reactions that can take place between the stored waste, the host rock and aqueous phases (groundwater) play an important role,
– quantitative description of the processes and scenario sequences with numerical models,
– Calculation of the radiological consequences and analysis of the consequences with regard to the protection objectives,
– Performing further analyses to evaluate uncertainties, robustness and sensitivity.

8.5.2 Scenarios Analysis—Scenarios Development

The aim of scenario development (scenario analysis) is to select from the multitude of conceivable development possibilities of the repository system in the post-closure phase those which represent the probable developments and the range of less probable developments. Concrete design steps of the repository system follow from the scenarios.

In this step of the long-term safety analysis, the so-called FEP—"Features", "Events", and "Processes" are first compiled. FEPs describe properties, events and processes that characterize the various scenarios, i.e. all development possibilities of the repository system. In this sense:

– "Properties": Conditions or conditions characterising a particular system or subsystem at any given time, such as the radionuclide inventory, available gas storage volume or backfilling permeability,[18]
– "Events" means natural events, spontaneous processes and changes, human intervention and seismic events in the area of the repository,
– "Processes": slow and long-lasting processes and changes such as corrosion, leaching, cavity convergence, diapirism or decay of long-lived radionuclides.

The selected FEPs are then combined into scenarios. A scenario results from the changes in the repository system and the barrier effect of its components in the post-closure phase under the influence of events and processes. Various international

---

[17]FEP: acronym formed from "Features", "Events", and "Processes" that is used in international literature; is used in the context of repository safety analyses.

[18]See also temperature dependence.

methods are used to combine the FEPs into scenarios, although a uniform methodology has not yet been established.

When compiling the FEPs, both site-specific information and generic databases such as the "NEA-FEP database" are used. The database contains a compilation of all FEPs from various international long-term safety analyses, which reduces the probability that relevant FEPs will be disregarded. This is followed by "FEP screening", NEA [50].[19]

The scenario catalogue determined in the scenario analysis is the starting point for the long-term safety analysis, in which the sequence of scenarios is simulated with the aid of calculation models. In particular, the models for the long-term safety analysis, in which the processes taking place in the repository are mapped, are constructed on the basis of the scenario analysis. The probability of occurrence is a quantitative or qualitative indication of the probability with which a risk event (scenario) occurs within a certain period of time. The complete specification of a scenario (risk) includes its effects (consequences) and its probability of occurrence p_E. These two parameters are determined as part of scenario analysis and serve as criteria for long-term safety analysis (scenario catalog). The scenarios can be classified as follows:

- "Reference scenarios" whose occurrence is certain or very likely,
- "Hypothetical scenarios" for which either low probabilities of occurrence or large uncertainties exist,
- "Practically excluded scenarios" with very low expected consequences or probabilities of occurrence or irrelevance for the concrete analysis and
- "Stylized scenarios", as a predefined (e.g. regulatory) development of the repository system or a subsystem in which the event sequences at the site to be considered are not taken into account in detail.

Reference scenarios are used to limit the high uncertainty of the input data and calculation models, because the reference case is based on the expected development, currently preferred hypotheses, the reference interpretation of the conceptual models (reference model and stylized models) and the following fundamental phenomena:

- Radioactive decay
- Period of complete containment of fuel elements and of high-level waste in the cask
- Inclusion in the waste matrices
- Geochemical immobilization and retention of released radionuclides by the waste matrices
- Limitation of the mobility of released radionuclides in the ECZ and during transport through natural and geotechnical barriers, dilution by regional aquifers and reduction of release in the biosphere.

It can be seen that reference scenarios depend on the location. Thus, reference scenarios for the different host rocks (salinar, crystalline, clay) are different.

---

[19]Nuclear Energy Agency (NEA) is an intergovernmental agency that is part of the Organisation for Economic Co-operation and Development (OECD).

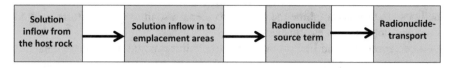

Fig. 8.11 Scenario "Solution inflow from a limited solution (brine) inclusion": Flowchart of the characteristic FEP's

Once a host rock has been selected, possible failures must also be investigated for long-term safe and stable storage, and from this, failure scenarios must be developed that examine the possible effects of a failure affecting the biosphere, see Fig. 8.11 Scenario solution inflow in the ECZ.

This means that compliance with long-term radiation protection targets must be demonstrated by means of a safety analysis for both the expected developments and the unlikely incidents.

A failure scenario that is to be expected with a probability of occurrence p_E in a period t should be able to be assigned a probability of failure p_f of the repository system in the period under consideration of 1 Million years, see equations for determination regulations. (Eqs. 8.13–8.17). The probability of failure within the observation period is not constant. It only makes sense to specify the probability of failure by specifying a time period. The specification of the probability density function would be mathematically correct, see Fig. 8.12.

The probability that an event at time t (e.g. earthquake) assumes a value between a and b (values on the Richter scale, or assigned accelerations) corresponds to the content of the area F_x (probability distribution). The graph at time t is the probability density function f_x, see also Eqs. (8.3) and (8.4).

In the case of assumed solution inflows in the ECZ, the probability of solution inflows in the ECZ must first be determined. According to the author's assessment, according to the classification described above, the scenario "solution inflows in the ECZ" is a hypothetical scenario with a very low probability of occurrence and

Fig. 8.12 Time-dependent probability density function $f_x(x,t)$

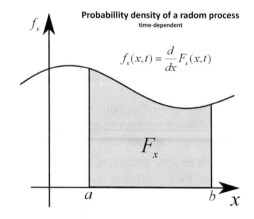

major uncertainties in the prognosis. Figure 8.11 shows the failure scenario with the following chain of events: solution inflow from the host rock into the emplacement areas (ECZ) → Formation of a radionuclide source term (radionuclides in (brine) solution), reaching the boundary of the ECZ and transition into the overburden, fluid-related radionuclide transport into the biosphere → Radiological Consequences → The final assessment is whether a violation of the protection target is to be expected or not, see Fig. 8.14.

Solution inflows to the emplacement areas may occur in the event of a disturbed repository development. How the solution inflows get to the emplacement areas depends on the host rock type, see Fig. 8.13.

This representation implies the following events (partial failures) which can then lead to the total failure of the repository system, with assigned time periods:

- Inflow of solution water from the host rock into the ECZ, failure of external barriers (geological barrier host rock/overburden, borehole or chamber seals, dam structures, shaft backfilling, shaft seals, etc., **prediction of the probability of the occurrence of the event and earliest possible inflow**);
- Earliest possible formation of a radionuclide source term in the ECZ, failure of internal barriers (backfill material, mineral mixtures such as bentonites, containers, waste matrix/immobilisates etc.)
- Fluid-related radionuclide transport from the ECZ into the host rock/overburden, **calculation of the probability of the occurrence of the event and earliest possible formation of the source term, determination, definition of the radionuclide source term for the near range**;
- Determination as early as possible of the solution with radionuclides entered in the solution from the emplacement areas to the limits of the ECZ;
- Fluid-related radionuclide transport with seepage waters into water-bearing horizons (deep aquifers), propagation of the radionuclide vector in the host rock/overburden; **determination of the propagation paths of the radionuclide source term of the far range and the time period of the earliest possible reaching of water-bearing horizons**;
- Fluid-related radionuclide transport in the overburden—transport period until radionuclides reach the biosphere, **calculation of the propagation of the source term in the overburden until reaching the biosphere, probability of the occurrence of the event, earliest possible reaching of the biosphere, see Fig.** 8.14. **Radiological consequences—use of scientifically recognised exposure models.**

It is well known that the proof of safety in the post closure phase of the repository cannot be carried out in a strictly scientific sense, since the potential consequences from the implementation of the repository elude a metrological check or verification due to the long observation period (proof period). Regarding the long-term safe, long-term stable decommissioning, this not only requires a continuous optimization process with periodic safety reviews, but it must also be pointed out here that there are possibilities to reduce the observation period in which e.g. long-lived radioactive isotopes are transformed into short-lived ones (partitioning and transmutation, acatech [51] or that the long-term safety assessment is carried out along time windows

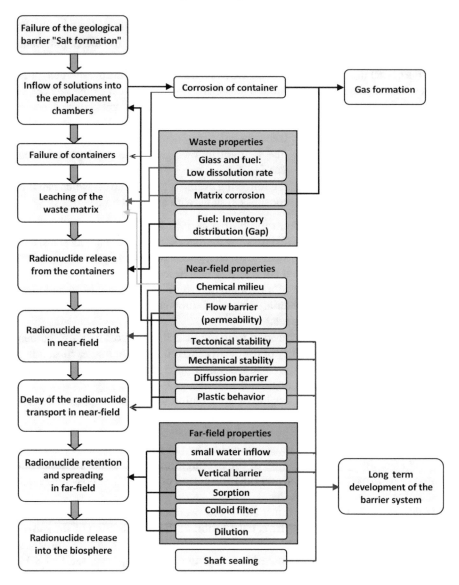

Fig. 8.13 Effectiveness of the repository system in salt formations in case of disturbed development

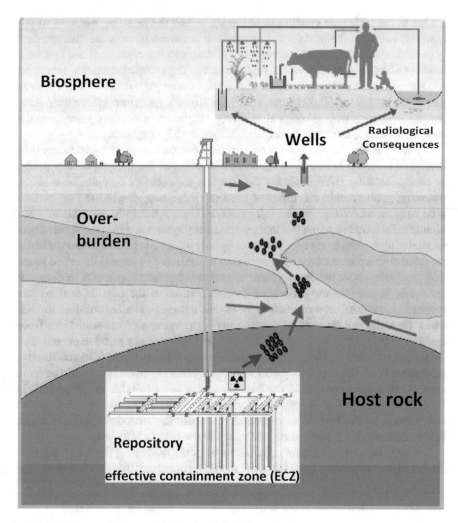

Fig. 8.14 Failure scenario—fluid-related radionuclide transport

with different uncertainties in the probability of occurrence of events and risks in the effects, without the safety requirements having to be reduced.

It only makes sense to specify the probability of occurrence by specifying a period for which this probability applies, as events with a very low probability density, considered across an unlimited period of time, will occur with certainty. The specification of the time-dependent probability density function would be mathematically correct, since it takes into account, for example, that a risk event (scenario) occurs more frequently at certain times than at others, see Fig. 8.12. The reference to the extent of damage is not included here.

Due to the high heat generation of the waste and the resulting mechanical and hydraulic stress on the host rock, FEPs, which dominate initially are characterised by geotechnical events or developments (e.g. the backfill compaction) and hydraulic events (e.g. the re-saturation of bentonite in seals). The group of scenarios, which are determined by climate-related FEPs (e.g. ice ages with glacial cover, warm periods after about 10,000 to 100,000 years, see Fig. 8.10) only gains importance later. After that, the scenarios with geological FEP (e.g. diapirism, subrosion, plate tectonics (continental drift), mountain formation), Mrugalla [52] dominate.

In principle, the protective objectives of final disposal are not subject to any time limit. It is assumed here that the general evaluation of the final repository design for heat-generating HAW includes the fact that the proof of compliance with the protection goals reaches the limits of the knowledge base due to the time periods to be considered. While, for carefully selected sites, it has to be possible to make scientifically sound statements about the repository system and its development over a period of about 1 million years, it is actually only possible to make a sound statement about the development of the biosphere over a maximum of several thousand years.

A prognosis about the development of man and human society with its demands and needs is possible for periods of 4–6 generations at the best. **It is therefore proposed here to select a repository system as a target-oriented solution in such a way that, during its normal development, the release of radionuclides from the effective-containment zone in the first 10^4 years is not to be expected. For the period of proof of 10^6 years, it must be ensured that releases of radioactive substances from the repository only marginally increase the risks resulting from natural radiation exposure in the long term. For less probable developments in the post-closure phase, it shall be demonstrated that the additional effective dose caused by the release of radionuclides from the stored radioactive waste does not exceed 0.1 millisieverts per year for the people affected**.

The long-term proof of safety could thus be performed along the following time windows (current state of knowledge). It should be noted that the indicated durations initially represent an expectation which is confirmed or not at the end of the proof:

- 0–500 years: Information about the repository is available/up to 500 years after the repository has been closed it is retrieved possible
- >500 years of human intrusion possible (Human Intrusion scenario)
- >50,000 years: earliest possible access of solution waters from the host rock
- >100,000 years: Earliest possible formation of a radionuclide source term
- >100,000 years: earliest possible reaching of a deep aquifer
- >1,000,000 years of radiation exposure of the population remains well below the applicable protection targets (0.1 mSv/a)
- Beyond 1,000,000 years: qualitative assessment.

8.5.3   Definition of Scenarios for Repository Development in the Post-closure Phase

The long-term safety of repositories for radioactive waste depends decisively on the possible effects of aqueous solutions on the waste, the release of radioactive substances and their migration through the rock strata of the host rock and of the overburden into the biosphere. For this reason, the long-term dryness of a repository is of utmost importance and the investigation of the failure scenario "solution inflows into the ECZ" is mandatory for the approval of the site. Within the process of optimising the repository concept, the inclusion of this failure scenario also has implications on the design of the multi-barrier system and thus the isolation potential of the repository system, and thus helps to improve the safety functions of the effective containment zone, Baltes et al. [16], as well as the technical measures. Requirements for waste are derived from the safety requirements in the operating phase and post-operation phase. It is the responsibility of the applicant to derive the final disposal and emplacement conditions in individual cases and to have them officially approved. This does not include the possibility of treating HAW with P&T.

In the following, the respective scenarios of undisturbed development are presented for the three host rock formations salt rock, clay rock and crystalline rock that are basically available in Germany for use in the final disposal of high-level radioactive heat-generating wastes, see also Fig. 8.12. Internationally, repositories for heat-generating HAW in these three host rock formations have already been approved, so that the general approval capability has been demonstrated. As a matter of course such approval has to be applied for and approved on a site-related basis again and again.

The host rock formations which are under consideration, contain chemically different aqueous media (highly saline solutions, clay-pore and granite water) and thus very different corrosion behaviours for the stored, site-specific waste containers have to be taken into account. Long-term stability of containers (waste matrices) >100,000 years, Raiko et al. [30], are dependent on the chemistry of the aqueous solutions. The radiological consequences of radionuclide leakage are determined using scientifically recognized exposure models.

Instead of the term "Scenario of undisturbed repository development", "Scenario of normal case—Normal Evolution Scenario" can also be used.

8.5.3.1   Scenario of the Undisturbed Development for a Repository in Salt Rock

The scenario of undisturbed development for a repository in salt rock can be described as follows: Within a period of up to a few hundred years, convergence will lead to the complete disappearance of the remaining cavities left after the closure of the repository. The porous salt breeze backfilling will be compacted to such an extent that it has a low permeability similar to the surrounding undisturbed salt rock. Until the waste packages are completely enclosed in the salt formation, no convective spread of pollutants will take place. According to current ideas, no long-term safety analytical dispersion calculation for this period is necessary. Since the natural salt rock and the backfill contain only very small amounts of water, anaerobic container corrosion, and thus gas formation, is limited. Prevention of leakage of containers is part of the undisturbed development of a repository.

Due to the tight containment of the corrosion gases, high local gas pressures may build up under certain circumstances. The effects and how to deal with them are currently undergoing a complex investigation.

8.5.3.2 Scenario of the Undisturbed Development for a Repository in Clay Rock (Clay)

In the case of clay formations, groundwater movement through the host rock to the repository area occurs in the reference case. However, due to the low hydraulic permeability of the clay rock, this movement is extremely slow. The galleries and cavities of the repository structure are generally backfilled with material that has similar properties to the host rock, such as bentonite or bentonite-sand mixtures. In the end zone, the host rock is loosened up by the tensile stresses occurring during excavation and, partially desaturated as a result of drying out by ventilation with fresh air in the operating phase. When water enters the near-field (emplacement areas), the loosened rock formation and the backfill in the sections and cavities are completely re-saturated again. Due to the re-saturation, the backfill swells, which in conjunction with the rock pressure leads to the healing of the cracks in the loosened zone, so that the low hydraulic permeabilities of the undisturbed host rock reappear there over time.

Depending on the type of waste container, the amount of incoming water may cause corrosion over time. The failure of the containers and the dissolution of the waste matrix can eventually lead to mobilisation of radionuclides.

Some of these radionuclides will precipitate again in the near-field when the solubility limits are exceeded and will thus be immobilised again.

The mobilized radionuclides are transported almost exclusively diffusively through the technical barriers, the host rock and further into the overburden. The advective transport through the technical barriers and the host rock is very slow and irrelevant due to their low permeability. A noticeable retention of the pollutants is achieved first by sorption on the backfill materials and subsequently effected to a great extent by sorption of the host rock. Radionuclide transport in the overburden is follows advective, with the groundwater. As a result of contamination of the groundwater occurs such that during its using as drinking water or for the production of foodstuffs it becomes a potential source of radiation exposure for the population.

8.5.3.3 Scenario for Undisturbed Development of a Repository in Crystalline Formations (E.g. Granites)

In crystalline rock formations the presence of fissures is to be expected. It is therefore assumed that water will enter the emplacement areas through the fissures and re-saturate the bentonite lining of the containers—the so-called bentonite buffer, see Chap. 7—in the near-field. "Bentonite buffer" is the term used to describe the embedding of the waste packages in the repository, which may consist of moulded bentonite blocks surrounding the waste, see Chap. 7. Since water ingress is not limited, it is possible that, depending on the type of waste container, corrosion may also occur in the waste containers. This does not necessarily lead to the mobilisation of the radionuclides and possibly only to a small portion of them. This is highly dependent on the geochemical environment in which the process takes place. In some cases, mobilized radionuclides in the near-field become immobile again, independent of the exceedance of the solubility limits (precipitation). When assessing this host rock type, attention should also be paid to the fact, that a release of radionuclides from the site-specific waste containers cannot be expected until well after over one hundred thousand years.

Once escaped from the waste container, the mobilized radionuclides can be diffusively transported through the bentonite buffer and released into fissures of the granite formation. In the fissures radionuclides can be transported by advection. The retention of radionuclides is activated by matrix diffusion and sorption in the rock matrix as well as on any fissure coatings or fillings.

If the gas penetration pressure for the bentonite is exceeded, gases formed in the storage areas can escape through the bentonite and be transported away through the fissures in the formation. The gas transport paths in the bentonite close again as soon as the gas pressure falls below the gas penetration pressure, so that there is no permanent damage to the bentonite buffer.

8.5.3.4 Alternative, Further Noteworthy Scenarios

These alternative developments can affect all areas of the repository system—near-field, far-field and biosphere. In these scenarios, developments of the repository system are described which deviate in one or more points from the expected course of the scenario of undisturbed development.

(a) Un intended development of a repository in rock salt

As described above, a repository in salt rock under undisturbed conditions, see Sect. 8.4.3.1, is not expected to allow water to enter the waste and release pollutants. However, disturbances may occur which result in alternative developments, such as access of water to the waste. It is assumed that water from the overburden and/or from solution inclusions in the rock inflows to an early stage after the closure of the repository, see also Chap. 6 (Asse II Mine). The solution inflow comes into contact with the waste packages, corrodes the containers, reaches the radioactive waste and thus will become contaminated. As soon as the remaining cavities in the repository are completely filled with liquid, the direction of the fluid flow reverses due to the convergence of the cavities. The contaminated solution is then pressed out of the repository into the overburden and can be transported from there via the water path into the biosphere. The radionuclide term is transported via aquifers into the biosphere with a laminar flow, see Fig. 8.14. The potential radiological consequences of this failure for the population are calculated in the long-term safety analysis. The use of contaminated groundwater by the people, which can lead to a potential radiation exposure of the population, is highlighted.

Due to which further events can aqueous solutions enter the ECZ although the selection of the site actually excludes them?

(b) **exceptional occurrences**:

- (b1) Terrestrial interventions from outside (drilling, exploration, etc.)—human intrusion scenarios,
- (b2) Temporal development of the climate (glacial and warm periods—scenario—Glacial and Interglacial periods), see Fig. 8.10 Cyclic occurrence of interglacial, l glacial and warm periods.

These lead to the loss or partial loss of the barrier effect of the multibarrier system of the repository with simultaneous changes in the condition of the overburden, which becomes fractured-porous in the long term and allows groundwater or overburden water to inflow the repository.

(b1) The possibility of human intrusion into the repository area (human intrusion scenario) and the consequences described is of particular interest. **For this reason, prevention should be provided in the selection of sites for the final disposal of HAW in deep geological formations by excluding as far as possible as site search areas where mineral deposits are already known. This could be made more effective by keeping the file (final documentation) in the registry of the respective responsible mining authority and retaining it there and maintaining the entire field of deep geological formations and their surface projection in the ownership of the federal state. There is a permanent obligation to mark the field above ground and a ban of trenches and test pitting by the responsible mining authority. This considerably reduces the probability of such intrusion events occurring**.

Pursuant to Section 3 of the StandAG, the Commission has excluded this important aspect from its recommendation for action, regarding to the documentation

(b2) Petit et al. [47] has demonstrated that within a time window $\geq 100,000$ years (interglacial period) with some probability of occurrence in the region under consideration, an ice age is to be expected which may lead to the weakening of the overburden and the multibarrier system described above.

In the case of heat-generating radioactive waste, depending on the storage conditions, we have to consider:

- temperature-related changes in rock properties;
- the thermally induced stresses which can lead to the formation of water pathways;
- the formation of secondary permeability under the influence of temperature;
- Migration of fluid inclusions in salt under temperature increase etc.

In the event of any failure scenario, solution inflow must be taken into account into the ECZ. The probability of occurrence during the observation period can be considerably reduced by the emplacement conditions and excluded with transmuted HAW.

8.6 Long-Term Safety Assessment

In Anon [53] a correlation between safety classes and annual failure probability is presented.

If this scheme is also followed in relation to the probability of occurrence of scenarios and is based on a probability of occurrence in the respective time window in accordance with Table 8.5, using the respective process equation systems or substance transport models, is it possible to determine the periods in which the events described above can be expected. This refers to the probabilities of occurrence for the scenarios:

- Scenario of undisturbed development for a repository of generic host rock formations
- Scenario of a repository in rock salt with a not being according as intended development
- Human Intrusion Scenario
- Glacial and warm periods—Scenario, interglacial
- Scenario: Loss or partial loss of the barrier effect of the multibarrier system of the repository, with simultaneous changes of state in the host rock and overburden.

If the chain of events in Fig. 8.11 was ordered according to physical chemical processes, submodels A to D would result, describing the entire process.

From Fig. 8.15, all necessary hydraulic calculations and radionuclide transport analyses can be derived for the description of the overall process: solution inflows

Table 8.5 Safety classes with annual failure probability acc. Anon [53]	Safety classes	Annual failure probability
	Low	$<10^{-3}$
	Medium	$<10^{-4}$
	High	$<10^{-5}$
	Unlikely	$<10^{-6}$

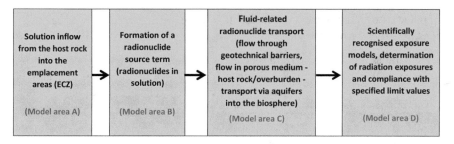

Fig. 8.15 Model areas of the scenario "solution inflows in the ECZ and as a consequence: transitions of radionuclides from the inventory of the repository to the biosphere"

in the ECZ and subsequent transition of radionuclides from the inventory of the repository into the biosphere. These can be divided as follows as follows:

- Verification of the barrier effect of the closure system (shaft and drift sealings, bentonite filling, compacted salt grit backfilling, etc.),
- Clarification as to whether there may be solution inflow into the emplacement areas,
- Analysis of the release dynamics and transport of dissolved and gaseous radionuclides,
- Analysis of the effect of gas pressures due to corrosion processes of the final disposal waste containers and -matrices,
- Assessment of radiological consequences using scientifically recognised exposure models.

In the following, the overall process is subdivided into sub-processes (model areas) and the fundamental relationships are shown for the model areas. The equations used can be replaced by more recent developments, but this does not result in a different verification scheme. In Germany, so far mainly generic safety analyses have been carried out for repositories for high-level radioactive waste, i.e. they were based on no specific site data. This approach is not pursued here because it does not do justice to the site selection process currently taking place.

The host rock formation rock salt is generally assumed here, because this is the area of greatest experience available in Germany. But, for other host rock formations the detection scheme is similar. This procedure also allows the determination of which verifications are to be carried out in each case. The events listed above are related to failure or partial failure of barriers of the multibarrier system, whereby failure can also be described as loss of function. I.e. in a repository concept, the lifetimes of the individual barriers in relation to the respective host rock must and can be coordinated in such a way that the primary protective objective of a permanent isolation of the radioactive material deposited is achieved with a probability bordering on safety.

8.6.1 Model Area A—Failure (Partial Failure) of the Sealing System

The design of the sealing system depends to a large extent on the choice of the host rock and thus on the choice of location. In rock salt, for example, the sealing system has to act as a fast-acting barrier by trapping the waste during the period, in which the backfilling material salt grit has not yet developed its sealing effect. The current German safety requirements state that the simultaneous failure of several independent technical components need not be assumed. For this reason, the current German repository design for the sealing system with shaft and drift sealings provides for two independent geotechnical barriers, arranged in a row.

The sealing system in the crystalline rock (granite) has a completely different structure and has hardly been researched in Germany. Due to the repository-relevant properties of crystalline rocks, see Table 8.3, the barrier function must therefore primarily be ensured by the repository container in combination with a bentonite backfill. The outer sheath, the double-jacketed waste container in the Swedish and Finnish repository concepts, see Fig. 8.7, consists of copper, see ESK [54].

Due to the potential fissurization (fractured crystalline rock) and the low radionuclide adsorption capacity of the granite, the cavities between the radioactive waste containers and the rock are still filled with an additional geotechnical barrier of bentonite (a swellable clay mineral). The barrier effect of the "sealing system" here is characterised by the swelling capacity of the bentonite backfill in connection with the temperature and the hydrological situation. A long-term barrier effect of currently more than 100,000 years can be expected. However, the Scandinavian repository concepts cannot be transferred to German conditions without adjustments. In Finland and Sweden, the safety of the repository must be proven for 100,000 years. In Germany, the verification period is 1 million years.

The exemplary analysis of a sealing system here assumes a repository in the host rock salt. Solution inflows from the host rock into the emplacement areas (ECZ) may occur if the sealing system fails or partial failure is present. In rock salt, the sealing system installed with the closure of the repository, consists of drift- and shaft sealing structures (dams and plugs) as well as chamber structures, see Fig. 8.14. An example of a sealing structure is shown in Fig. 8.5.

According to the state of the art, shaft sealings are composed of several functional elements. These can be divided into the following main groups:

- Abutments
- Sealing elements
- Filling column

can be subdivided.

As essential elements of the safety proof for technical barriers, according to the state of the art, the following individual proofs can be described, according to Kreienmeyer et al. [55] and Herold and Müller-Hoeppe [56]:

- Proof of sufficient hydraulic resistance (tightness proof)
- Proof of load-bearing capacity
- Proof of deformation limitation
- Proof of crack limitation
- Proof of durability.

These proofs to be carried out are decisive for the proof of function of a shaft sealing. The proof of manufacturability with the corresponding quality assurance certificate must be anchored in the repository concept.

8.6.2 Model Area B—Formation of a Radionuclide Source Term in the ECZ

With respect to the time windows, the probabilities of occurrence (occurrence of events) of a failure scenario could be derived for solution inflow to the ECZ, see Fig. 8.13:

- Earliest possible inflow of aqueous solutions into the ECZ, flows through or around the geotechnical barriers, P_{E1};
- Earliest possible time of decay of the waste matrix—start of formation of radionuclide source term in the ECZ—chemical (corrosion) process—probability of occurrence, P_{E2};
- Earliest possible starting time of radionuclide discharge into the host rock/overburden—radionuclide transport (spreading) in the ECZ (process: spreading and leakage of the radionuclide term from the ECZ—mass transport)—probability of occurrence, P_{E3};
- Earliest possible entry of the radionuclide source term into a deep aquifer—radionuclide transport (spreading) in the overburden (process: mass transport in porous media)—probability of occurrence, P_{E4};
- Earliest possible exit of the radionuclide source term into the biosphere—radionuclide transport (spreading) in the overburden—laminar flow process—probability of occurrence, P_{E5};
- Radiological consequences—Use of scientifically recognised exposure models—Probability of occurrence, P_{E6};

For the independent basis variables (random variables): $X_1, \ldots, X_n$ the probability of occurrence can be determined with

$$P_E = P(X_1, X_2, \ldots, X_n) \tag{8.6}$$

And the probability density function with

$$f(x_1, \ldots x_n) = f_1(x_1), \ldots, f_n(x_n) \tag{8.7}$$

Can be determined. Where f_i is the real probability density of X_i.

If the time periods t_i are determined and the associated probabilities of occurrence p_{Ei} are known, these are entered into a probabilistic model of the overall process and the overall probability of failure P_f is determined from this with Eq. (8.3). The long-term stability of the technical barrier of the waste container and thus the probability of occurrence P_{E2} can be understood as a variable controlling the process. It makes it clear that technical and geotechnical barriers can be compared with the natural barriers to determine the insulation capacity of the repository system.

The relevance (consequences) in the event of events is an essential measure of the precautions to be taken, see Table 8.6.

The overall probability of occurrence PE is called conditional probability. Definition: If A and B are events from the same sampling area and $P(B) \neq 0$, then the conditional probability of occurrence of A is under the condition.

$$P(A|B) := \frac{P(A \cap B)}{P(B)} \qquad (8.8)$$

An important conclusion from the definition of the conditional probability is the multiplication theorem:

$$P(A|B) = P(B)P(A|B) \qquad (8.9)$$

This theorem can be generalised for any events A1, A2, A3, ..., An; Krucker [57].

According to Eq. (8.9), the probability of the occurrence of the failure can be calculated, when

$$p_E = p_{E1} \cdot p_{E2} \cdots \cdots p_{En} \qquad (8.10)$$

Table 8.6 Assessment of relevance (consequences) in case of occurrence of events

Designation	Relevance/consequence in case of occurrence of event
Desaster risk	When the event occurs, the repository fails, causing mass discharge of radio-toxic contaminants into the biosphere. In addition to the occurrence of life-threatening radiation exposure, all aquifers in the far-area of the repository will be contaminated. In case of such an occurrence evacuation is mandatory. The choice of a different disposal channel in the event (recovery of radioactive inventory) of an accident is imperative but risky for the personnel
High risk	The occurrence of the event forces the evacuation (partial recovery of radioactive inventory) and choice of another disposal channel
Medium risk	The occurrence of the event forces partial evacuations until the elimination of defects/faults (if possible)
Small risk	The occurrence of the risk forces to make repairs/improvements, even temporarily
Minor risk	The occurrence of the event has no consequences, but increased environmental monitoring in the surroundings area of the repository should be implemented

It must now be shown that the probability of occurrence for the failure within the period under consideration (detection period) is so small, here 1 million years, that this event can be excluded or a release of radioactivity in this failure remains so low, that it is within the tolerable range.

8.6.2.1 Solution Transport Inside the Effective Containment Zone (ECZ)

The considerations here refer to a repository in rock salt, see Rübel and Mönig [58]. In the host rock formation of salt, the container has the function of acting as a retention barrier until a dense enclosure is produced by the convergence of the rock and the host rock assumes the retention function as a geological barrier.

The basis for this proof is the solution transport in the repository structure (close range). The solution transport is determined by the driving forces and the flow resistances in the mine works. If the driving forces are defined as the differences in the solution pressure, then the corresponding transport process is advection, and if the driving forces are defined as density differences, then the transport process is convection. Density differences can be caused by differences in solution composition, solution temperature or gas saturation.

Advective Solution Transport

The advective solution flow Q results from the hydraulic pressure difference p and the permeability k of the medium flowed through according to Darcy's law, Eq. (8.11):

$$Q = \frac{k \cdot A}{\mu} \nabla p \qquad (8.11)$$

A Cross-sectional area of the flowing through area
μ dynamic viscosity of the solution

In with salt grit backfilled mine workings, the permeability of the backfill depends on its porosity, Eq. (8.12).

$$k = f_p' \cdot c \cdot \varphi^q \qquad (8.12)$$

φ Porosity of salt grit
c; q constants from experiments with salt grit.

Processes of dissolving and reprecipitating (redissolution) and settling phenomena can lead to higher permeabilities in a drift, in this case the factor f_p is taken into account.

Convective Solution Transport

With so-called free convection, the solution circulates within the repository mine works. The driving mechanisms for this solution circulation are density gradients in the solution. Density differences in the solution can be caused by inhomogeneities in the chemical composition of the dissolved rock salt, by corrosion products of the containers or the waste matrix and by temperature differences. The dissolved radionuclides hardly influence the density differences and are neglected. According to Storck [59]; the solution exchange flow $Q_{T,C}$ can be calculated as follows, Eq. (8.13).

$$Q_{T,C} = \frac{g \cdot k \cdot \beta}{8 \cdot v} H^2 B \frac{\Delta T}{\Delta L} - \frac{g \cdot k}{8\mu} H^2 B \frac{\Delta \rho_C}{\Delta L} \qquad (8.13)$$

With

B, H, L width, height, length of the distance
ΔT temperature difference
$\Delta \rho_C$ Difference of the density of the solution due to the concentration of dissolved substances
g   acceleration due to gravity
k   Permeability
μ dynamic viscosity of the solution
v kinematic viscosity
β Thermal expansion coefficient.

The exchange currents by temperature gradients and concentration gradients add up and can be represented as a total exchange flow of solution. Convective solution flows are generally excluded for radionuclide transport, since in all cases considered here a release of radionuclides from the waste into the solution only occurs at a time when the temperature increase in the repository due to the heat production of the waste has already decreased.

8.6.2.2   Release of Radionuclides from the Waste Container

The radionuclides can only enter the solution when water enters, if there is a partial or complete failure of the waste package. The release as such only leads to solution transport if the radionuclides are mobilised. This depends on the one hand on the radionuclide itself and on the other hand on the geochemical environment that is formed. In addition, the timing of release is important. It is important whether the release occurs after 10,000, after 100,000 years or even later. This makes it clear that the retention properties of the waste packages are of outstanding importance. Final storage concepts for argillaceous rock and granite are based on significantly different conditions as far as those compared with salt with regard to the significance of the retention function of the container, the backfill, the technical sealings and the geological host rock formation. The backfilling and sealing concepts for clay and granite therefore differ significantly from those for salt. In Germany, there is a great need for development in this area, not only because one has so far concentrated on a repository in rock salt, but also because one has not paid much attention to container development. The following statements are based on the findings available internationally.

The German repository concept assumes the following behaviour: The contact of the waste containers with water leads to their corrosion and can lead to a failure of the containers if there is sufficient water inflow and correspondingly across long periods of time. The corrosion rate increases significantly with temperature. At approx. 175 °C, a corrosion rate of approx. 0.3 mm/a was determined. If a shielding Pollux container with a wall thickness of 435 mm steel/gray cast iron is used as a basis for the storage of the tracks (see Table 8.4), then at an average corrosion rate of 0.05 mm (corresponds to an average temperature of approx. 50 °C), the container would corrode after 8700 years and the solution could reach the matrix. The matrix itself has a retaining effect, so that radionuclides do not immediately go into solution and if so, by no means completely then. This has not been investigated in Germany. However, in its 380th meeting on 13.03.2005, the Reactor Safety Commission dealt with the retention effect of the corrosion products of Pollux containers and stated: *"The corrosion of iron creates a reduced environment which is advantageous for the immobilisation of radionuclides. The significantly larger amounts of iron in the case of a POLLUX container therefore also have a significantly higher buffering effect with regard to the "redox properties".*

Both the restraining effect of the matrix and the immobilizing effect of the milieu play no role in the German repository concept. It is assumed that the mobilization takes into account the solution-induced conversion of the waste matrix to corrosion products and the diffusion of radionuclides from the waste matrix. Depending on the type of waste, these sub-processes are described in detail or as a whole, e.g. by a constant rate. It is assumed that the radionuclides distributed in the matrix are initially completely transferred to the solution during mobilisation. A possible subsequent precipitation or sorption on the corrosion products and on the offset material is not treated in the mobilization models, but in the models for the radionuclide transport. This does not adequately reflect the radionuclide transport that may actually take place in the mine workings and leads to a considerable overestimation of the radionuclide discharge into the biosphere.

- **Loss of activity of the radionuclide inventory in the repository over time**

Radioactive decay leads to a reduction in the radionuclide inventory in the repository over time. The activity of a particular radionuclide i can be calculated on the assumption that there is exactly one daughter nuclide for each parent nuclide and for a known initial inventory $A_{i,0}$ at any time according to the Eq. (8.14):

$$A_i(t) = A_{i,0}e^{-\lambda_i t} + \lambda_i e^{-\lambda_i t} \sum_j \int_o^t A_j(t)e^{\lambda t}\, dt \qquad (8.14)$$

determine. And there are:

λ_i the decay constant of the radionuclide i, where $\lambda = \ln(2)/T_{\frac{1}{2}}$ with $T_{1/2}$ the half-life of the radionuclide and

A_j the activity of the parent nuclide j.

The relative decline of the activity of the radioactivity inventory in relation to the initial activity is plotted by spent fuel elements as a function of the decay time in Chap. 2, Fig. 2.4.

- **Failure of the waste containers**

When modelling the container failure, consider for each time point τ from the relative number N_B of containers which failed at this time the proportion

$$n_B(\tau) = \frac{N_B(\tau)}{N_B(0)} \tag{8.15}$$

calculated as a distribution function of the container life. In addition, the relative increase in failed containers per time unit $\partial n_B(\tau)/\partial t$ is determined as the distribution density of the container service life, whereby this quantity is referred to below as the container failure rate. The relative values refer in each case to the total number of containers $N_B(0)$ in the emplacement location under consideration. To simplify matters, the addition "relative" is always omitted in the following, since all the sizes given are relative sizes.

For an evenly distributed service life of the containers, the proportion of failed containers $n_B(\tau)$ increases linearly, from "zero" at the beginning of the container failure to "one" at the maximum container service life, and so results

$$n_B(\tau) = \begin{cases} \tau/\tau_B & \text{for } 0 \le \tau \le \tau_B \\ 1 & \text{for } \tau > \tau_B \end{cases} \tag{8.16}$$

With uniform distribution, the maximum service life τ_B is twice as long as the average service life of the containers. For times longer than the maximum service life, the failure rate is zero. The CSD-V and CSD-C vitrified canisters have only a small wall thickness of approx. 5 mm. At an average corrosion rate of 0.05 mm, these would be corroded 100 years after solution inflow. A short average service life can therefore be assumed in contrast to Pollux containers.

- **Mobilization model: "Glass Matrix" (of vitrified waste canisters)**

In the mobilization model, the glass matrix is considered to be proportional to the surface of the glass. The effective surface is assumed to be about ten times the geometric surface. For the **surface-related reaction rate j(t)** applies:

$$j(t) = j_r \cdot exp\left(-\frac{Q_G}{R}\left(\frac{1}{T(t)} - \frac{1}{T_r}\right)\right) \tag{8.17}$$

with

j_r Surface-related reaction rate at the reference temperature,
Q_G Activation energy for glass corrosion,
R   general gas constant,
T(t)   Temperature at the current time,
T_r Reference temperature.

● **Mobilisation model: Fuel elements**

The model for the description of the mobilisation of radionuclides from spent fuel elements is based on work carried out by the Research Centre Karlsruhe [60] GRA 98 within the framework of a research contract of the Federal Office for Radiation Protection (BfS). The activity release rate $\mathbf{R_i(t)}$ of the radionuclide $\mathbf{i}$ is calculated as follows:

$$R_i(t) = n_B(t) \sum m_X \cdot c_{X,i}(t) \cdot r_X \tag{8.18}$$

with

n_B Proportion of failed containers,
m_X Initial mass in the area $\mathbf{x}$,
$c_{X,i}$ Activity concentration of the radionuclide $\mathbf{i}$ in the area $\mathbf{x}$,
r_X Release rate from the area $\mathbf{x}$.

To calculate the radionuclide flow from a container, the waste is divided into three areas:

Area 1: Metal parts,
Area 2: Gas space in the fuel rods,
Area 3: Fuel matrix.

8.6.2.3   Transport of Radionuclides in Solution in the Mine Works

The radionuclides dissolved in the solution can be transported in the mine works through three different effects. These are solution transport by pressure gradients, exchange processes, dispersion and diffusion. The first two points have already been discussed during solution transport. The activity stream of a radionuclide by a forced solution movement results from the product of the solution stream and the concentration of the radionuclide in the area from which the solution stream comes. A solution movement also results in the dispersion, which is due to the heterogeneous distribution of the flow velocities in a porous medium and leads to a balancing of the radionuclide concentration in the water. The diffusive radionuclide stream $\mathbf{J_D}$ is determined by Fick's first law. In addition, a diffuse radionuclide transport takes place, which is driven by existing concentration gradients of the radionuclides. The diffusive radionuclide stream $\mathbf{J_D}$ is determined by the first Fick's law Eq. (8.19):

$$J_D = -D(T) \cdot A \cdot \varphi \cdot \nabla c \tag{8.19}$$

with

c   amount of substance concentration in mol m^{-3};
∇c Concentration gradient of radionuclides
D   Diffusion coefficient at the temperature $\mathbf{T}$,
A   Cross section plane and
φ Porosity. The unit J_D for example $[J_D] = $ mol m^{-2} s^{-1}.

The total radionuclide transport results from superposition of all previously discussed processes (transport by forced solution movement, convective exchange processes and diffusion), Eqs. (8.11), (8.13) and (8.19). These processes are very different from each other, so that an analytical treatment of the superposition is not possible. Since exchange processes are generally of minor importance for radionuclide transport, a simple approximation for the entire radionuclide transport is obtained by summing the individual contributions to the activity stream. By summing the activity currents from exchange processes and from the enforced solution movements the total current may be overestimated. This approach is conservative with respect to radionuclide release.

8.6.2.4   Retention of Radionuclides in the Effective Containment Zone

Radionuclides released from the waste matrix by mobilisation can be retained in the repository by sorption and precipitation, i.e. they no longer participate in the radionuclide transport. Sorption is the process of retention of dissolved radionuclides by interaction with immobile materials in the repository, e.g. the backfill or corrosion products. For most radionuclides, the salt grit backfill shows only a very low interaction. Otherwise, no other substances in larger quantities are introduced into the final repository in rock salt where a significant sorption of radionuclides is to be expected. The retention of radionuclides by sorption on corrosion products of the waste packages is known as a process, but is currently still little investigated in Germany and not sufficiently quantifiable. For this reason, such sorption processes are currently not considered in the context of a German repository project. When solubility limits are reached, radionuclides are precipitated in solid phases. The extent of these processes depends on the geochemical environment of the solution. In radionuclide transport, this process is taken into account by considering solubility limits for different geochemical milieus in a repository. The solubility limits are to be understood in terms of maximum radionuclide concentrations in the solution. However, with corresponding site-specific waste containers,, achieving maximum radionuclide concentrations is almost impossible and is only mentioned internationally for the sake of completeness. Within the framework of a German repository concept, differences in the geochemical milieu within the mine workings are currently not considered. In the German repository concept for rock salt, precipitation of radionuclides within the mine works is only considered for the waste containers. If the radionuclides are released from the repository into the overburden, they are strongly diluted by the groundwater flow in the overburden, so that precipitation of radionuclides is not expected there either, see model area C.

8.6.2.5   Processes of Dissolving and Reprecipitating (Redissolution)[20]

In Germany, hydraulically setting salt concrete (Sorel concrete) with the following composition is mostly used in the manufacture of sealing structures in the ECZ in rock salt: binder (cement), concrete additives (rock powder, coal fly ash, etc.), aggregates (salt grit, quartz sand, etc.), mixing liquid (water, salt solutions, etc.), see Chap. 6. It is stable against NaCl solution. When the solution containing magnesium interacts with the binder, calcium hydroxide is dissolved and magnesium hydroxide precipitates. This increases the pore space in the salt concrete and the permeability increases. After complete consumption of the magnesium in the solution, the reaction with the salt concrete comes to a standstill. For a complete conversion of the salt concrete and a corresponding permeability increase to the maximum, a certain amount of magnesium-containing solution is therefore required. The availability of the solution and its magnesium content thus determines the redissolution rate and thus the rate of progress of the permeability increase of the salt concrete. The modelling of the redissolution of the seals has been derived from the procedure in the Morsleben Radioactive Waste Repository (ERAM), see Becker [61]. The permeability k of the partially corroded seal thus results as a function of the quantity of solution V_L flowing through to

$$k = k_0 \frac{1}{1 - \frac{V_L}{V_p} \frac{n \cdot \kappa}{(1-n)+n \cdot \kappa}(1 - 10^{-\varepsilon})} < 10^{-\varepsilon} \qquad (8.20)$$

with

κ   Redissolution capacity
ε   Number of decades
k_0 Permeability of the non-corroded seal
V_P Pore volume
n   Porosity.

The integrity of the structure, here of sealing structures, is determined by the design. In the case of chemical actions, a direct relationship can be established with the proof of durability, in which the corrosion resistance of the structure is evaluated, and in the case of mechanical actions with the proof of bearing capacity, in which the load-bearing capacity is evaluated. Equation 8.20 is part of a proof of the long- term durability. The corrosion of materials with cement or Sorel phases is initiated via the initial conditions, and is then considered part of the process description. In summary,

---

[20]Redissolution: Interaction of unsaturated solutions with rocks by solution and crystallization of individual components.

it can be stated that the closure system exhibits sufficient resistance to corrosion-induced access to the radioactive waste from surface-, overburden- and formation water when an additional 10 m thick salt-concrete sacrificial layer is added. The corrosion that occurs then concentrates on the sacrificial layer. For comprehensive investigations to uncertainty and sensitivity analysis for the corrosion of salt concrete by saline solutions see [62].

8.6.2.6   Gas Transport

In a repository for radioactive waste, gases can be produced by various processes in the presence of water. During the design and construction of a repository in a salt formation, the aim is to reliably prevent significant quantities of water from entering the repository mine works from outside. However, even if the repository system is developed undisturbed, there is always a certain amount of water available in the repository mine works so that the processes can run at least partially, so that not all of the available amount of material is converted. The gas-forming processes are

– Radiolysis,
– The corrosion of metals and
– The microbial degradation of organic substances.

   The presence of water is a prerequisite for the gas formation processes. The scenario of repository development considered in each case determines the quantities of water to be considered. The water can be fed into the repository from different sources, whereby some sources are only relevant in the case of disturbed developments of the repository system:

• Natural moisture on the layer boundaries of the salt crystals in the host rock ("pore water"): Rock salt has a natural water content of 0.1–1‰.
• Microscopic solution inclusions in salt rock: The individual salt crystals of the host rock may contain the smallest solution inclusions.
• Moisture introduced with the backfilling: Depending on the degree of drying, a quantity of water of 2 (dry) to 20 (humid) kilograms per cubic metre of salt grit backfill is introduced with the backfilling.
• Inflow from macroscopic solution inclusions in the host rock: In salt formations there are usually solution inclusions whose volumes can be up to several hundred cubic meters. If these solution inclusions are not detected during the exploration work, they can release their solution content into the repository.
• Inflow from the overburden via the shaft or other pathways.

   The spreading of the gas into the salt rock has to be considered especially against the background of the concept of the complete encapsulation of the radionuclides in the host rock formation. This includes the question whether the residual pore space in the host rock, together with the gas storage capacity of the host rock, is sufficient to store all the gases formed and how extensive the area impregnated with gases is, or whether gases are released from the host rock into the overburden. This would compromise the integrity of the host rock formation.

Just like the natural barrier of the host rock, the function of geotechnical sealing structures can also be impaired by the occurrence of high gas pressures. On the one hand, this affects the permeability of the geotechnical sealing structures themselves, as well as their coupling to the host rock. Possible disturbance zones around the sealing structures could lead to peripheral pathways at the fluid movement. This impairs the integrity of the closure structures.

The gas produced in the mine workings is transported in the pore space of the backfill. The transport mechanisms are diffusion due to concentration differences of the gases and gas pressure induced advection. In the case of the normal development of the dry repository, this is a single-phase flow whose advective flow is determined by the pressure gradient and the gas permeability of the materials flowed through. For very impermeable materials, such as seals, a certain threshold pressure must be exceeded before the gas flow can begin. This threshold pressure is also called the gas penetration pressure. The gas pressure in the repository is determined on the one hand by the quantity of gas from enclosed air and produced gases and on the other hand by the available cavity, which is successively reduced by convergence. The gas pressure thus increases due to two processes, the progressive gas production and the reduction of the cavities due to convergence. As soon as the gas pressure in the repository reaches the lithostatic pressure in the repository depth, the further convergence of the cavities ends. If gas formation continues at this point, this would in principle lead to a further increase in pressure above the rock pressure. On the other hand, when the rock pressure is reached, the gases formed can penetrate the rock salt mountains and be transported away. This overcomes the local strength of the salt rock and increases the gas permeability of the rock. This process is reversible, so that when the gas pressure drops, the original permeability of the salt rock is restored Popp [63].

The gases formed in a repository can have an influence on the geochemical environment and thus on processes and properties in the repository that are dependent on the geochemical milieu. This concerns in particular the solubility of the radionuclides and their sorption behaviour in the mine works as well as the corrosion rate of the fuel or glass matrix and thus the release rates of the radionuclides from the waste matrix.

The gases influencing the geochemical milieu that can be formed in the repository are mainly hydrogen from iron corrosion and carbon dioxide from the possible degradation of organic components. With increasing hydrogen partial pressure, the geochemical milieu becomes increasingly reducing and thus influences the states of redox-sensitive radionuclides such as uranium, plutonium, neptunium, selenium and technetium.

8.6.3 Model Area C—Mass Transport Modelling for the Far-Field

According to the scenario considered, pollutants from the near-field reach the far-field and are transported from there with the groundwater into the biosphere. In the case of repositories in crystalline formations (e.g. granite) or in clayey rock, far-field is understood to mean the area outside the technical bentonite barriers, i.e. host rock and overburden, whereas in the case of repositories in salt rock this will only be the overburden and that part of the host rock regarded as far-field. In the scenario of the undisturbed development of a repository in salt rock, a flow through the far field is assumed, but due to the tightness of the salt rock the groundwater is not contaminated and thus no radionuclide transport takes place. It is only in the case of failure scenarios that radionuclides are released and transported through the far field. The release of radionuclides from a repository in rock salt can be caused by convergence and gas production, in which case the contaminated solution is also pressed from the repository into the overburden. In this case, radionuclides from a repository reach deep groundwater aquifers and can be transported from there via the water path into the biosphere.

In the case of a repository in clay rock, it is assumed that radionuclide transport in the host rock itself is purely diffusive; only within the overburden can advective transport take place. In contrast, in a repository in crystalline rocks it is assumed that advective radionuclide transport always takes place via fissures. The structure and properties of the overburden must be described in detail. For the Gorleben site, see Klinge [64]. When radionuclides are released from the ECZ into the overburden, the radionuclide concentration is diluted in the groundwater flow in the overburden. The transport of the radionuclides in the overburden takes place by advection, diffusion and dispersion, as well as retention by sorption.

As already mentioned above, a sufficiently precise description of the groundwater flow is a prerequisite for transport modelling. The following model can be regarded as sufficiently reliable. The groundwater flow is assumed to be laminar. For the modelling, the fluid mass is balanced locally (i.e. within a volume element) and for the entire model area, Eq. (8.21). Density-driven flows will be additionally simultaneously balanced: the mass of the dissolved salt, Eq. (8.22), and/or the amount of heat. In case of a non-density-driven flow, Eq. (8.22) is omitted, see potential flow, see [65].

$$\frac{\partial}{\partial t}(n\rho_f) + \vec{\nabla} \cdot (\vec{q}\rho_f) = 0 \tag{8.21}$$

$$\frac{\partial}{\partial t}(n\rho_f \chi_S) + \vec{\nabla} \cdot \left(\vec{q}\rho_f \vec{\nabla}\chi_S - D_z\rho_f \vec{\nabla}\chi_S\right) = 0 \tag{8.22}$$

n effective porosity (-);
ρ_f density of fluid (kgm^{-3});
$\vec{q}$ filtration (Darcy-) velocity (m s^{-1});

χ_S- mass fraction of dissolved salt (kg kg^{-1});
D_S diffusion-dispersion tensor for a salt solution (m^{2}s^{-1}).

In the case of laminar flow, the **linear Darcy approach for streamline motions** or the **extended Darcy approach for density currents** applies to both the porous and the fissured areas:

$$\vec{q} = -K\vec{\nabla}h = -\frac{k}{\mu}\left(\vec{\nabla}p + \rho\vec{g}\right)$$ (8.23)

K Permeability tensor (m s^{-1});
$\vec{\nabla}h$ Hydraulic gradient (-);
k Permeability (matrix) (m^2);
μ dynamical viscosity (Pa s);
$\vec{\nabla}p$ pressure gradient (Pa m^{-1});
ρ Density of fluid (kg m^{-3});
$\vec{g}$ local acceleration of gravity $= 9.81$ (m s^{-2}).

To simulate the groundwater flow in porous media, Eqs. (8.21)–(8.23) are solved numerically simultaneously for two or three dimensions with corresponding boundary and initial conditions. For fissured-porous media the corresponding equations can be taken from e.g. Kolditz [66]. The simulations of the flow field are generally carried out in order to be able to reproduce the temporal development of the flow system (historical matching). Since one is mainly interested in the description of the flow in the deep underground, one generally assumes tense groundwater conditions. There the piezometric heads are specified as pressure and the groundwater recharge results from the calculation. This often leads to discrepancies between measurement and simulation for groundwater recharge and/or the measured and assumed permeabilities in the uppermost layers of the model area. This procedure is acceptable because of the long periods of time considered and because of the interest in the flow in the deep subsurface, especially since the computational effort is considerably reduced as a result.

8.6.4 Model Area D—Transition of Radionuclides into the Biosphere; Possible Radiological Effects—Radiation Exposures, Recognised Exposure Models

In order to determine radiological effects in the biosphere, long-term safety analyses must firstly model the transfer of radionuclides within the biosphere. In particular, a person or a group of persons with their lifestyle must be defined for which the effects are to be determined.

Different objectives can be set for the modelling. The models can follow the following objectives to illustrate the effects of radionuclide spreading in the biosphere

- as realistic as possible,
- in the sense of mean values,
- covering, with a certain percentile (e.g. 95% percentile),
- basically covering.

For long-term safety analyses, it is usually necessary to use models and values that cover basically the relevant parameters or to carry out comprehensive uncertainty analyses and sensitivity analyses with regard to the parameter values, for example in order to record the effects of climate changes.

In Germany, reference is generally made to a "reference person" when determining the radiation exposure of persons in the population. This person is defined in the Radiation Protection Ordinance (StrlSchV) with certain characteristics and habits for use in the determination of radiation exposure, through discharges with air and water from nuclear facilities; this process is mandatory. The model for the determination of radiation exposure in the AVV is further specified in Section 45 StrlSchV, BA [67]. The radioecological model specified in the AVV is, by is characterized by the following basic defined facts:

- Real possibilities of use in the surroundings are to be taken as a basis, i.e. not the currently available types of use such as residential area, agriculture, etc.
- The determination of the spreading into the environment is carried out as realistically as possible by using site-specific statistical data sets from long-term meteorological measurements.
- Many parameter values are not to be regarded as broad, but rather as an average of values, some of which scatter considerably (e.g. transfer factors soil plant, dose coefficients).
- Food consumption rates are based on statistical data to cover the 95% percentile; see [68, 69].
- Some assumptions are covering a great extent (e.g. year-round outdoor exposure of the reference person, all food obtained by the reference person from the most unfavourable site).
- Overall, the result should be wide ranging, which is achieved in particular by the latter assumptions.

In the long-term safety analyses carried out so far, probabilistic exposure assessments have already been carried out for several scenarios, see also VSG [14]. Thus, a scenario "normal development" was also investigated. It was assumed that container corrosion increasingly releases radionuclides into the mine works, so that they reach water-bearing horizons with seepage water and are used as groundwater 18 km away. Within the framework of this probabilistic analysis, the dose curves for this scenario as well as the mean and median, the upper and lower limits of the 90% confidence interval for the radiation exposure of all single simulations and the regulatory limit values (0.15 mSv/a in the first 10,000 years of the decommissioning phase and thereafter 3.5 mSv/a) were determined and presented. This proof-sharing, described by the author as scientifically verifiable, is also used in the USA for the long-term proof of safety for the approval of the central geological repository at Yucca Mountain. This divides the observation period into a term of up to 10,000 years for which reliable scientific data and findings are available and a term of 10,000–1,000,000 years for which the statements are associated with considerable uncertainties with regard to data, processes and models. A temporal extrapolation of obtained data, derived process descriptions and models is afflicted with large errors, so that this procedure seems scientifically justified because of the very long period. Whether the regulatory limit values at 0.15 mSv/a in the first 10,000 years of the decommissioning phase and thereafter 3.5 mSv/a should be set in this way is initially a secondary consideration, see EPA [70].

Since 1991, the "Working Group on Principles and Criteria for Radioactive Waste Disposal" (until 1995 "INWAC Subgroup on Principles and Criteria for Radioactive Waste Disposal") of the IAEA, which is assigned to the "International Radioactive Waste Management Advisory Committee (INWAC)", has been dealing with certain issues relating to the disposal of radioactive waste. In its fourth report, IAEA [71], the working group dealt with the "critical group" and the modelling of the biosphere in connection with dose determination in the long-term safety demonstration of a repository.

The BIOMASS programme initiated by the IAEA was launched in October 1996. According to IAEA [72], the first step in the determination of a representative biosphere for dose estimation is the definition of the context for this estimation. This step includes:

- Definition of the task of estimation and the assessment variables to be considered (e.g. individual dose, collective dose, risk),
- Recording of the connection between repository and site,
- Determination of the source term, the work on with the interface of geosphere and biosphere,
- Determination of the time horizon of the calculations as well as
- Definition of the underlying philosophy (in particular the degree of conservatism sought).

The examples of reference biospheres developed in IAEA [72] serve on the one hand to demonstrate the procedure, but on the other hand should also be suitable for practical use in many contexts. Three reference biospheres were developed, which refer to a moderate climate and constant conditions in the biosphere:

- Extraction of drinking water from the contaminated aquifer via a well,
- Irrigation of agriculturally used areas with water from the contaminated aquifer via a well,
- Natural discharge from a contaminated aquifer into various habitats, including arable land, pastures, semi-natural wetlands, inland lakes.

In addition, in IAEA [72] three reference biospheres were developed for demonstration purposes, which contain changing biospheres. Two real sites for which detailed information was available (Harwell/South England and Äspö/Sweden) and one generic site were investigated.

8.7 Environmental Monitoring/Preserve Evidence

The designation of long-term safety, which should ultimately lead to the approval of a repository at site X, does not rule out the possibility of failures occurring and radionuclide transition into the biosphere. In the interests of transparency for the citizens and public safety, it is imperative that the repository site be monitored on a large scale. The task of environmental monitoring will be to ascertain the initial state of the environment at the site of a future repository over a period of at least 10 years, then to monitor the development during the construction phase, the subsequent operating phase and over an appropriate period also during the post-closure phase, over least 500 years. Thus, even the smallest impact of the repository on the environment can be determined. The proposed monitoring is independent of the type of repository—deep geological repository or near surface. Sampling should include meteorological observations as well as the air, water and soil paths. For this purpose, permanent environmental monitoring stations must be set up both with sensors and with conventional measurement technology coupled with appropriate analytical technology. Inside of this area more detailed studies will be carried out in a reference area of more than 200 km^2.

The investigation system to be set up should contain several hundred stationary observation points as well as data from satellite and aerial photographs, experimental plots and permanent monitoring stations (in the forest, in agricultural areas, air monitoring stations and waters monitoring stations). The measuring network shall follow the monitoring progress. Every year, flora and fauna will be monitored at more than 1000 points, agricultural production at more than 100 points and more than one tonne of samples will be taken and analysed to monitor the physico-chemical and biological quality of the soil. The 10,000 data points will be collected in accordance with best practice.

In order to ensure the traceability and sustainability of the data collected by an independent authority. A public "eco-data registry" should be established and maintained, see Chap. 9, where samples from the local agricultural food chain (milk, cheese, maize, vegetables, fruit…), from forest ecosystems (leaves, mushrooms, wood, game …) and from the aquatic environment (water, fish …) can be stored. The "eco-data registry" is to be operated for at least 500 years in order to be able to follow the changes in the environment during the phases of the repository. In national and international partnerships, scientific studies can be carried out using a monitoring system (IAEA, EURATOM). A quality management system is being developed for this purpose, which is to be certified by an independent authority within reasonable periods of time. The explanations on repository monitoring can be found in Chap. 9.

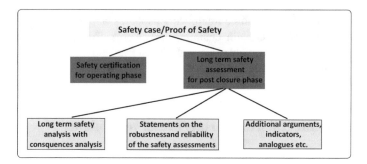

Fig. 8.16 Relationship between safety case, long-term safety assessment and long-term safety analysis

8.8 Safety Case

The "Safety case" for a repository project describes the project and demonstrates that the repository with the underlying safety concept fulfils the safety requirements for both the operation and a long period after closure of the repository, see also [73]. In each case, the safety case represents the central set of documents for the approval or decision for the individual development steps, see Fig. 8.16.

8.9 Outlook

The disadvantage of final disposal in deep geological formations is that—measured by the longevity of the waste—observation of the processes taking place in the repository, such as geochemical interactions, is only possible for a relatively short time and then only to a limited extent. The input values determined in the sensitivity analysis, to which the result reacts most sensitively, and the supplementary uncertainty analysis are further aids for limiting the range of possible solutions. However, they cannot completely rule out misjudgements. Lasting disturbances only become recognizable at a very long time interval after the occurrence of the disturbance, if at all. Even a wrong location decision would then no longer be correctable and repair measures in the repository itself would no longer be practically possible. The precautions for reversibility, the retrieval, recoverability and retrievability are immense and it is not certain that they will be maintained over a very long period of time.

A decision for final disposal in deep geological strata places high demands on the site selection procedure, on the proof of suitability for the repository and, in particular, on the procedure for the long-term safety proof. It must be taken into account that the essential basics for the long-term safety verification and the achievable prognostic reliability are already set out with the site determination. The procedures must therefore be methodically correct and coherent, fulfil the legal and social framework conditions in an appropriate and binding manner, and be comprehensible to technical outsiders. This also includes communicating the scientific and technical fundamentals and the international state of the art in science and technology.

A prognosis about the development of man and human society with its demands is at best possible for periods of 4–6 generations. It is therefore advisable to select a repository system in such a way that its normal development does not lead to a release of radionuclides from the effective-inclusion mountain range in the first 10^4 years. For the verification period of 10^6 years, the release rate of radionuclides should be small compared to the naturally occurring radionuclide concentrations, see also EPA [70].

It is therefore worthwhile to examine extensively and carefully which alternatives open up for final disposal in deep geological strata and whether these are acceptable to society. The careful examination of the treatment of heat-generating radioactive waste with the P&T process towards more short-lived and less heat-generating waste should not be omitted lightly, also against the background of international cooperation and programmes, which Germany also supports and finances. Possibly a pilot repository should be set up from which important detailed findings can be obtained. As there is no hurry to build a repository for heat generating HAWs, there are several ways to develop a German repository for heat generating HAWs:

- Since a worldwide phase-out of nuclear energy is not to be expected, one can currently speak of a 1st generation of repositories. At least in other countries such as France, Sweden, Finland, the USA, etc. more repositories are needed than those which are currently under construction or be in the approval phase, so that this will be accompanied by further development of disposal options and procedures. As things stand at present, it is to be expected that Germany will not meet the planned commissioning date in 2050 for a repository for heat-generating HAW. The approval phase for the Konrad shaft alone took approx. 25 years. It can rather be assumed that the commissioning will not take place before 2075.
- In such a long period of time, considerable knowledge gains and further development of procedures and methods can be expected, so that it seems advisable to involve the next generation(s) in the decision-making process. This may also include the inclusion of partitioning & transmutation in the process. Strictly speaking, however, this procedure is not currently available. On the one hand because the process is not yet technically mature and on the other hand because reprocessing has been banned in Germany since July 2005. It violates the "harmless recovery" required in § 9a (1) of the Atomic Energy Act; § 9a (1): *"Whoever erects, operates, otherwise holds, essentially modifies, decommissions or disposes installations in which nuclear fuel is handled, or handles radioactive material outside such installations, or operates installations for the generation of ionising radiation, shall make provisions to assure that residual radioactive material as well as disassembled or dismantled radioactive components are utilised without detrimental effects in conformity with the purposes referred to in § 1, subparas. 2 to 4, or are disposed of as radioactive waste (direct disposal) in controlled manner. The delivery of irradiated nuclear fuel originating from the operation of installations for the fission of nuclear fuel for the commercial generation of electricity to*

an installation for the reprocessing of irradiated nuclear fuel for the purposes of non-detrimental utilisation shall become unlawful as of 1 July 2005".
In the long term, however, a regional scenario in Europe could be considered with shared facilities and costs, for example: shared use of fuel cycle facilities (fuel production and processing, transmutation reactors).

• Until a final disposal facility for heat-generating HAW is commissioned, all operating permits for the existing interim storage facilities in Germany will expire. In its statement of 11 July 2002, the RSK stated that the safety of German interim storage facilities for irradiated fuel assemblies in storage casks is ensured in case of a deliberate crash of large aircraft, as "… the transport and storage casks ensure the essential protective function of the safe confinement of the radioactive material due to their design under mechanical and thermal loading even in the case of a deliberate crash of a large aircraft". Nevertheless, the inclusion of interim storage facilities in the waste management concept in Germany requires that the safety requirements be raised so that they are brought up to the state of the art. The establishment of 3–5 near-surface interim storage facilities in caverns or bunkers is suggested, whereby a pilot project for near-surface final disposal could be carried out simultaneously.

• It seems inappropriate to exclude near-surface final disposal for the following generations and for all time, especially as reversibility of the waste containers should be guaranteed for more than 500 years. With appropriately equipped interim storage facilities, essential information and experience on final disposal can be gained in order to ensure the best possible transition to final disposal. Appropriate instrumentation, data acquisition and evaluation are required. The use of sensor technology offers previously untapped possibilities [74].

• The establishment of a long-term monitoring system with comprehensive environmental monitoring, which provides information on protracted processes, is imperative.
The task of environmental monitoring will be to ascertain the initial state of the environment at the site of a future repository, then monitor its development during the construction phase, the subsequent operating phase and, over a reasonable period of time, the post-closure phase, at least over 500 years. This will make it possible to demonstrate even the smallest impact of the repository on the environment.

• This considerable effort is also necessary in order to transfer suitable procedures and methods as well as extensive data material to the next generation, constantly improving the security of the site and to further develop repository research. The current generation is doing pioneering practical work in repository research.

The repository concept, which should be based as a multibarrier system of natural and technical/geotechnical barriers, is generally considered safe. The fact that the technical/geotechnical barriers will in particular contribute to the required isolation capacity is to be emphasised once again here. It will be possible in the future to produce containers with waste matrices that hold and fix the enclosed radionuclides immobile for up to 500,000 years, even in the case of water ingress. It therefore seems appropriate that in Germany the period of observation for the post-closure phase to determine long-term security (proof-sharing) should be divided as follows:

- into a period of 10,000 years, for which reliable scientific data and findings are available, with a regulatory limit value for the local dose rate of 0.15 mSv/a (release of radionuclides from the ECZ not to be expected in this period) and
- into a period of 10,000–1,000,000 years, in which the statements are associated with considerable uncertainties in data, processes and models, with a regulatory limit value of 3.5 mSv/a (release rate of radionuclides in this period small compared to the naturally occurring radionuclide concentrations).

Technical barriers in a repository system that have a guaranteed insulation capacity of up to 500,000 years are also a sign, that near-surface repositories are by all mean justified, especially as they would be controlled during the post-closure phase and accessible for aftercare.

The safety case should also include the scenarios: Human intrusion and climate change; cyclic occurrence of glacial and warm periods.

The following activities are proposed:

- the selection of sites for the final disposal of heat developing HAW in deep geological formations, to exclude as far as possible already known promising raw material areas (minerals, fossil fuels…etc.),
- that the file (final documentation) is kept and remains in the registry of the responsible mining office,
- the entire field of deep geological formations and their surface projection remains permanently the property of the federal state,
- there is a permanent above-ground labelling obligation and a ban on test pitting and/or drilling by the responsible mining authority.

The human intrusion scenario practically requires a "guardian" concept, as indicated above and as excluded in the current repository philosophy.

References

1. Reinhold, Klaus; Sönnke, Jürgen: Methodenentwicklung und Anwendung eines Sicherheitskonzeptes für einen generischen HAW-Endlagerstandort in Tonstein (AnSicht), BGR, Hannover/Berlin; Juni 2012; https://www.bgr.bund.de/DE/Themen/Endlagerung/Downloads/Lanzeitsicherheit/2_Langzeitprognosen_Szenarienanalysen/2012-06-11_AnSichT_Geologische_Referenzprofile.pdf?__blob=publicationFile&v=3
2. Methods for Safety Assessment of Geological Disposal Facilities for Radioactive Waste Outcomes of the NEA MeSA Initiative; ISBN 978-92-64-99190-3; © OECD 2012 NEA No. 6923
3. Ulrich Noseck at. al.: Assessment of the long-term safety of repositories; Scientific basis; GRS - 237 ISBN 978-3-939355-11-3; December 2008
4. Bonano, E.J.; Baca, R.G.: Review of Scenario Selection Approaches for Performance Assessments of High-Level Waste Repositories and Related Issues. - NUREG/CR-6333 (CNWRA 94-020), 1995

5. Emissions- und Immissionsüberwachung sowie sanierungsbegleitende behördliche Kontrollmessungen für die Niederlassungen der Wismut GmbH; Freistaat Sachsen; Staatliche Umweltbetriebsgesellschaft Geschäftsbereich Umweltradioaktivität; Juni 2006
6. Jan-Martin Hertzsch, Zur Wahrscheinlichkeit der Beeinträchtigung der Integrität untertägiger Endlager durch Impaktereignisse, BGR - Hannover, 2013
7. Abschlussbericht der Kommission Lagerung hochradioaktiver Abfallstoffe „Verantwortung für die Zukunft", Deutscher Bundestag (2016)
8. Sicherheitsanforderungen an die Endlagerung wärmeentwickelnder radioaktiver Abfälle, Stand 30. September 2010, BMU (2010)
9. Gesetz zur Suche und Auswahl eines Standortes für ein Endlager für Wärme entwickelnde radioaktive Abfälle (Standortauswahlgesetz - StandAG); vom 23. Juli 2013
10. Planfeststellungsbeschluss für die Errichtung und den Betrieb des Bergwerkes Konrad in Salzgitter als Anlage zur Endlagerung fester oder verfestigter radioaktiver Abfälle mit vernachlässigbarer Wärmeentwicklung vom 22. Mai 2002; Niedersächsisches Umweltministerium (2002)
11. Endlager Morsleben (ERAM)-betriebliche radioaktive Abfälle, Bundesamt für Strahlenschutz (BfS), 2009
12. Stilllegung der Industriellen Absetzanlagen (IAA) der SDAG Wismut – BBergG, VOAS, HaldAO, StrlSchV, AtG etc.
13. Brennecke, P., Anforderungen an endzulagernde Abfälle (Endlagerungsbedingungen: Stand Dezember 2014), Schacht Konrad; Bundesamt für Strahlenschutz – Fachbereich Sicherheit nuklearer Entsorgung, Salzgitter, 2014
14. GRS „Vorläufige Sicherheitsanalyse Gorleben (VSG)", Köln 2013
15. B. Baltes,; K.-J. Röhlig; A. Kindt; Sicherheitsanforderungen an die Endlagerung hochradioaktiver Abfälle in tiefen geologischen Formationen - Entwurf der GRS - Auftrags-Nr.: 854752; Köln; Januar 2007
16. Bruno Baltes et. al., Auswahlverfahren Endlagerstandorte, Empfehlungen des AkEnd – Arbeitskreis, Köln 2002
17. Evaluation der Rand- und Rahmenbedingungen, Bewertungsgrundsätze sowie der Kriterien des Arbeitskreises Auswahlverfahren Endlagerstandorte (AkEnd), Diskussionspapier der Entsorgungskommission vom 10.12.2015
18. Physikalisch-technische Bundesanstalt: „Einleitung eines Planfeststellungsverfahrens gemäß § 9 Atomgesetz (AtG) für die Schachtanlage Konrad als Endlager für radioaktive Abfälle, 31.08.1982
19. http://www.posiva.fi/en/final_disposal/final_disposal_facility#.Wtw3Ti6uzcs
20. Forschungsvorhaben; Förderkennzeichen: 02E10921/02E10931: Schachtverschlüsse für Endlager für hochradioaktive Abfälle – ELSA Teil 1; Freiberg, Peine, den 30.04.2013
21. Bundesamt für Strahlenschutz (BfS): Grundsätze für die sichere Endlagerung radioaktiver Abfälle - Die Sicherheitsphilosophie des BfS; 2. Entwurf; Dezember 2004
22. BGR; Endlagerung radioaktiver Abfälle in Deutschland Untersuchung und Bewertung von Regionen mit potenziell geeigneten Wirtsgesteinsformationen; Hannover/Berlin, April 2007
23. Jörg Mönig et. al.: Grundzüge des Sicherheits- und Nachweiskonzeptes Bericht zum Arbeitspaket 4- Vorläufige Sicherheitsanalyse für den Standort Gorleben; GRS – 271 ISBN 978-3-939355-47-2; Dezember 2012
24. Gerhard Spaethe: Die Berechnung der Versagenswahrscheinlichkeit einfacher Tragwerke unter statischer Belastung, Technische Mechanik 1 (1980), Heft 1
25. Dimitri Suchard, Silvio Sperbeck, GRS – 329 Entwicklung eines PSA- Bewertungsansatzes zur Zuverlässigkeit baulicher Anlagen, Juni 2014
26. Sitz, P. et al.: Entwicklung eines Grundkonzeptes für langzeitstabile Streckenverschlussbauwerke für UTD im Salinar, Bau und Test eines Versuchsverschlussbauwerkes unter realen Bedingungen. Abschlussbericht zum BMBF-geförderten Vorhaben 02C05472 und 02C0902, 2003
27. Nina Müller-Hoeppe (DBETEC) et. al.: Integrität geotechnischer Barrieren, Teil 2; Vertiefte Nachweisführung Bericht zum Arbeitspaket 9.2, Vorläufige Sicherheitsanalyse für den Standort Gorleben; GRS – 288 ISBN 978-3-939355-64-9

28. Ingo Kock (GRS) et. al.: Integritätsanalyse der geologischen Barriere; Bericht zum Arbeitspaket 9.1; Vorläufige Sicherheitsanalyse für den Standort Gorleben; GRS – 286 ISBN 978-3-939355-62-5; 2012
29. Wilhelm Bollingfehr (DBE TEC) et. al.: Endlagerkonzepte; Bericht zum Arbeitspaket 5;Vorläufige Sicherheitsanalyse für den Standort Gorleben; GRS – 272 ISBN 978-3-939355-48-9; Juli 2011 mit Corrigendum Dezember 2011
30. Raiko H. et. al. 2010; Design analysis report for the canister. Stockholm, Sweden: Swedish Nuclear Fuel and Waste Management Co. (SKB). Technical Report TR-10-28. 79 p. ISSN 1404-0344
31. U.S. Department of Energy.: "Stabilize high salt content waste using polysiloxane stabilization", DOE/EM-0474, U.S. Department of Energy, Office of Environmental Managment, Office of Science and Technology, 1999
32. Pawel Kucharczyk, Qualifizierung von Polysiioxanen für die langzeitstabile Konditionierung radioaktiver Abfälle, Berichte des Forschungszentrums Jülich; 4195 ISSN 0944-2952, Dezember 2005
33. Wilhelm Bollingerfehr et. al., Entwicklung und Umsetzung von technischen Konzepten für Endlager in tiefen geologischen Formationen in unterschiedlichen Wirtsgesteinen (EUGENIA) –Synthesebericht, Förderkennzeichen 02E 10346, Peine, 2011
34. CROP, "Swiss Country Annex for WP1 Design and construction of engineered barriers"
35. [NAGRAb] Lager HAA, http://www.nagra.ch
36. Pettersson, S.: Spent Nuclear Fuel Management Strategies in Sweden and the deep repository for spent fuel. ESDRED Workshop "Technology Related to Deep Geological Disposal of High Level Long Lived Radioactive waste", University Politehnica of Bucharest, 8–9 November 2006
37. Regulation for the safe transport of radioactive Material; IAEA SAFETY STANDARDS SERIES No. TS-R-1; 2009 Edition
38. Sources and effects of ionizing radiation; United Nations Scientific Committee on the Effects of Atomic Radiation; UNSCEAR 2008; Report to the General Assembly with Scientific Annexes Volume I; United Nations; New York 2010
39. GNS Gesellschaft für Nuklear-Service mbH, www.gns.de, Produktinfo: CASTOR® V/19 Transport- und Lagerbehälter für Brennelemente (DWR)
40. W. König, Zwischenlager im Entsorgungskonzept für Deutschland, Deutsches Atomforum e.V. Wintertagung, Berlin 2001
41. BGR „Literaturstudie Wärmeentwicklung- Gesteinsverträglichkeit", Beratung der Endlagerkommission gemäß § 3 Standortauswahlgesetz, K-MAT 55, Hannover, März 2016
42. Jost Halfmann; Klaus-Peter Japp (Hrsg.); Risiko Entscheidungen und Katastrophenpotentiale – Elemente einer soziologischen Risikoforschung, Westdeutscher Verlag GmbH, Opladen, 1990; ISBN 978-3-531-12216-8
43. Scope of Radiological Protection Control Measures; ICRP Publication 104: ICRP 37 (5), 2007
44. Radiation protection recommendations as applied to the disposal of long-lived solid radioactive waste; ICRP Publication 81: ICRP 28 (4), 1998
45. Dr. Odile Mekel; Evalution von Standards und Modellen zur probabilistischen Expositionsabschätzung, Abschlussbericht, im Auftrage des BMU, Bielefeld 2007
46. "Leitlinie zum menschlichen Eindringen in ein Endlager für radioaktive Abfälle" Recommendation of the Nuclear Waste Management Commission (ESK), 26.04.2012
47. PETIT, J. R. et. al. 1999: Climate and atmospheric history of the past 420,000 years from the Vostok ice core, Antarctica. In: Nature, Heft 399, S. 429–436, (1999)
48. Barnola, J.-M., D. Raynaud, C. Lorius, and N.I. Barkov. 2003. Historical CO_2 record from the Vostok ice core. In Trends: A Compendium of Data on Global Change. Carbon Dioxide Information Analysis Center, Oak Ridge National Laboratory, U.S. Department of Energy, Oak Ridge, Tenn., U.S.A.
49. Ingo Müller-Lyda et. al. Endlagerung wärmeentwickelnder radioaktiver Abfälle in Deutschland, Anhang Langzeitsicherheitsanalyse Die Methodik zur Durchführung von Langzeitsicherheitsanalysen für geologische Endlager, GRS – 247 ISBN 978-3-939355-22-9, Köln, September 2008

50. Mazurek, M.; Pearson, F.J.; Volckaert, G.; Bock, H.: Features, Events and Processes Evaluation - Catalogue for Argillaceous Media. - OECD-Nuclear Energy Agency, NEA 4437, 376 pp., Paris, 2003

51. acatech STUDIE, Forschung – Entwicklung –Gesellschaftliche Implikationen, Partitionierung und Transmutation, Zentrum für Interdisziplinäre Risiko- und Innovationsforschung (ZIRIUS) der Universität Stuttgart, Herausgeber: Prof. Dr. Dr. hc Ortwin Renn, acatech – DEUTSCHE AKADEMIE DER TECHNIKWISSENSCHAFTEN, 2014

52. Mrugalla, S.: Geowissenschaftliche Langzeitprognose. Bericht zum Arbeitspaket 2, Vorläufige Sicherheitsanalyse für den Standort Gorleben, GRS-275, ISBN 978-3-939355-51-9, Gesellschaft für Anlagen- und Reaktorsicherheit (GRS) mbH: Köln, Juli 2011

53. Anon: Offshore Standard OS-F101, Submarine Pipeline Systems Det Norske Veritas (DNV); 2000

54. Empfehlung der Entsorgungskommission: Anforderungen an Endlagergebinde zur Endlagerung Wärme entwickelnder radioaktiver Abfälle; K-MAT 56, 17.03.2016

55. Kreienmeyer, M. et. al.: Überprüfung und Bewertung des Instrumentariums für eine sicherheitliche Bewertung von Endlagern für HAW: ISIBEL; AP 5: Nachweiskonzept zur Integrität der einschluss-wirksamen technischen Barrieren; DBE Technology; TEC-15-2008-AP FKZ 02 E 10065; April 2008

56. Philipp Herold & Nina Müller-Hoeppe: Safety demonstration and verification concept - Principle and application examples – Peine, Germany, December 2013

57. G. Krucker; Wahrscheinlichkeitsrechnung, Hochschule für Technik und Architektur Bern Informatik und angewandte Mathematik, Ausgabe 1996/98, http://www.krucker.ch/skripten-uebungen/IAMSkript/IAMKap9.pdf

58. Rübel, A.; Mönig, J.: Prozesse Modellkonzepte und sicherheitsanalytische Rechnungen für ein Endlager im Salz; GRS mbH; Okt. 2010

59. Storck, R.; Buhmann, D.; Hirsekorn, R.-P.; Kühle, T.; Lührmann, L.: Das Programmpaket EMOS zur Analyse der Langzeitsicherheit eines Endlagers für radioaktive Abfälle. Version 5. Gesellschaft für Anlagen- und Reaktorsicherheit (GRS) mbH, GRS-122, Braunschweig, 1996

60. Grambow, B.: Vorläufiger Quellterm LWR-Brennstoff zur Beschreibung der Korrosion im integrierten Nahfeldmodell (Rev.). Forschungszentrum Karlsruhe, Institut für Nukleare Entsorgung, Karlsruhe, 15.08.98

61. Becker, D.-A.; et al: Endlager Morsleben, Sicherheitsanalyse für das verfüllte und verschlossene Endlager mit dem Programmpaket EMOS. Planfeststellungsverfahren zur Stilllegung des Endlagers für radioaktive Abfälle Morsleben. Verfahrensunterlage P 278, Bundesamt für Strahlenschutz (BfS), Salzgitter, 2009

62. Hagemann, S., Xie, M., Herbert, H.-J.: Unsicherheits- und Sensitivitätsanalyse zur Korrosion von Salzbeton durch salinare Lösungen. Projekt Morsleben PSP-Element 9 M 232 100 51, Endlager Morsleben, GRS-A-3458, Gesellschaft für Anlagen- und Reaktorsicherheit (GRS) mbH: Braunschweig, März 2009

63. Popp, T.: Bewertung der Gasproblematik in einem Endlager im Salz auf Basis von In-situ-Tests. In: Rübel A.; Mönig, J. (Hrsg.): Gase im Endlager im Salz. Bericht über einen Workshop der GRS und des PTKA-WTE am 17.-18. April 2007 in Berlin

64. Klinge, H.; et al: Standortbeschreibung Gorleben: Die Hydrogeologie des Deckgebirges des Salzstocks Gorleben. Geologisches Jahrbuch, Reihe C, Band 71, E. Schweizerbart'sche Verlagsbuchhandlung, Stuttgart, 2007

65. Fein, E.; Müller-Lyda; I. Rübel, A.: Endlagerung wärmeentwickelnder radioaktiver Abfälle in Deutschland; Anhang Langzeitsicherheitsanalyse: Die Methodik zur Durchführung von Langzeitsicherheitsanalysen für geologische Endlager; Anhang zu GRS-247; Sept. 2008

66. Kolditz, O.: Strömung, Stoff- und Wärmetransport im Kluftgestein. – Gebrüder Borntraeger, Berlin, Stuttgart, 1997

67. Allgemeine Verwaltungsvorschrift (AVV) zu § 45 Strahlenschutzverordnung: Ermittlung der Strahlenexposition durch die Ableitung radioaktiver Stoffe aus kerntechnischen Anlagen oder Einrichtungen vom 21. Februar 1990. - Bundesanzeiger Jahrgang 42, Nr. 64a, 31.3.1990

68. BfS - Radioaktivität in Lebensmitteln –Verzehrsraten; http://www.bfs.de/DE/themen/ion/umwelt/lebensmittel/lebensmittel_node.html

69. Gesetz zur Neuordnung der Organisationsstruktur im Bereich der Endlagerung, vom 26. Juli 2016; Bundesgesetzblatt Jahrgang 2016 Teil I Nr. 37; S. 1843 ff

70. Environmental Protection Agency (EPA): EPA's Proposed Public Health and Environmental Radiation Protection Standards for Yucca Mountain. - EPA Yucca Mountain Fact Sheet #2, EPA 402-F-05-026, 2005

71. International Atomic Energy Agency (IAEA): Critical groups and biospheres in the context of radioactive waste disposal. - Fourth report of the Working Group on Principles and Criteria for Radioactive Waste Disposal, IAEATECDOC-1077, Vienna, 1999

72. International Atomic Energy Agency (IAEA): "Reference Biospheres" for solid radioactive waste disposal. - Report of BIOMASS Theme 1 of the Biosphere Modelling and Assessment (BIOMASS) Programme, Part of the IAEA Co-ordinated Research Project on Biosphere Modelling and Assessment (BIOMASS). - IAEA-BIOMASS-6, Vienna, 2003

73. W. Bollingerfehr et. al., Überprüfung und Bewertung des Instrumentariums für eine sicherheitliche Bewertung von Endlagern für HAW – ISIBEL, Peine, April 2008

74. Offenlegungsschrift, DE 10 2011 100 731 A1, DPMA 06.05.2012

Chapter 9
Environmental Monitoring

Environmental monitoring is an integral part of the repository concept for all geotechnical environmental structures used for the final disposal of radioactive residues and waste. In this context, reference can be made to the explanations given in Chap. 5, where such environmental monitoring has been presented for the safe long-term storage of uranium tailings ponds. Therefore, the statements made here are initially limited to the final disposal of radioactive waste, in particular highly radioactive, heat-generating waste (HAW-HG).

It is through carrying out monitoring that the operators and regulators can check that the safety-relevant requirements for a repository are being met. This applies to both the operating phase and the post-closure phase. In the following sections, the requirements for a monitoring programme are presented in detail as an integral part of a repository concept.

9.1 Repository Concept and Monitoring

The concept of maintenance-free, direct final disposal of high-level radioactive, heat-generating waste (HAW-HG), which has long been favoured by many countries, has been seriously questioned in recent years. Some countries such as France, Sweden, Finland, Switzerland amongst others, are moving away from the notion of maintenance-free final disposal. They are considering new concepts, which are usually in the form of a gradual approach with the retaining of options for targeted retrieval and recovery of waste. It is therefore assumed that private or public operators have to monitor both the mine structure and the deposited waste over significant periods of time. If this is done in a knowledge-based, systematic and risk-conscious manner, it is subsumed under "monitoring" of a repository system. It is possible that monitoring may shift the point in time, where the waste has to be completely transferred to waste disposal technology, well into the future Berkhout [1]. In Germany, this cannot be ruled out although there is currently no need for timely final disposal of high-level radioactive, heat-generating waste.

© Springer Nature Switzerland AG 2020
M. Lersow and P. Waggitt, *Disposal of All Forms of Radioactive Waste and Residues*, https://doi.org/10.1007/978-3-030-32910-5_9

In connection with the disposal of radioactive waste and residues, the term "monitoring" is usually used to refer to the technical collection of data on the development of a repository and its surroundings, see also Hocke et al. [2]. This implies the repetition of geotechnical and technical measurements over long periods of time. The repeated measurement of the data serves here as a basis for decisions for or against an option. However, since decisions in the field of tension between technology and geology always have effects on the social environment, Hocke et al. [2] suggest a differentiation of the term monitoring between social and technical monitoring. In contrast to technical monitoring, social monitoring does not focus on the collection of data, but considers the social processes that are necessary to transform technical data into options for action [2].

With the help of social monitoring, it is possible to identify suspected social problems to be assumed in connection with the choice of location, construction, operation and permanent safe closure of a repository, from which solutions can be derived. Social monitoring usually uses the index procedure with selected indicators (attention indicators) in order to classify and evaluate the social problems in different social status- and dynamic classes. Indicators in connection with repository sites are, for example, the presumed risk and the presumed disadvantages of the repository site.

In 2016, the "Act on the Reorganisation of the Organisational Structure in the Field of Final disposal", [3], came into force. In accordance with § 8(1) of this Act, a pluralistically composed National Supporting Body (NBG) was set up to support the site selection procedure in Germany in the public interest. According to § 8 (2), the *"central tasks of the National Supporting Body are the mediating and independent monitoring of the site selection procedure, in particular also the implementation of public participation in the site selection procedure ..."*. The NBG is thus part of the social monitoring in the site selection phase. One will have to think how this task can be further defined through the various phases of the repository construction.

In order to develop and implement a repository programme for radioactive waste and residues, it is therefore important that, in addition to engineering competence and a sound safety strategy, there is also a growing understanding that social aspects such as acceptance, participation and trust in of the regions, in society and also on the part of interest groups, should be included on an equal footing. Monitoring, especially long-term monitoring is the key if engineering competence and a consistent safety strategy are to be combined; on the one hand with social matters and ethical issues on the other. Monitoring can be used as an important instrument for public communication, which on the one hand proves the technical safety strategy, the safety of the technology and its service life, and on the other hand promotes public understanding of processes in operation and later in the decommissioning phase; and it contributes to building confidence in the predicted repository development in the post-operational phase.

The monitoring in a repository concept reacts to the task of ensuring that certain protection goals with regard to radiation exposure of man and nature, but also to ensuring compliance with binding safety provisions, see BMU [4] *"Safety Requirements for the Disposal of Heat-Generating Radioactive Waste"*. In the operating and decommissioning phase, this also includes the monitoring of personnel as persons exposed to radiation.

If deviations from these objectives are detected, this may have different consequences depending on the status of the procedure. The phase in which the repository is located is a decisive factor here. If the repository is not yet in operation, decisions that led to the potentially unsafe situation can be reversed. This also includes the abandonment of the selected site or the termination of the selection procedure. In the phases of operation, decommissioning or after closure of the repository, the reversibility of decisions becomes more and more complex and thus more difficult. Technical and social monitoring should therefore interlock in a specific way and prepare problem-solving decisions.

The prerequisite, however, is that the monitoring provides the overseeing authorities[1] with data that allow them to review security and integrate assessment procedures into governance[2] processes [6], involving the small circle of the Government authorised people, formally responsible officers, institutions, etc., as well as supervisory authorities and others who guarantee or accompany public control.

The focus of presentation here is on technical monitoring.

In this context, "monitoring" is a technological sub-concept that is currently being continually and that can help to shape, depending on its design and integration into the respective political overall concept of waste management, the possible development paths and the social interaction with uncertainties. What is certain is that monitoring can always offer assistance in making qualified decisions based on reliable information regarding the continuation, transformation or abandonment of the processes of nuclear disposal above and below ground, especially in connection with reversibility and retrievability. Since there is no uniform definition of the terms, reference is made here to the broad definition of reversibility provided by the Nuclear Energy Agency (NEA) of the Organisation for Economic Cooperation and Development (OECD).

Definition NEA-OECD [7][3]: "Reversibility means the ability to reconsider and reverse the course during the development and implementation (also progressive implementation) of a disposal facility. Reversal is the actual action of going back on (changing) a previous decision, either by changing direction, or perhaps even by restoring the situation that existed prior to that decision. Reversibility implies making provisions in order to allow reversal, should it be required."; see Fig. 9.1.

This definition also explicitly includes decisions in the post-operational phase. The NEA stresses that retrievability becomes more and more complex the longer

---

[1] In Germany, the Federal Office for Safety of Nuclear Waste Management (BfE), see [5].

[2] Since the 1990s, the term "governance" has been used in EU research, including repository research, to describe new, non-hierarchical forms of political control and "network governance". It is contrast to traditional forms of governance ("government"), for a coordination and integration of political decision-making levels "Multi-Level-Governance". Nuclear waste governance can be considered a generic term here.

[3] There are, however, differences, among countries, in the use of terms and in the application of the underlying concepts during the different repository development and implementation steps.

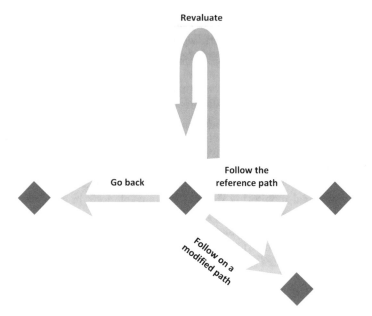

Fig. 9.1 Definition of reversibility, according to NEA [7]

the post-operational phase lasts, NEA [7]. The task of monitoring in this phase must, among other objectives, prepare the existing decision-making options on the basis of measurement results, weighing them against each other and sharpening the decision-making options. The current difficulties in obtaining relevant measurement results in the post-closure phase should be pointed out. Against this background, it is particularly difficult to weigh up whether retrieval at a specific point in time makes sense or not. In addition to retrieval, there are other possibilities. For hazardous situations, emergency planning tailored to the situation is always part of the scenarios In this context, reference is made to the Asse II mine facility, see Chap. 6.

There is still a considerable need for research into the provision of relevant technical, geotechnical data, especially in the post-closure phase. On the one hand, the installation of a suitable technical monitoring system must be carried out in such a way that the integrity of the multi-barrier system is not violated; on the other hand, wireless transmission systems with the associated power supply are not yet sufficiently developed to be able to supply data from the sealed repository system from the underground to the surface over a long period of time. Which technical possibilities for the procurement of meaningful data (also for emergency scenarios) are basically available or must be developed and for which "observation periods" can be realized in the post-closure phase is currently an open field. At present, it is also unclear which facts are to be empirically recorded in the post-closure phase, e.g. which parameters allow reliable conclusions to be drawn on heat pulses, geological deformations and stress redistributions in the host rock.

Germany has decided to guarantee a retrieval time of 500 years, France only wants to carry out the final closure of the Cigéo deep final storage facility after the end of the monitoring period (> 100 years), see Sect. 7.4.2, Finland and Sweden provide for an operating phase of around 100 years and other countries, including Switzerland, have not yet adopted any binding declarations. On the question of the "duration" of the observation phase after emplacement operation, there is currently no internationally uniform position. But there is still enough time to reliably provide solutions for the issues and developments that have been raised.

The task of monitoring to prepare existing decision possibilities on the basis of measurement results, to weigh them up against each other and to refine decision possibilities can be transferred to the site search and the repository concept. A strategic monitoring concept forms both the basis for a comparative site selection procedure with a high degree of transparency and appropriate participation as basic building blocks, and for the optimisation of a repository concept, from the planning to the closure of a repository. Chapter 2 proposes the use of the Conceptional Site Model (CSM), see also [8]. This includes a multi-stage procedure, see also Flüeler [9], in which so-called branching points are introduced, at which stages the model is checked again and a decision is made as to whether the reference path should be pursued further or modified, or whether there should again be a step back in the procedure by one or more stages, see also Fig. 9.1.

When optimizing a repository concept, the formulation of optimization goals is of particular importance. Optimization goals are, see also Chap. 2:

- Radiation protection in the operating phase
- Long-term safety
- Operational safety of the repository
- Reliability and quality of long-term containment of waste
- Safety management
- Technical and financial feasibility.

The benchmarking procedure is particularly suitable for achieving optimization goals. Here, actual conditions and targets, methods and procedures for achieving the target field, etc. are compared in order to develop iterative solution strategies. Benchmarking in the optimization of a repository concept is supported by a strategic monitoring concept. In the course of geotechnical and technical measurements, enormous amounts of data are generated over long periods of time. With the application of data mining, the position of the actual conditions in the target area can be determined. Data mining is therefore part of benchmarking. For this purpose, the huge amount of data is processed using computer-aided methods and used to optimize the repository concept. Data mining should therefore be understood here as the systematic application of statistical methods to large databases with the aim of identifying new cross-connections and trends. Data mining is carried out in all phases of the repository project (site selection, construction, operation, decommissioning and post-closure phase). The term data mining covers the entire process of the so-called "Knowledge Discovery in Databases", which also includes steps such as pre-processing and evaluation, see Fayyad [10].

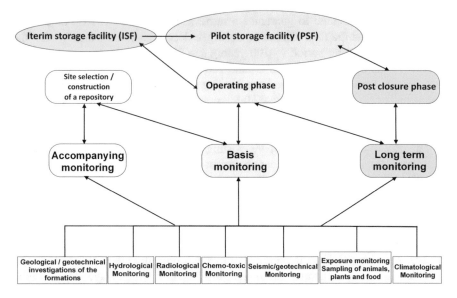

Fig. 9.2 Environmental monitoring for the disposal of radioactive waste in the various development phases of the repository concept

Chapter 5 shows an environmental monitoring system used for the long-term safe and stable storage of uranium tailings in ponds, see Chap. 5, Fig. 5.27. Based on this, the following scheme for monitoring during the phases of a repository for HAW-HG can be proposed, see Fig. 9.2.

In addition to the integration of a monitoring as a component of a repository concept, data management is of central importance for technical monitoring. In addition to the application of data mining, appropriate evaluation routines must be provided and fed back into monitoring. The question of what to measure, how to measure and which routines are to be used to evaluate the resulting data volumes has not yet been conclusively answered. The submission of a generally recognized long-term safety assessment is absolutely necessary for this purpose, see Chap. 8. Thus, it seems initially to make sense to classify the monitoring sub-concepts according to: site selection, construction of repositories, operation and decommissioning as well as the post-closure phase.

9.2   Monitoring in Different Phases of Final Disposal

For the development of a monitoring programme as an integral part of a repository concept, it makes sense to note that significant differences exist in whether the repository is unsealed and the accessibility to the emplaced waste still exists or whether the repository is sealed and the stored waste can only be reached after significant interventions into the barrier system. The following remarks are arranged accordingly.

9.2.1 Monitoring of the Unsealed Repository

The monitoring of the unsealed repository is divided into:

- Site search and selection procedure;
- Preservation of evidence before and during construction, environmental monitoring;
- Monitoring during the operating phase

  - Acceptance conditions; declaration; storage scheme
  - Shielding of repository containers; exposure in the repository- room air monitoring, personnel monitoring (radiation passport (§ 40 para. 2 Radiation Protection Ordinance))
  - Temperature; rock behaviour;
  - Environmental monitoring; food chain.

- Monitoring during the decommissioning phase;

  - geotechnical and technical barriers; rock behaviour
  - Shielding of repository containers; exposure in the repository- room air monitoring-, personnel monitoring (radiation passport (§ 40 Para. 2 Radiation Protection Ordinance))
  - Environmental monitoring; food chain.

9.2.1.1 Monitoring the Site Search- and -Selection Procedure

For the site search- and -selection procedure in Germany the Act on the Further Development of the Act on the Search and Selection of a Site for a Repository for Heat-Generating Radioactive Waste and other laws has been put to the vote, StandAG [11]. This is also to regulate the investigation and selection criteria according to which a preselection of the investigation areas, and ultimately the site, are to be preselected. The Act integrates technical and social monitoring, in particular through public participation, a National Monitoring Committee and the Council of Regions. It is already foreseeable that the StandAG will not be able to solve the task set. It requires not only an update but also a underpinning with directives and regulations, which should ultimately form a body of rules. The body of rules should include the basis for decision-making:

– general safety requirements for storage, in particular HG-HAW,
– geoscientific, water management and spatial planning exclusion criteria and minimum requirements with regard to the suitability of geological formations for final
  disposal,
– host rock-specific exclusion and selection criteria for the possible host rocks: salt
  rock, clay rock and crystalline rock,
– host rock independent balancing criteria and
– a methodology for carrying out the preliminary safety investigations.

As long as no site-specific properties have been determined, this can be done in
an index procedure for the consideration of the site selection, e.g. for the evaluation
of the barrier properties. The indexing can be classified as favourable, conditionally
favourable or less favourable; or 3 points, 2 points, 1 point etc. and be summarised in
an evaluation matrix, see Table 7.21 Chap. 7. In StandAG [11], Annex 1 to Sect. 24
Paragraph 3, are the evaluation-relevant properties for the criterion: transport of
radioactive substances by groundwater movements and diffusion in the effective
containment zone, given:

– the groundwater flow prevailing in the effective containment zone,
– the available groundwater and
– the diffusion rate.

According to Table 9.1.
A national site search and selection is subject to regional restrictions on the one
hand due to the geological formations and hydrological conditions found regionally
at sites available for selection and on the other hand due to the climatic zones in
which potential sites can be found. In Germany the focus will be on the host rock
formations salt rock, clay stone and crystalline rock (granite etc.). In other regions
this geological diversity is not found. This can be compensated by final disposal
containers which are adapted to the host rock formation. For the individual host
rock formations, differing designs of final disposal containers have to be provided.
Ultimately, however, a repository, regardless of the type of host rock, must meet the
long-term safety requirements and be capable of being approved. The observation
period, currently 1 million years in Germany, is also set independent of the host rock
type. According to the current state of affairs, the process of national site search and
selection can be arranged in accordance with Fig. 9.3.
In particular, the geoscientific evaluation criteria listed in StandAG [11] do not
do justice to the developments in repository research, see also Table 9.1. The fixing
of the evaluation-relevant properties and their admissibility ranges includes at least
the measurement methods and the methodology with which these properties are to
be determined.
From this point of view, the exclusion criterion must be that repository sites in
currently known promising raw material areas (minerals, fossil fuels … etc.) are
excluded, as in such locations an external intervention initiated by humans, e.g.
through exploration, over the long observation period is very likely. This does not
prevent possible later development of deposits found at repository sites. However,

Table 9.1 Groundwater movements and diffusion in the effective containment zone; evaluation-relevant properties; see StandAG [11], Annex 1 to Sect. 24 Paragraph 3

Valuation-relevant property of the criterion	Evaluation parameter or indicator of the criterion	Ranking-group		
		Favourable	Relatively favourable	Less favourable
Groundwater flow	Groundwater flow velocity [mm/a]	<0.1	0.1–1	>1
Available groundwater	Characteristic rock permeability of the rock type [m/s]	$<10^{-12}$	$10^{-12} - 10^{-10}$	$>10^{-10*}$
Diffusion rate	Characteristic effective diffusion coefficient of the rock type for tritiated water (HTO) at 25 °C [m^2/s]	$<10^{-11}$	$10^{-11} - 10^{-10}$	$>10^{-10}$
Diffusion rate in clay stone	Absolute porosity	<20%	20–40%	>40%
	Material consistence	Clay stone	Firm clay	Semi-firm clay

*For repository systems that are essentially based on geological barriers, sites with a rock permeability of more than 10^{-10} m/s are to be excluded from the procedure as unsuitable, according to StandAG Sect. 23 paragraph 4 number 1

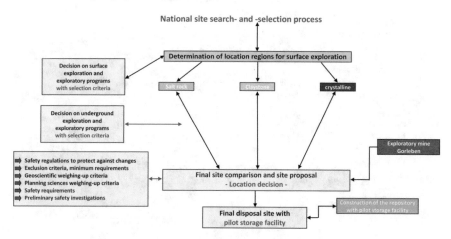

Fig. 9.3 Process of national site search and -selection

this is not affected by the principle of intent, as the Federal Mining Act currently grants protection of deposits and guarantees the supply of raw materials in § 1, amongst others.

9.2.1.2   Monitoring During Repository Construction—Pilot Waste Storage; Specified Normal Operation; Decommissioning

The concept of monitored long term disposal of HAW-HG in deep geological formations, as currently being established in some national repository sites (Finland, Sweden, France, etc., see Chap. 7), consists of a test waste storage facility, a pilot waste storage facility and a main waste storage facility. In the staged procedure, the test waste storage facility is erected as the first part of the overall facility, see Fig. 9.4. It is used for site-related investigations and carrying out the safety verifications for the main waste storage facility. The existence of a generally recognised long-term safety assessment is absolutely necessary for this purpose. The required proofs are listed in Chap. 8. The assessment criteria, safety investigations etc., compiled in Fig. 9.3, can be used for this purpose.

A pilot waste storage facility (transitional waste storage facility) was to be integrated into a repository concept. The construction of a pilot waste storage facility

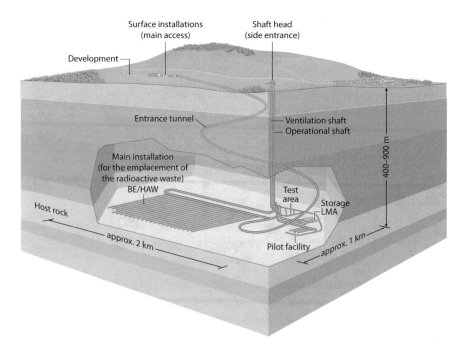

Fig. 9.4   Test waste storage facility; pilot waste storage facility; main waste storage facility—Nagra concept *source* Nagra [12]; designed according to Infel AG Claudio Köppel

(PL) for an observation phase to be defined also opens up the possibility of optimising the safety of the repository. A representative part of the planned total inventory is stored and transferred to the pilot waste storage facility at a relatively early stage. Systematic monitoring will be implemented. The technical and geotechnical barriers are monitored beyond the operating time of the main waste storage facility. For this purpose, the storage sections and the final storage containers will be surrounded by an extensive and complex measuring and monitoring system (wired probes, sensors, etc.). This measuring and observation system should be designed in such a way that the integrity of the chambers and or sections is not compromised. Thus, the behaviour of the safety barriers after closure can be analysed and results determined in an observation phase which is currently not specified in more detail. Such actions could allow conclusions to be drawn about the functioning of the main waste storage facility and thus provide early optimization and empirical evidence for its safety, as well as suggest emergency measures (such as retrievals) that would be necessary if central conceptual weaknesses were to become apparent in the pilot waste storage facility.

In addition, the knowledge forms the basis for the routine decisions as to whether the repository should be finally closed or further monitored, whether technical barriers in the main repository should be checked or whether the waste should be retrieved. For the operation of a pilot storage facility in a German repository concept, an observation phase of about 500 years is recommended, during which retrievability is to be guaranteed; because reversibility of disposal is in contrast to the retrievability of a final sealed final storage. At present, however, not all technical possibilities for the generation of meaningful data (also for emergency scenarios) have yet been explored in principle. However, the need for research is largely outlined. This approach is also sustainable, because the repositories for heat-generating HAWs which are built now, are first-generation repositories. Future generations of repositories will certainly have to be further developed if nuclear energy is to be used for energy supply in future.

The vast majority of radioactive waste is stored in the main repository after it has been stored in the pilot repository. It is backfilled section by section and sealed as soon as the waste is stored. These measures are carried out in such a way that it remains relatively easy to retrieve the waste. The access and service tunnels remain open. No measures should impair the passive safety barriers. The specified normal operation of a repository requires in particular the monitoring of personnel exposure, personal dosimetry and cyclical evaluation, as well as the monitoring of the acceptance and storage of the waste inventory (declaration and localization), see Chap. 2. From both the approval conditions for the specified normal operation and the monitoring results of a pilot storage facility as well from the safety requirements a basic equipment list of the instrumentation can be prepared. This will enable, considerable amounts of data to be obtained during specified normal operation, with which conclusions can be drawn on the long-term behavior or the data sets as well as supplementing the data sets of the pilot repository.

At the French repository Cigéo, the independent French public waste management company ANDRA has presented a monitoring programme, see [12], which is to be constructed with the equipment for the repository facility. During the entire operating phase, all safety-relevant parameters of the repository (speed of the machines, hydrogen concentration, performance of the highly effective air filters of the ventilation system, etc.) are to be monitored. In addition to monitoring in the narrower sense, parameters are monitored which may influence the development of the repository in the medium and long term: Ambient air temperature, expansion of the waste facility (meeting of the galleries), concrete durability, steel corrosion, etc. For example, individual monitoring of control containers is carried out in separate rooms, which allows easy access to each control container and its regular inspection. In the first phase of construction of the Cigéo deep storage facility, representative control structures of the various storage components (containment, chambers...) are constructed, which are used only for observation and monitoring. These control structures are extensively instrumented in order to follow their behaviour and evolution over time in every detail. This means that tens of thousands of sensors are installed in the Cigéo deep storage facility. The methods of investigation envisaged are based on proven sensors already used in the French nuclear industry and in civil engineering, for which a great deal of experience has been gained (several decades of use in nuclear power stations, dams, etc.), as well as on innovative means currently being developed within the framework of R&D programmes.

The chambers for MAW (SMA chambers), see Nagra [13], will be set up during the operating period of the Cigéo deep storage facility. The first chambers will then be fully equipped with instruments for detailed observation. In addition, ANDRA proposes to close one of the control chambers a few years after the storage of the waste containers and then to continue the observations in a closed configuration in order to draw conclusions from these observations that can be used for the actual decommissioning of the repository. The following monitoring areas are described in detail for the Cigéo deep repository.

(a)   Monitoring of the environment of the Cigéo deep storage facility

Already in 2011 in "Departement Meuse" near "Houdelainecourt" an air monitoring station was set up, with a 120 m high mast which is equipped with meteorological sensors and analytical measuring devices on the ground together with connected air samplers.

In 2007, ANDRA set up a permanent environmental observation station (Observatoire pérenne de l'environnement—OPE), see [12]. Its task is to survey the initial state of the environment at the site of the future repository over a period of 10 years and then to monitor its development during the construction of the Cigéo repository and its subsequent operation. Thus, even the smallest impact of the deep geological repository on the environment can be evaluated.

The area monitored by the environmental monitoring station covers 900 km^2 around the area where the Cigéo deep geological repository could be located. Within this area, more detailed studies will be carried out in a reference area of about 240 km^2.

The established inspection system is based on several hundred stationary observation points, satellite and aerial data, experimental plots and permanent monitoring stations (in forests, agricultural areas, air monitoring stations and waters monitoring stations). The monitoring network is adaptable. Every year, flora and fauna are monitored at about 2000 points, agricultural production at about 100 points, physical-chemical and biological soil quality is monitored—more than a ton of samples are taken and analysed, and more than 85,000 data points are collected and logged according to best practice.

In order to ensure the traceability and sustainability of the data collected by the OPE, ANDRA is setting up an "eco-data registry", which went into operation in 2013. Here samples from the local agricultural food chain (milk, cheese, maize, vegetables, fruit…), from forest ecosystems (leaves, mushrooms, wood, game…) and from the aquatic environment (water, fish…) can be stored. The "eco-data registry" is to be operated for at least 100 years in order to monitor the changes in the environment during the operation of the Cigéo deep geological repository. In national and international partnerships, scientific studies are carried out within this framework using a monitoring system with a quality certification. For further details, see [12].

(b) Health surveillance

The local population has repeatedly expressed the wish for health monitoring in the area of the repository. In the meantime, a group of experts has been set up to define the technical conditions for this monitoring. ANDRA has approached its supervising ministries to clarify the management and organisation of such an institution, see monitoring concept in [12].

Whether and how the repository (main waste storage facility) is to be decommissioned and permanently closed will also be decided by the monitoring results of the pilot storage facility. In addition, a repository concept will stipulate whether or not underground monitoring is to be carried out in the post-closure phase. If this is planned, the instrumentation must be installed during decommissioning and closure of the repository. The highest principle is the integrity of the effective containment zone.

Final repository monitoring in the post-closure phase is currently regarded as critical. Firstly, because of the measurement technology currently available. In sealed repository areas, wire-based sensors cannot be used because they would impair the sealing effect of the barriers and there are currently considerable limits to signal transmission and power supply of wireless-based sensors. Other concepts have to be developed. Data transmissions with magnetic waves at frequencies of 500 and 2500 Hz are being tested, and autonomous power supplies with thermoelectric isotope generators or beta voltaic batteries are being discussed. Furthermore, a seismic tomography technique is being developed which could be used to follow the development in a repository area from a distance. Neutrino spectroscopy has also come into focus, as neutrinos or antineutrinos are released in a repository at every beta-plus or beta-minus decay. Neutrino spectroscopy, which has already been used for geophysical investigations of natural radioactivity, could potentially also be applied here. All this is not yet fully developed and there is a considerable need for research. However, it is to be expected that, within the available time frame, solutions will be added which have not yet being considered.

On the other hand, monitoring in the post-closure phase is more likely to be seen as an additional security risk. This is also not completely unfounded, because the evaluation of information must not allow for different interpretations because there is no longer any possibility of visual inspection. This raises the legitimate question of how to deal with unexpected monitoring results which also have nothing to do with the repository. It is also assumed that this can then lead to politically motivated dangerous reactions arising from ignorance.

From the author's point of view, these questions are justified, but the closure of a repository in Germany will take at least 150 years. It will take 2–3 generations until a HAW-HG repository is closed. So it does not seem to make much sense today to speculate about the technologies available at that time. Monitoring for the post-closure phase may well be operational by then and this will considerably increase the long-term safety of a geotechnical environmental structure—the repository structure. Conveying this message is certainly an important contribution to transparency in dealing with the question of final storage, see Chapman [14].

The strategy currently favoured is the errection of a pilot storage facility parallel to the main storage facility and its operation beyond the time of closure of the main storage facility, combined with the assessment of comprehensive long-term safety, justified on the current state of the art and the long-term safety and continuous optimisation of the repository system up to the time of closure.

Environmental monitoring is unaffected by this. It should cover at least the observation period of the pilot storage facility.

In addition, the following options are pursued worldwide, see [15]:

- Final disposal of HAW-HG in deep geological formations without monitoring and precautions for retrieval (repository without monitoring—"ELo"); in accordance with repository planning in Germany valid until 2010. This has been supplemented by general requirements on retrievability, recoverability and monitoring ("control and evidence preservation programme").
- Final disposal of LAW/MAW and HAW-HG in deep geological formations with monitoring and precautions for retrieval (repository with monitoring—"Elm");
- Final disposal of HAW-HG in structures near the surface disposal, disposal option (long-term storage near the surface—"LzLo"); see Dutch interim storage facility (COVRA).

9.2.2 Monitoring the Sealed Repository

Little concrete work has so far been carried out on long-term monitoring as a whole and on the post-closure phase. The generally accepted thesis is that underground monitoring in the post-closure phase is not technically feasible at present. In StandAG [11], Sect. 26 (2) 3. stipulates for a German repository for HAW-HG: "*It shall be ensured that the stored waste can be retrieved during the operational phase and that sufficient provision is made for the possible recovery of the waste for a period of 500 years after the planned closure of the repository.*"

According to current understanding, the post-operational phase begins after the complete closure of the repository structure, whereby the repository is transferred to the passively safe state. After closure of an optimized repository (ELo b*)[4] in deep geological formations, however, the reversibility of the decisions becomes very complex and difficult and is not possible without intervention in the barrier system of the repository.

However, for a monitoring concept to be developed at the beginning of the post-closure phase, the monitoring measures, some of which were already used in the operating phase, can be started or continued in a modified form and with modified objectives (e.g. preserve evidence, environmental monitoring). So far, it can still be assumed that even after closure of the repository structure, above-ground monitoring of the surroundings of the facility will take place, how long and with which firm concept is not finally discussed. The only thing that is certain is that monitoring should be carried out for the air, water and earth pathways, and that a possible spread of radioactivity via the food chain must also be considered and also the requirements of the population living in the region for a health surveillance must not be forgotten. Monitoring concepts are available at the Federal Office for Radiation Protection. The basis for further consideration of the monitoring concept in the post-closure phase is the protection of future generations and the environment against ionising radiation; this is to be implemented in the long-term safety proof. There is international agreement that calculated or estimated risks or doses in this phase may only be interpreted as indicators of the level of protection to be achieved with final disposal. For a German repository concept for HAW-HG, the following evidence has to be provided, see BMUB [4]:

– Proof of the integrity of the effective containment zone in the post-closure phase;
– For probable developments in the post-closure phase, it shall be demonstrated that the release of radionuclides originating from the stored radioactive waste can only result in an additional effective dose in the range of 10 microsieverts per year[5] for individuals in the population. Here, it shall to be consider individuals with a present-day life expectancy who are exposed during the entire lifetime.
– For less probable developments in the post-closure phase, it shall be demonstrated that the additional effective dose caused by the release of radionuclides from the radioactive waste stored does not exceed 0.1 millisieverts per year[6] for the people affected. Here also individuals with a present-day life expectancy who are exposed during their entire lifetime should also be considered. Higher releases of radioactive substances are permitted for such developments, as the occurrence of such developments is less likely.

The long-term safety proof, see Chap. 8, must therefore include the following modules

• Integrity proof of the effective containment zone and

---

[4]See [15]—Elo b*—Repository with retrievability and recovery.
[5]See ICRP 104 [16] (triviale Dosis).
[6]See [17] according to ICRP 81 (risk less than 10^{-5}/a), see Chap. 8 Table 5.

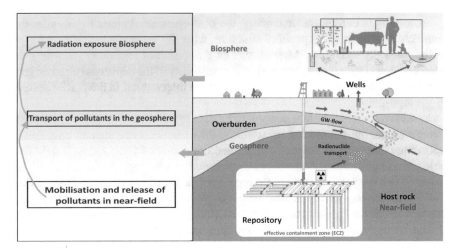

Fig. 9.5 Radionuclide spreading from the effective containment zone (near-range) into the biosphere; spreading spheres (monitoring areas)

• The spread of radionuclides in the spheres:

  • Repository (or near-range);
  • Geosphere or overburden and
  • Biosphere,

See Fig. 9.5.

Since different propagation processes take place in the three spheres, they are edited with different numerical methods. The three modules are largely independent. In a modelling of the entire repository system with the aim of calculating the radiation exposure, the modules for the three spheres are processed sequentially. If monitoring of the repository is to be installed in the post-closure phase, monitoring of the three spheres, which require completely different instrumentation, would also have to take place.

Radionuclide transport in the near-field of a repository begins at the time of container failure and mobilisation of the radionuclides. The radionuclides can be transported in dissolved form via the water phase ("transport via the water path") or via the gas phase ("transport via the gas path"). If radionuclides reach a near-surface aquifer from the stored waste, they can be expected to reach the biosphere. The limit value for the effective dose for the protection of individuals of the population is 1 millisievert per year in accordance with Sect. 46 of the Radiation Protection Ordinance. According to Sect. 47 of the Radiation Protection Ordinance, the radiation exposure from a single facility (here referred to as the repository) via the waste water and exhaust air exposure paths must not exceed 0.3 millisieverts per year. First of all, this limit value for the general population must also be complied with in the post closure phase. Evidence can only be provided if permanent measurement is carried out in any inflows and outflows to and from the repository area, the local dose rate is measured at several relevant locations, and the soil and the food chain are monitored on a random basis. It remains to be seen which part of the detection period of 1 million years would be prescribed for this activity.

Considering the monitoring of the biosphere, there is a risk of being "surprised" by a large-scale radioactivity leakage into the biosphere that is difficult to locate. Consideration must be given, for example, to constructing deep wells at relevant points in the facility, through which a near-surface aquifer of the water dispersion path could be sampled in a suitable manner and through which the radionuclide-polluted waters can even be drawn off. These could then be cleaned in a water treatment plant and discharged from there into the receiving water in a downstream area. All the technical possibilities are already available to implement such a project, see also Chap. 5. However, these are not currently part of the spectrum of measures for the post-closure phase of a repository. If the unplanned release of radionuclides from the stored waste cannot be restricted or stopped, and the limit value for the effective dose for the protection of individuals of the population is approached, then emergency measures must be initiated, which could also include the recovery of the waste or part of it.

The author does not see a need to monitor the near-range in the post-closure phase and to carry out integrity checks. Monitoring in the post-closure area could be designed according to Fig. 9.6.

Of course, the great concern with a direct monitoring measure is that the overburden is additionally perforated, thereby facilitating the spread of radionuclides. This concern must be countered if radionuclides from the stored waste have found their way into this aquifer and if there is a very high probability that they will also find their way into the biosphere or food chain. This represents an irresponsible risk.

With the monitoring of a relevant near-surface aquifer, on the one hand the early discovery of a dispersal path and thus the limitation of the emission of radioactivity

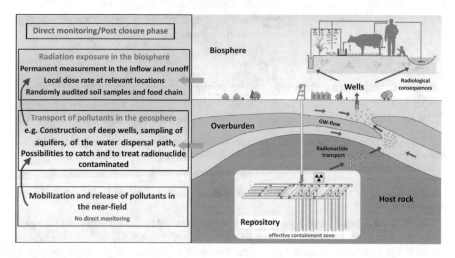

Fig. 9.6 Principle—direct monitoring in the post-closure phase (biosphere and geosphere)

is possible, on the other hand there is the possibility of isolating and removing the radioactivity. In addition, the potential gain resulting from the controllability of the repository behaviour outweighs the expense. For the author, this early detection measure results in a responsible risk and is also robust, with regard to the unambiguity of the results. This extension of direct monitoring would represent a fundamental change in the configuration compared with the previous conceptual planning.

As an alternative, and in addition to the direct monitoring of the repository structure, a pilot storage facility would be an option, see Lux et al. [18]. A pilot storage facility implemented in the repository concept could be arranged as a monitoring floor above the emplacement floor, also known as the main storage facility, which is kept open even after offsetting and closure of the emplacement floor and remains connected to the emplacement floor via observation—or measuring boreholes. This results in a two-level disposal mine structure. The controllability of the repository behaviour described here results in confidence in the reliability of the repository design and execution as well as in the possibility of a reliably well founded error correction. However, direct monitoring also requires that appropriate analysis routines be available for these investigations. This results in the generalized sequence of the HAW-HG disposal described here, see Fig. 9.7.

Fast TracKer (FTK)[7] simulators are currently being developed and used in order to obtain reliable results for the developments and processes of the effective containment zone (near-range) under repository-relevant effects. FTK simulators used in Germany consist of a coupling of the known and established simulators FLAC3D[8] for the mapping of thermomechanical processes and TOUGH2[9] for the mapping

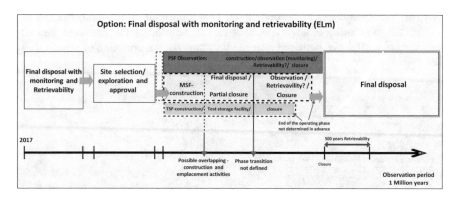

Fig. 9.7 Currently emerging disposal option for HAW-HG in Germany according to [19] MSF—Main Storage Facility; TSF—Test Storage Facility; PSF—Pilot Storage Facility, see [15]

---

[7]Fast TracKer (FTK)—A trigger and data acquisition system is equipped with a Fast Tracker coprocessor.

[8]FLAC3D is a numerical modeling software for 3D geotechnical analysis of loose and hard rock, groundwater, engineering design, risk analysis, etc.

[9]TOUGH2 is a universal numerical simulation program for multiphase liquid and heat flows in porous and fissured media.

of thermohydraulic processes, see Lux et al. [20]. The rock behaviour is simulated in combination with the filling behaviour under mechanical, hydraulic and thermal influences. With a FTK simulation it is also possible to study large-scale 3D structures with regard to their thermomechanical—thermohydraulic—behaviour under repository-relevant impacts such as cavity excavation, waste emplacement and residual cavity displacement, heat development and gas development for various host rocks and in particular to investigate statements on the fluid dynamic processes in the sealed repository for HAW-HG also in the post-closure phase with more probable and less probable development. Currently, an international benchmarking cooperation is taking place, so that the whole range of possible host rocks can be investigated.

For the development of a monitoring system as an integrated part of a national repository concept, which also includes the post-closure phase, it is advisable to remember that a German repository for HAW-HGT, according to current plans, will be closed at the earliest in 150 years and for a period of 500 years after the intended closure of the repository, provision must be made for possible recovery of the waste. Based on these facts alone, it can be expected that the scientific and technical development will make it possible to present a monitoring concept for the post-closure phase of a repository in deep geological formations that will satisfy all safety requirements. This will also rely on direct monitoring of the geosphere. However, it is likely that a considerable number of countries will need further repositories for their HAW-HG waste in the future because it cannot be expected at present that nuclear energy will be removed from their energy mix. The reach of uranium/thorium resources extends far into the future, see Lersow [21] and is not an limitation on the long term use of nuclear power.

9.2.3   Monitoring for Near-Surface Final Disposal

In the consideration of retrievability and recovery as well as the monitoring of the HAW-HG repository in the post-closure phase, storage in deep geological formations receives negative assessments. For this reason, near-surface disposal in caverns and bunkers would be an option. Even though the final disposal of HAW-HG in deep geological formations is currently favoured, it cannot be ruled out that near-surface repositories in caverns and bunkers will become the focus of final disposal in the future. Here, some elements and experiences from the long-term monitoring of uranium tailings ponds could be used or modified, see Chap. 5 and 9.3. In particular, this concerns the transition of radioactive contaminants into the biosphere, their containment and safe disposal. The procedures and methods described are of particular interest for contingency planning at deep geological repositories.

In all evaluations of near-surface final disposal in caverns and bunkers, it is noticeable that acceptance of this method by the affected population is not to be expected currently. This is due to the justification (assumptions) from theoretical comparisons that a direct release of radioactivity into the biosphere is possible from such near-surface repository structures. There is no evidence that the storage of HAW-HG in caverns and near-surface bunkers poses a greater risk than in deep geological formations, since the geotechnical and technical barriers tailored to near-surface final disposal would also have to be assessed for this purpose, but they do not yet exist. A first step in this direction could be the establishment of near-surface interim storage facilities. In this context, however, it should be noted that the operating licences of all existing interim storage facilities in Germany are not valid for a sufficiently long time until a national repository is available. Here, too, considerable amounts of data are generated.

Due to the fact that the existing interim storage facilities do not currently meet the requirements of a long-term storage facility, an interesting project has been started. In the Netherlands, in Vlissingen-Oost at the mouth of the Westerschelde, COVRA operates a long-term interim storage facility on behalf of the Dutch Ministry of Finance with 1.7 m thick external walls, earthquake-proof up to an earthquake of magnitude 6.5 on the Richter scale, safe against the impact of a fighter jet, safe against water levels of ten metres above sea level, etc. with an operating permit for 100 years, which can be extended. However, the current amount of HAW-HGT in the Netherlands is very small at around 100 m^3 and is increasing only slowly due to the low annual amount of waste produced. The containers with the HAW-HGT are located in concrete pipes, which are checked at regular intervals for radioactivity leaks, see [22]. Although the Dutch waste management concept provides for the search for a repository site for all types of radioactive waste in deep geological strata, this long-term interim storage facility will provide significant insights into handling, monitoring, central interim storage, container tightness, perhaps also acceptance, etc. might be performed. This leaves the question of a near-surface final storage in caverns and bunkers under discussion, also against the background of the expected time sequences and the expected scientific-technical solution possibilities. Long-term interim storage therefore means that in the foreseeable future no decision will be made or expected on the final handling of high-level radioactive waste. Long-term interim storage is therefore a planned condition. It follows from this that the overall system must be taken into account when designing the long-term interim storage facility (LzL) with an operating time of several hundred years. Due to their long-term nature, the focus is on potential developments that are relevant to the design and construction of nuclear facilities and have not played a role in the design of nuclear facilities to date. these considerations are noted here in connection with the transitional storage facilities discussed in Chaps. 5 and 7.

9.2.4 Central Database

One problem should be tackled very quickly: the establishment of a central database, in which all data sets determined and compiled so far should be stored. This should include data on the host rock formations that are suitable for site search in Germany, on the repository containers, parameters for determining radiation exposures, hydrological data of aquifers, parameters for determining stability in the repository structure, etc.,. This should also be available to the interested public. This creates transparency and trust and ultimately the confidence to be able to organise a participation process in which the interest group feels its views have been adequately taken into account.

9.3 Long-Term Monitoring of Uranium Mill Tailings Ponds and in Landfill Construction

The long-term monitoring for the geotechnical environmental structures for the long-term safe storage of uranium mill tailings ponds is described extensively in Chap. 5, so that the brief summary here emphasises the permitted differences in the effects on the biosphere.

First of all, for the geotechnical environmental structures for the storage of the uranium tailings ponds of the SDAG Wismut, the description "guardian principle" can be used, because the (direct) long-term monitoring also indicates the long-term safety, necessary aftercare measures are identified and arranged and because the 1 millisievert in the calendar year—criterion according to § 46 Radiation Protection Ordinance—is also to be enforced here. On the one hand, radon exhalations are monitored. This is monitored by Wismut GmbH with a limitation of the external radiation exposure (ground radiation) of ≤ 0.15 μSv/h. In addition there are environmental quality standards for the assessment of the condition of the watercourses around the uranium tailings pond sites used by Wismut GmbH (discharge of mining water), in accordance with the European Water Framework Directive (Directive 2000/60/EC).

In former uranium mining areas of Saxony and Thuringia, water-specific quality standards were agreed with the water authorities, which generally range from 10 to 20 μg/l Uranium. Now the Federal Environment Agency has submitted a proposal for an environmental quality standard (UQN)—for uranium (U-238) in surface waters with a target value of 3 μg/L; UBA [22]. In Schmidt [23] has established the connection between the remediation of the uranium ore mining legacies in Saxony and Thuringia and the general radiation protection criterion, also valid for the final disposal of radioactive waste, see Fig. 9.8. This path also seems to be expedient because an UQN for uranium (U-238) is not to be expected in the medium term and in some cases in the former uranium mining areas of Saxony and Thuringia existing guide values are far exceeded.

This shows that considerable differences are permitted in the assessment and treatment of radioactivity and their influence on the biosphere. What may be impossible under the Atomic Energy Act in Germany is permitted on areas and objects which fall under mining law. Through the European Directive 2013/59/Euratom, [3], the conceptual separation of the regulatory regime for naturally occurring radionuclides and radioactive substances whose ionising radiation or nuclear properties are used, is abolished as far as possible; this comes too late for the remediation of the remains of uranium mining in Saxony and Thuringia and has hardly any effect. The effects

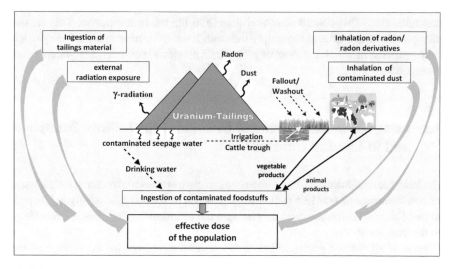

Fig. 9.8 Culminating radioactivity effects—1 millisievert/calendar year—criterion for former uranium mining areas, according to [23]

are documented using examples in Chap. 5 for the water path and in Chap. 6.1 for the treatment residues (immobilisates) from water treatment plants.

The statements on the monitoring of landfills in accordance with the Closed Substance Cycle Waste Management Act and the Landfill Ordinance can be dispensed with here because no new monitoring aspects need to be listed, including decommissioning and aftercare, see Chap. 4.

References

1. Berkhout, Frans (1991) Radioactive Waste: Politics and Technology. London; Routledge
2. Hocke, P.; Bergmans, A.; Kuppler, S., 2013: Einführung in den Schwerpunkt. In: Technikfolgenabschätzung– Theorie und Praxis 21/3 (2012)
3. RICHTLINIE 2013/59/EURATOM DES RATES vom 5. Dezember 2013: zur Festlegung grundlegender Sicherheitsnormen für den Schutz vor den Gefahren einer Exposition gegenüber ionisierender Strahlung und zur Aufhebung der Richtlinien 89/618/Euratom, 90/641/Euratom, 6/29/Euratom, 97/43/Euratom und 2003/122/Euratom
4. BMUB 2010: Sicherheitsanforderungen an die Endlagerung wärmeentwickelnder radioaktiver Abfälle, Stand 30. September 2010
5. Gesetz zur Neuordnung der Organisationsstruktur im Bereich der Endlagerung, vom 26. Juli 2016
6. Brunnengräber, Achim [Hrsg.]: Problemfalle Endlager: Gesellschaftliche Herausforderungen im Umgang mit Atommüll; Nomos; Berlin 2016
7. NEA No. 7085: Reversibility of Decisions and Retrievability of Radioactive Waste - Considerations for National Geological Disposal Programmes © OECD 2012
8. Conceptual Site Model for Disposal of Depleted Uranium at the Clive Facility - Clive DU PA Model v1.4; 5 November 2015; Prepared by NEPTUNE AND COMPANY, INC.; Los Alamos

9. Flüeler, T.: Decision making for complex socio-technical systems. Robustness from lessons learned in long-term radioactive waste governance. Series Environment & Policy, Vol. 42. Springer, Dordrecht NL, 2006

10. Usama M. Fayyad, Gregory Piatetsky-Shapiro, Padhraic Smyth: From Data Mining to Knowledge Discovery in Databases; AI Magazine Volume 17 Number 3 (1996), p. 37–54

11. StandAG: Gesetz zur Fortentwicklung des Gesetzes zur Suche und Auswahl eines Standortes für ein Endlager für Wärme entwickelnde radioaktive Abfälle und anderer Gesetze; vom 05. Mai 2017

12. http://www.andra.fr/pages/fr/menu1/l-andra/nos-missions/realiser-l-inventaire-42.html

13. Projektkonzepte für die Lagerkammern und Versiegelungsstrecken und deren Bewertung Arbeitsbericht NAB 16–45, Nationale Genossenschaft für die Lagerung radioaktiver Abfälle (nagra), Juli 2016

14. Chapman, N.,C. McCombie. (2003). Principles and Standards for the Disposal of Long-lived Radioactive Wastes; Imprint: Pergamon; 07. Oct. 2005

15. Appelt, D. et. al.: ENTRIA-Bericht-2015-01; Darstellung von Entsorgungsoptionen; K-MAT 40; Sept. 2015

16. ICRP Publication 81; Radiation protection recommendations as applied to the disposal of long-lived solid radioactive waste; Ann. ICRP 28 (4), 1998

17. https://www.oecd-nea.org/rwm/profiles/Netherlands_report_web.pdf

18. Karl-Heinz Lux, Ralf Wolters und Juan Zhao: Auf dem langen Weg zu einem Endlager für hochradioaktive, Wärme entwickelnde Abfälle; atw - International Journal for Nuclear Power Vol. 62 (2017), Issue 4 (April)

19. http://www.nagra.ch/de/wasentsorgen.htm

20. Lux, K.-H., Wolters, R., Zhao, J., Rutenberg, M., Feierabend, J. & Pan, T. (2017): TH2 M-basierte multiphysikalische Modellierung und Simulation von Referenz-Endlagersystemen im Salinar- und Tonsteingebirge ohne bzw. mit Implementierung einer Möglichkeit für ein direktes Monitoring des längerfristigen Systemverhaltens auch noch nach Verschluss der Einlagerungssohle. Ein Beitrag zur Verbesserung der Robustheit von Sicherheitsfunktionen mit sehr hoher Relevanz im Hinblick auf die Entwicklung von Bewertungsgrundlagen zum Vergleich von Entsorgungsoptionen. Hannover/ Clausthal-Zellerfeld, ENTRIA-Arbeitsbericht-07, ISSN (Print): 2367-3532, ISSN (Online): 2367-3540

21. Lersow, Michael: Energy Source Uranium – Resources, Production and Adequacy; Mining Report 153 (2017) No. 3, p. 286–302

22. Wenzel, A.; Schlich, K.; Shemotyuk, L. und Nendza, M.: Revision der Umweltqualitätsnormen der Bundes-Oberflächengewässerverordnung nach Ende der Übergangsfrist für Richtlinie 2006/11/EG und Fortschreibung der europäischen Umweltqualitätsziele für prioritäre Stoffe; Umweltbundesamt, Dessau-Roßlau; ISSN 1862-4804; Texte 47/2015, Juni 2015

23. ICRP Publication 104; Scope of Radiological Protection Control Measures; Ann. ICRP 37 (5), 2007

24. Stellungnahme zur zukünftigen Bewirtschaftung der von der Wismut GmbH beeinflussten Oberflächenwasserkörper in Thüringen in Umsetzung der EU-WRRL - Bewirtschaftungszeitraum 2015 bis 2021; Paul, M. et al., Wismut GmbH, Chemnitz, Mai 2014

Chapter 10
Summary/Outlook

In the preceding chapters, the author has attempted to discuss the various problems arising from the need to dispose of respective classes of radioactive waste and residues and consequently the associated geotechnical environmental structures—repository structures—without able to fulfil a necessary licensing practice. This task initially requires reference to a legal and regulatory framework as this is the only way geotechnical environmental structures can be compared. The author has decided to refer to the German legal and regulatory framework. Where it appeared necessary, a comparison was made with the legal and regulatory framework of the USA. On the one hand because the USA has the greatest worldwide experience in dealing with radioactivity; on the other hand because it is in the USA that by far the largest quantities of radioactive waste and residues are generated annually, due to both the peaceful and military uses of nuclear energy. In addition, because the USA uses a different classification of radioactive waste compared to that in use in Germany. This is due in part to tradition and in part because in the USA the need for classification arose earlier, before the IAEA had published a classification system.

The fact that some technical terms are interpreted differently in the individual countries has not been taken into account. This only results in minor differences in the repository concepts of the individual countries. To enable the readers to make a suitable comparison, examples of geotechnical environmental structures from different countries were presented in Chaps. 4–7 and compared with the geotechnical environmental structures erected in Germany.

The author wanted to appeal to a broad readership and not to hide the fact that a whole series of problems are still unsolved, this concerns in particular the final disposal of highly radioactive, heat-generating waste; but also to be aware that the safety structures of the existing geotechnical environmental structures—repository structures—can often not be justified with the stored radioactive inventories, but rather with whether the mining law or the Atomic Energy Act was used as the basis for the geotechnical environmental structure design and the associated long-term safety assessments. The reader is always in a position to make detailed comparisons and receive suggestions, both with regard to site selection and design and creation

© Springer Nature Switzerland AG 2020
M. Lersow and P. Waggitt, *Disposal of All Forms of Radioactive Waste and Residues*, https://doi.org/10.1007/978-3-030-32910-5_10

of repository structures, and with regard to the radio-toxic inventories to be stored in each case. A glossary is attached for clarification of the terms used in different applications.

In order to enable comparison of the radioactive inventories stored or to be stored in the various geotechnical environmental structures, irrespective of their location, a dimensionless activity index has been defined. On the basis of this dimensionless activity index, with which the activity of the respective radionuclide is related to its exemption limit according to Radiation Protection Ordinance [StrlSchV—Annex 3, Table 1, Column 2], see Chap. 7, the radiotoxicity of the various radioactive inventories can be compared with each other irrespective of location.

Efforts are being made worldwide, including in international benchmarking co-operations, to solve outstanding problems, in particular in the final disposal of high-level radioactive, heat-generating waste; also, the acute pressure for action appears to be only moderate due to the amount of high-level radioactive, heat-generating waste generated to date, in particular in Germany, and the existing interim storage capacities. A considerable pressure to act could arise in Germany if the operating licences were also to be questioned due to safety concerns at the existing interim storage facilities. In the meantime in Germany the interim storage facilities for high-level radioactive waste have been taken over by the federally owned BGZ Gesellschaft für Zwischenlagerung mbH,[1] based in Essen, in accordance with [1]. This does not change the duration of the operating permits for the interim storage facilities.

According to Nuclear Waste Management [2] and others, 12,000 metric tons of high-level radioactive waste are produced worldwide every year. As a result of Germany's announced withdrawal from the peaceful use of nuclear energy, the amount of high-level radioactive waste is known precisely and after 2022 only extremely small amounts will be added each year. According to the US Department of Energy (US DOE), approximately 90,000 tonnes of HAW are currently stored at 80 sites in 35 states in the USA, including 14,000 tonnes arising from the US government's nuclear weapons program. Approximately 3000 tonnes are added to this total annually.

The total amount of Heat-Generating, High Active Waste in the USA is expected to increase to about 140,000 tonnes over the next several decades.

Also in the USA are $1.57E+05 \, m^3$ of transuranic (TRU) waste, with a total activity of approximately 176 PBq (4.76E+06 Ci), see DOE [3].

According to IAEA estimates [4] 1.3 million m^3 of LLW (LAW) and ILW (MAW) are produced annually worldwide.

In addition the corresponding worldwide production of uranium mill tailings is between 100 and 200 million tonnes, containing 4.0 PBq of Ra-226; 4.0 PBq of Th-230, see [5]. These are large quantities of radioactive waste and residues, which have to be stored safely and remain stable over long periods of time, as described in this publication.

Globally recognized practice is to rely on national disposal concepts. This is tacitly justified by the Treaty on the Non-Proliferation of Nuclear Weapons (Nuclear Non-Proliferation Treaty), which is intended to prevent misuse of fissile materials

---

[1]BGZ Company for Interim Storage mbH.

becoming possible, including for terrorist purposes,. In general, the handling of high-level radioactive waste for final disposal, transmutation or reuse is seen as an important task for mankind. It should be noted here that internationally, in addition to final disposal, the reuse (reprocessing) and transmutation of high-level radioactive, heat-generating waste are also components of national disposal concepts. Both are prohibited by law in Germany., No plants for reprocessing such materials exist in Germany and transport to reprocessing plants in France and Great Britain has been forbidden by law since 1 July 2005. Also since the agreement between the Federal Government and the energy supply companies of 15 June 2000 (nuclear energy consensus) a ban on new construction for the construction of new nuclear power plants applies (no licences). As a result, the construction of transmutation plants in Germany is currently ruled out. The components of reprocessing and transmutation are missing from Germany's national waste management programme, while they are included in the French waste management programme. This means, that regardless of the German phase-out scenario, further international work is being carried out on both reprocessing and transmutation.

It should be noted that generally although international recommendations and guidelines by the IAEA exist, the disposal of radioactive waste is ultimately subject to national legal requirements.

The IAEA is discussing an international storage facility for high-level waste. However, the fact that an international disposal centre under IAEA control could be a possible solution has met with little positive international response, let alone acceptance. From the point of view of most states, the safety concept presented is not sufficient to deal with the high risks associated with the handling of radioactive materials. The safety precautions deemed necessary for this purpose, as they may be found in various countries, are far from being taken into account sufficiently. In conclusion, however, it can be stated that the establishment of an international waste disposal centre cannot be ruled out for all time, and that it would then be subject to both national legislation and international supervision. The extent to which this can be agreed has not been clarified conclusively.

In the preceding chapters, various national disposal concepts are presented and compared. Of course, the German disposal concept with the various existing sites and geotechnical environmental structures, which are dealt with in detail in Chaps. 4–7, has been emphasised here. The examples include the closure concepts of Wismut GmbH, the repository for radioactive waste in Morsleben and the Asse II mine facility. Also included are the conversion of the Konrad mine facility into a repository for low-level and intermediate-level radioactive waste—which is currently in the implementation phase—and the initial work on the development of a repository concept for high-level radioactive, heat-developing waste. Not only repository structures but also the radioactive inventories to be stored are presented. The comparison drawn on the basis of the dimensionless activity index shows that the differences in the inventories to be stored worldwide are small, i.e. location-independent, and therefore the requirements for the repository concepts are similar, see Chap. 7. Nevertheless, in the specific case of application, this is a site-dependent repository concept, which arises in particular from the geological and hydrogeological conditions encountered situate each site.

In Olkiluoto/Finland, the first repository for heat-generating, high-level radioactive waste is expected to be operational in 2022. The storage sections will run 400–450 m below ground level in a granite formation. The final storage containers are mainly made of copper and will be surrounded by a Betonite buffer. The concept of the repositories is adapted to the host rock formation. The ONKALO underground laboratory is directly integrated into the repository as a pilot plant. Sweden and France will follow with the commissioning of repository structures. This will also have an influence on the development of the repository concept in Germany, see Chap. 7. Many of the objections previously raised can now no longer be upheld.

It is interesting to note in this context that in the USA, Yucca Mountain/Nevada, which essentially consists of Miocene volcanic tuffs, was selected as the host rock formation for the civil repository for heat-generating HAW. Miocene volcanic tuffs differ significantly in their retention capacity from other previously favoured geological formations such as saltstone and claystone. The focus of the selection was that the transition conditions for radionuclides into the biosphere overall—into deeper groundwater horizons and to the surface—in relation to a multi-barrier system are many times lower than at comparable sites, especially because of the arid climate there. In accordance with the American disposal concept and the submitted long-term safety concept, it is certainly possible to obtain a permit. There are similarities to the Swedish and Finnish repository concepts. The USA has thus made a considerable contribution to repository research. With the submitted work the author has consistently shown that the earth has not been created for a repository. The individual host rock formations also exhibit more or less large defects, which must be compensated for by a multi-barrier concept. In the case of approval, repositories in different host rock formations are comparable on the basis of safety.

In Germany, the search for a site for a repository for high-level radioactive, heat-generating waste will focus on the host rock formations of salt rock, clay rock and crystalline rock (granite, etc.) with open-ended results. In other regions of Europe and beyond this geological diversity is not found. For each individual host rock formation different repository containers have to be provided. In this way, shortcomings in the host rock formation can be compensated for by the development of repository containers adapted to the host rock formation, as shown by the Finnish and Swedish repository concepts. Ultimately, however, a repository, regardless of the type of host rock, must meet the long-term safety requirements and be capable of being approved. The Finns and the Swedes have achieved this and, by optimising their repositories during the construction, operation and decommissioning phases, will not only meet the national protection targets for the long-term safe and stable disposal of their high-level radioactive, heat-generating waste, but will also make a significant contribution to gaining further knowledge about the long-term behaviour of the repository structures, see Chaps. 7 and 8.

It is therefore by no means absurd that an engineering, geotechnical structure of a repository close to the surface with only a small overlap could also be suitable in principle. It is possible that such a geotechnical environmental structure can be proven to be safe in the long term. Because the period of recovery is left at 500 years in Germany, a repository will meet this requirement in full. Also, the requirement for an installation of a long-term monitoring network in this type of storage of HAW-HG could not only be realized, but the system could be operated for long periods in the post closure phase. If international cooperation in the research and design of repositories is further intensified, future generations will be able to benefit greatly from the repository development of the present generation. It cannot be ruled out that near-surface storage of HAW-HG will be realised in the future.

The final development of a German repository concept for high-level radioactive, heat-generating waste started anew in 2017 with the passing of the "Act on the Further Development of the Act on the Search and Selection of a Site for a Repository for Heat-Generating Radioactive Waste …", StandAG [6]. According to current, realistic planning, the operating phase will not be able to begin before 2070, so that the repository structure will be sealed at the earliest in 150 years and that sufficient precautions will be taken for a possible recovery of the waste for a period of 500 years after the planned sealing of the repository. Based on these facts alone, it can be expected that scientific and technical developments will make it possible to solve problems that are still unresolved today. This also includes a monitoring concept for the post-closure phase of a repository in deep geological formations, see Chap. 9. Monitoring can be used as an important instrument for public communication which, on the one hand, documents the technical safety strategy, the safety of the technology and its service life, and, on the other hand, promotes public understanding of processes during operation and later in the decommissioning phase. This should contribute to building public confidence in the predicted repository development for the post-operation phase.

It is likely that a number of countries will also need further repositories in the future for their high-level radioactive, heat-generating waste, because nuclear energy cannot be expected to be removed from their energy mix. The useable amount of uranium/thorium resources extends far into the future, see Lersow [7], and is not an obstacle to long term use of nuclear power. Thus the present generation also contributes considerably to the increase in knowledge about the long-term behaviour of repository structures for future generations. What can be stated in that in Germany at present the interim storage of radioactive waste, both for low-level and medium-level radioactive waste with negligible heat generation and for high-level radioactive, heat-generating waste is being undertaken. The Konrad mine facility will not be completed before 2022, see Chap. 6 and the long road to a repository for high-level radioactive, heat-generating waste has only just begun, see Chaps. 7 and 8. The disposal option currently emerging for high-level radioactive heat-generating waste in Germany is summarised in Fig. 10.1.

The repository concept for high-level radioactive, heat-generating waste is referred to as final disposal with monitoring and retrievability. How the development will continue is open. A monitoring concept for the post-closure phase of a repository in deep geological formations is being worked on intensively. If near-surface

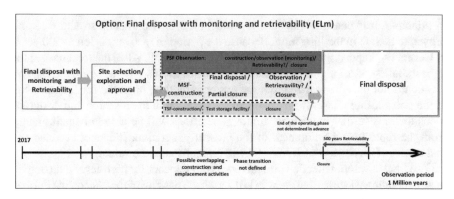

Fig. 10.1 The currently emerging disposal option for HAW-HG in Germany, see [8]

disposal in caverns or bunkers becomes the focus of consideration or if transmutation is applied, this will also have an impact on the requirements for a monitoring concept, see Chap. 7.

Another subject of intensive investigations is the development of a general long-term scientifically-based safety plan that can be adapted to the respective site and shows the remaining risks and expected probabilities of occurrence. The basic principles are summarised in [9, 10].

If a risk assessment with probabilities of occurrence for the period under review of one million years is not quantifiable, the long-term safety calculated on this basis is subject to a high degree of uncertainty. It is suggested at least to assume a shorter observation period of e.g. 10,000 years, for which the probability of occurrence of processes and events can be quantified. From the long-term safety thus determined, it is then possible to extrapolate to longer observation periods. The long-term safety factors determined in this way are then assigned uncertainties which are equivalent to the reliability with which the long-term safety factor was verified.

To this end, a methodology could be pursued which aims to present the proof of long-term safety as a scientifically proven prognosis of the long-term safety risk emanating from the repository structure and of the probabilities of occurrence of various scenarios, while presenting possible, necessary measures to avert danger; in particular for high-level radioactive, heat-generating waste and for the post-closure phase of the repository structure. On this basis, a comparison of different repository structures and the selection procedures have been undertaken, see Chap. 8.

In a matrix, Table 10.1, the radionuclide species are evaluated and compared with the safety-relevant functional elements. The safety level of the effective containment zone can be designed independently of the depth. The overburden is not considered in some safety analyses. It is assumed here that the overburden certainly contributes to the long-term safety of the geotechnical environmental structure. The degree of dryness in the geotechnical environmental structure depends on the overburden. In addition, the transition of the small quantities of substances escaping, into the biosphere is delayed. A dilution effect is also achieved in the different groundwater

Table 10.1 Evaluation matrix for various repository systems (Geotechnical Environmental Structures) for HG-HAW

Repository system (Geotechnical Environmental structure)/functional element	EBW	TBL	KuB
Effective containment zone (ECZ)	+	+	+
Overburden	+	+	+[a]
Technical und geotechnical barriers	Balanced	Only containers, waste matrix and sealing	Predominantly existing, overburden is missed
Recovery possibility of radioactive waste	~	−	+
Reversibility	~	−	+
Costs	Very high	Comparably low	Comparably low
Total/review			

[a]Since the overburden has a decisive influence on the degree of dryness in geotechnical environmental structures, even a thin layer of overburden can be advantageous in this respect if it is appropriately located and designed

Rating: + fully achievable (4); ~ sufficient (2);—not achievable (0)

EBW—repository structure; TBL—deep boreholes; KuB—cavern construction

levels. The overburden has a considerable safety function in the event of external influences such as earthquakes, aircraft crashes on the repository system or attempted external attacks (penetration of overburden), which can/should lead to damage in the effective containment zone.

For all possible events, probabilities of occurrence (PD) can be determined for the site. According to current knowledge, the probability of occurrence of events that have a negative impact on the effective containment zone has been divided into 3 categories:

- probable occurrence, significant influence on the functionality of the geotechnical environmental structure, $EW > 10^{-2}$;
- less probable occurrence; influence on the functionality of the geotechnical environmental structure not to be neglected, $10^{-3} < EW \leq 10^{-2}$;
- unlikely occurrence; not significant, negligible influence on the functionality of the geotechnical environmental structure, $EW \leq 10^{-3}$.

If this standard is applied, then it must be proven in a long-term safety demonstration that the probability of failure (AW) of the geotechnical environmental structure at location X (loss of functionality) is achieved in the verification period $NZ < 10{-3}$. Loss of functionality is understood to mean that quantities of substances leave the effective containment zone, resulting in radiation exposures in the biosphere, which lie outside the socially accepted limits. The geotechnical environmental structure (repository system), in such a case, has lost its required protective function of "permanently protecting man and the environment from the consequences of ionising radiation and the toxic effects of the radioactive waste absorbed".

Thus, the long-term safety proof for the geotechnical environmental structure (repository system) for the permanent protection of man and the environment against the consequences of ionising radiation and the toxic effects of the radioactive waste absorbed is initially broken down into two partial proofs:

– Long-term safety demonstration for the effective containment zone: demonstrates that with a failure probability of only (VW) $< 10^{-3}$ (this corresponds to an annual failure probability of 10^{-6}) the radio-toxic waste in the ECZ remains within the demonstration period—demonstration of internal safety.
– Evidence that, during the period under consideration, no external impacts can occur that violate the effective containment zone to such an extent that it can no longer fulfil its protective function, EW of $< 10^{-3}$—Evidence of safety against external impacts.

If external security cannot be demonstrated, the location is not considered. If internal safety cannot be demonstrated, this may have to lead to a different concept of the geotechnical environmental structure—repository structure. This can have a significant impact on the costs for the safe and stable long-term storage of radioactive waste, see Table 9.1.

The long-term safety record for the geotechnical environmental structure for the final disposal of radio-toxic waste must therefore take account of this:

– The safe confinement of radio-toxic waste
– Isolation of radio-toxic wastes
– The obstruction or delay of significant radionuclide releases or delay in
– Radionuclide migrations
– The guarantee that any radionuclide releases which cannot be avoided in the long term will not lead to a significantly increased risk for humans and the environment
– Precautions to prevent the possibility of unintentional human intrusion into the storage areas of waste and its associated impacts (violation integrity of ECZ).

The long-term safety certificate shall be preceded by a long-term safety analysis. Parts of the long-term analysis may already be part of the site selection. A long-term safety analysis, see BMUB [11], is understood to mean: *"The analysis of the long-term behaviour of the repository after decommissioning. The central aspect of consideration is the analysis of the containment capacity of the repository system and its reliability. It includes, for example, the development of conceptual models, scenario development, consequence analysis, uncertainty analysis as well as the comparison of the results with specified safety principles, protection criteria and other verification requirements".*

In a licensing procedure for the operation and decommissioning of a repository, the formal addressees of the long-term safety certificate are the Federal Office for

Nuclear Safety of Disposal (BfE)[2] and, in the case of a repository mine facility, the competent mining authority with the participating authorities (water, radiation protection, etc.). They have to examine the evidence submitted by the applicant. They can request additions and confirm the evidence as acceptable or refuse to confirm it. Strictly speaking, this is also a Conceptual Site Model, which is contained in the permit. However, the term model of the (site) repository system has prevailed. The preparation of a long-term safety record can also be useful outside an approval procedure to support a decision-making process (e.g. in a legislative procedure, public participation or approval procedure with municipalities). The Atomic Energy Act (AtG) requires the implementation of a licensing procedure with an environmental impact assessment (EIA). A transboundary consideration for plants close to a border is not only mandatory but can be also provide trust and confidence for others. According to Article 37 of the EURATOM Treaty, each EU Member State is obliged to provide the Commission with information on any plan for the disposal of radioactive waste; "…. as will enable the Commission to determine whether the implementation of such plans is likely to involve radioactive contamination of the water, soil or airspace of another Member State." Section 8 of the Environmental Impact Assessment Act (UVPG) contains provisions on cross-border participation by authorities in the assessment of environmental impacts; BMU [12].

The safeguarding of data records for the repository site, the repository register, is currently not yet conclusively regulated in Germany. In the case of repository mine facilities, the responsible registry ot the State Mining Agency would be the best solution, in which the data records (mine plans) for a repository mine would be permanently stored anyway.

The purpose of this presentation is to present and compare the necessary geotechnical environmental structures for the long-term safe and stable storage of the vast majority of all radioactive residues and waste. And this as far as possible as neutral scientists who cannot be assigned to one of the competing groups. Ultimately, all these geotechnical environmental structures have the overarching objective of guaranteeing the isolation of radioactive waste and residues so that they are shielded and encapsulated in such a way that the personnel and the environment (e.g. through transport) do not suffer any damage when handling the waste as well as ensuring that any radio-toxic substances cannot escape from the repository into the biosphere during the observation period (detection period), or any escapes are only within the socially and legally accepted limits. For this purpose, the repository structures are equipped with a multi-barrier system. A multibarrier system can consist of coordinated technical, geotechnical and geological barriers. The requirements, and thus the approval criteria, for geotechnical environmental structures are handled very differently. While the uranium mill tailings ponds of SDAG Wismut were approved by the authorities without base sealing, see Chap. 5, the supervision authority of Asse 2 mine facility is considering clearing the entire nuclear facilities in accordance with Section 9 of the Atomic Energy Act, and taking the recovered radioactive waste and

---

[2]The BfE is not only responsible for the long term safety assessment, also for the safety during the construction, operation and sealing of the repository.

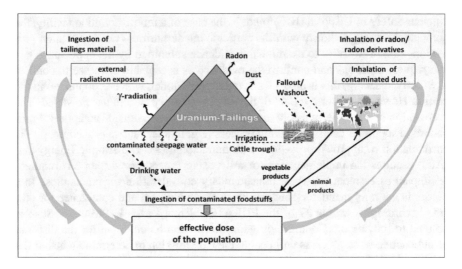

Fig. 10.2 Culminating radioactivity effects—1 millisievert/year—criterion for former uranium ore mining areas, according to [11]

transferring it to a repository, which is currently not available. A robust estimation of the real exposures of the workers and the real and potential exposures of the population during the retrieval of the radioactive waste from the Asse 2 mine facility is not possible on the basis of the current state of knowledge. see Chap. 6.

European Directive 2013/59/Euratom [13] repeals as far as possible the conceptual separation of regulatory regimes for naturally occurring radionuclides and radioactive substances whose ionising radiation or nuclear properties are used. Germany transposed this Directive into national law in 2019. For the remediation of the immense legacies of uranium ore mining in Saxony and Thuringia, this regulation comes too late and thus has hardly any regulatory effect. Chapter 5 provides extensive evidence of this fact with examples of the water path in the vicinity of the uranium mill tailings ponds and the shipment of the water treatment residues of Wismut GmbH, see also Chap. 6. The resulting radiation exposure at the respective sites requires permanent monitoring. The associated concept of safekeeping is also referred to as the "guardian principle". Chapter 5 explains in detail that, especially for the uranium mill tailings ponds of SDAG Wismut, the monitoring of contamination and radiation exposures in connection with the management of mining-based water at the former processing sites in Saxony and Thuringia still has to take place over an extremely long period of time.

The limit value for the effective dose for the protection of individuals of the population is 1 millisievert per year for all geotechnical environmental structures considered here in accordance with § 46 of the Radiation Protection Ordinance. Figure 10.2 shows an example of the uranium mill tailings at SDAG Wismut.

It is definitely recommended that the results of the handling of large quantities of radioactive material, the collection and safe disposal of leaked radioactive radionuclides in the biosphere and of the established long-term monitoring and its maintenance as well as data management from the remediation of the uranium tailings ponds of SDAG Wismut are to be included in the consideration of repository planning for radioactive waste. Especially in relation to any emergency planning. Some elements of the existing emergency planning could already be sensibly incorporated or adapted site-specifically to the emergency planning for the Asse II mine facility.

One problem deserving special mention is the establishment of a central database in which all the data records determined and compiled so far for the host rock formations that are suitable for site search in Germany can be accessed: for the repository containers, parameters for determining radiation exposures, hydrological data from aquifers, parameters for determining stability in the repository structure, etc. To link this database with the databases of the responsible mining authorities seems to be reasonable, because among other things those records also currently list known host areas of mineral deposits.

From this point of view, the exclusion criterion must be that repository sites in currently known host areas of mineral deposits are excluded, as in such locations an external intervention initiated by humans, e.g. through exploration, is very likely over the long period under consideration. Although this does not prevent mineral deposits from being found at repository sites in the later stages of development, it does not affect the principle of intent, as the Federal Mining Act in § 1, currently serves to protect deposits and secure the supply of raw materials, among other things.

A central database should also be available to the interested public. This creates transparency and trust and ultimately the confidence to be able to organise a participation process in which the interest groups feel their opinions have been adequately taken into account.

References

1. "Gesetz zur Neuordnung der Verantwortung in der kerntechnischen Entsorgung"; vom 27. January 2017 (BGBl. I S. 114, 1222).
2. https://wipp.energy.gov/library/TRUwaste/DOE-TRU-18-3425_Rev_0.pdf.
3. DOE/TRU-18-3425; November 2018.
4. ESTIMATION OF GLOBAL INVENTORIES OF RADIOACTIVE WASTE AND OTHER RADIOACTIVE MATERIALS IAEA, VIENNA, 2008 IAEA-TECDOC-1591 ISBN 978–92–0–105608–5 ISSN 1011–4289.
5. ANNUAL TRANSURANIC WASTE INVENTORY REPORT – 2018 (Data Cutoff Date 12/31/2017).
6. StandAG: Gesetz zur Fortentwicklung des Gesetzes zur Suche und Auswahl eines Standortes für ein Endlager für Wärme entwickelnde radioaktive Abfälle und anderer Gesetze; vom 05. Mai 2017.
7. Lersow, M.: Energy source uranium – resources, production and adequacy. Glueckauf Min Reporter 153(3):286–302; June 2017
8. Appel, D. et. al.: ENTRIA-Bericht-2015-01; Darstellung von Entsorgungsoptionen; K-MAT 40; Sept. 2015.

9. Methods for Safety Assessment of Geological Disposal Facilities for Radioactive Waste; ©
    OECD 2012; NEA No. 6923[8] Ulrich Noseck et. al.: Assessment of the long-term safety of
    repositories - Scientific basis; December 2008; GRS - 237 ISBN 978-3-939355-11-3.
10. Claire Corkhill; Neil Hyatt: Nuclear Waste Management; Physics World Discovery; https://
    doi.org/10.1088/978-0-7503-1638-5; isbn: 978-0-7503-1638-5; p. 1–18; January 2018.
11. Peter Schmidt, Sanierung der Hinterlassenschaften der Uranerzförderung und –verarbeitung in
    Sachsen und Thüringen durch die Wismut GmbH, Technisches Seminar, DESY Zeuthen 13.
    Febr. 2007.
12. BMU: Sicherheitsanforderungen an die Endlagerung wärmeentwickelnder radioaktiver
    Abfälle, Stand 30. September 2010.
13. RICHTLINIE 2013/59/EURATOM DES RATES vom 5. Dezember 2013: zur Festlegung
    grundlegender Sicherheitsnormen für den Schutz vor den Gefahren einer Exposition gegenüber
    ionisierender Strahlung und zur Aufhebung der Richtlinien 89/618/Euratom, 90/641/Euratom,
    6/29/Euratom, 97/43/Euratom und 2003/122/Euratom.

Glossary

(Radioactive Waste Management/Nuclear Terms—Geological Repository/Mining Terms)

Absorbed dose—is a measure of the energy deposited in a medium by ionising radiation. It is equal to the energy deposited per unit mass of a medium, and so has the unit joules (J) per kilogram (kg), with the adopted name of gray (Gy) where 1 $Gy = 1\ Jkg^{-1}$. The absorbed dose is not a good indicator of the likely biological effect. 1 Gy of alpha radiation would be much more biologically damaging than 1 Gy of photon radiation for example. Appropriate weighting factors can be applied reflecting the different relative biological effects to find the equivalent dose.

Actinide—series encompasses the 15 metallic chemical elements with atomic numbers from 89 to 103. It does include: Actinium, Thorium, Uranium and the transuranics: Neptunium, Plutonium, Americium, Curium, Berkelium, Californium, Einsteinium, Fermium, Mendelevium, Nobelium and Lawrencium.

Active safety—Safety, in particular long-term safety, of a storage or disposal facility which relies substantially or solely on the technical barriers of the storage system and measures to ensure its protective function for man and the environment (monitoring, maintenance, repair).

Activity—Number of nuclear transformations occurring in a radioactive substance per unit of time.
The unit of measurement of activity is Becquerel (Bq) or Curie (Ci). $1\ Ci = 3.7 \times 10^{10}\ Bq = 37\ GBq$.

Advective transport—see transport.

Assessment[1]—The process, and the result, of analysing systematically and evaluating the hazards associated with facilities and activities, and associated protection and safety measures.
Assessment is often aimed at quantifying performance measures for comparison with criteria.

---

[1] Full text see [1].

© Springer Nature Switzerland AG 2020
M. Lersow and P. Waggitt, *Disposal of All Forms of Radioactive Waste and Residues*, https://doi.org/10.1007/978-3-030-32910-5

In IAEA publications, assessment should be distinguished from analysis. Assessment is aimed at providing information that forms the basis of a decision on whether or not something is satisfactory. Various kinds of analysis may be used as tools in doing this. Hence an assessment may include a number of analyses.

Varieties of assessment include: Consequence assessment; Risk assessment; Dose assessment; Exposure assessment; Hazard assessment; Performance assessment; Radiological environmental impact assessment; Safety assessment etc.

- **Safety assessment**: Assessment of all aspects of facilities and activities that are relevant to protection and safety; for an authorized facility, this includes siting, design and operation of the facility.
  The systematic process that is carried out throughout the design process (and throughout the lifetime of the facility or the activity) to ensure that all the relevant safety requirements are met by the proposed (or actual) design.
- **Consequence assessment**: Assessment of the radiological consequences (e.g. doses, activity concentrations) of normal operation and possible accidents associated with an authorized facility or part thereof.
  ! Care should be taken in discussing 'consequences' in this context to distinguish between radiological consequences of events causing exposure, such as doses, and health consequences, such as cancers, that could result from doses. 'Consequences' of the former type generally imply a probability of experiencing 'consequences' of the latter type.
- **Radiological environmental impact assessment**: Assessment of the potential radiological impacts of facilities and activities on the environment for the purposes of protection of the public and protection of the environment against radiation risks.
- **Dose assessment**: Assessment of the dose(s) to an individual or group of people. For example, assessment of the dose received or committed by an individual on the basis of results from workplace monitoring or bioassay.
  The term exposure assessment is also sometimes used.
- **Hazard assessment**: Assessment of hazards associated with facilities, activities or sources within or beyond the borders of a State in order to identify:
  (a) those events and the associated areas for which protective actions and other emergency response actions may be required within the State;
  (b) actions that would be effective in mitigating the consequences of such events.

Atomic numbers—Atomic numbers indicate the number of protons in an atomic nucleus. Atoms with the same number of protons but different neutron numbers are called isotopes.

Barrier—A barrier is a natural or technical component of the final disposal (repository) system, which completely or partially prevents or delays the transport of radionuclides or other substances into the biosphere. A distinction is made between technical, geological and geotechnical barriers. Such barriers are, for example, the waste matrices, the waste containers (canisters), the chamber- and shaft gate structures, the effective containment zone (ECZ) and the ECZ surrounding or overlapping geological strata.

Bentonite—is an absorbent aluminium phyllosilicate clay consisting mostly of montmorillonite. The different types of bentonite are each named after the respective dominant element, such as potassium (K), sodium (Na), calcium (Ca), and aluminium (Al). For example in final disposal structures (Finland, Sweden) the space around the canisters will be filled with bentonite clay.

Big Bag (Wismut GmbH)—flexible bulk material container with glued inner foil lining and 4 lifting loops with the dimensions 90 cm × 90 cm × 125 cm.

Biosphere—is the layer of the planet Earth, the earth's surface and the atmosphere, inhabited by living organisms and forming a variety of habitats.

Blind shafts—Shafts that have no connection with the surface.

Backfilling—The placing of backfill material (usually relatively inert) in the mine structures to reduce the remaining underground volume.

Backfill material—Is often associated with the final disposal; the filling of cavities among others to enclose the stored repository containers in storage areas with a backfill material whose properties correspond as far as possible to those of the host rock (salt rock: salt grit, clay rock: bentonite).

Cooling-down period—Cooling-down period, in the context of radioactive waste is the period in which spent fuel must be stored in a cooling pond until, due to lower heat output, it can be packed in containers (canisters) for interim storage. With regard to the disposal of highly radioactive waste, the term also describes the cooling-down period from which the heat release of the waste has diminished to such an extent that it can be disposed of.

Castor- container (canister)—A transport and storage container for highly radioactive waste.

Cladding tubes—External enclosure of fuel rods in fuel elements made of zircaloy (alloy based on zirconium and tin).

Chemical speciation—in chemistry means either (a) the process (i.e. the formation) of chemical characteristic forms of a substance as a result of an adjusting reaction equilibrium or (b) the adjusting equilibrium state itself.

Curie—Non SI unit of activity of a radioactive substance has been replaced by the SI unit Bq. $1Ci = 3.7\ E+10\ Bq$.

Coefficient of permeability—Parameter for describing water permeability, in particular for rocks and mineral seals (bentonite); unit: meters per second; ms^{-1}.

Containment—Primary safety function of the repository system. Radioactive waste is isolated in a underground area in such a way that it essentially could remain at the place of storage and at most small amounts of material leave this storage area. Containment is the primary way of preventing contamination being released into the environment or coming into contact or being ingested by humans. See also Effective Containment Zone (ECZ).

Cavern—an (manmade) underground cavity.

Crevices—Surfaces or gaps through rock masses that are formed by tectonic movements, among other processes.

Coquilles (also Vitrified waste containers (canisters))—In nuclear engineering, the term used to describe a highly active waste from reprocessing that has been consolidated by vitrification, including its gas-tight welded metal casing made of corrosion-resistant steel.

Conditioning—designates to the conversion of radioactive waste into a chemically stable state, so that it may become water insoluble or difficult to dissolve.

Convergence, Convergence of cavity—Natural process of volume reduction of underground cavities due to deformation or loosening due to the rock pressure.

Contamination—Contamination is the presence of an unwanted constituent, contaminant or impurity in a material, physical body, natural environment, persons, workplace, etc. Contaminants are biological, chemical, physical or radiological substances, here especially a radiological substance.

Crystalline rocks—Term for rocks that were transformed during a regional metamorphism. The mineral elements of the rocks were completely crystallized. A distinction is made between fine crystalline (e.g. marble or gneiss) and coarse crystalline (e.g. many granites).

Criticality—The state of a nuclear reactor in which a self-sustaining chain reaction takes place. Subcriticality is the state in which no chain reaction can be maintained.

Closure and repository sealing—It includes all measures taken after cessation of storage, including closure of the repository to ensure a long-term maintenance-free condition.

The closure and repository sealing related to the cessation of waste disposal operations at a site and the backfilling and sealing of access tunnels and shafts. The intention of repository sealing is to prevent human access to the wastes. Sealing should also promote a return to pre-excavation hydrogeological conditions. Individual sections of a repository may be closed in sequence, but: closure usually refers to final closure of the whole repository, and will probably include removal of surface installations.

Claystone (Mudstone)—is formed by settling the finest particles (diameter max 2 μm). It differs from plastic clays by its higher strength and its lower water content. These properties are the result of pressure and compaction through superposed layers.

Decay chain—the decay chain refers to a series of radioactive decays of different radioactive decay products as a sequential series of transformations. It is also known as a "radioactive cascade". Most radioisotopes do not decay directly to a stable state, but rather undergo a series of decays until eventually a stable isotope is reached.

(Residual) Decay heat—The heat generated by the further decay of the fission products in the fuel elements, after the reactor has been shut down.

Decontamination—Removal of chemical, biological or radioactive contaminants (contaminations) from persons, objects or unprotected areas.

Detection period—The long-term safety must be demonstrated throughout the whole detection period.

Diapirism—is a type of geologic intrusion in which a more mobile and ductility deformable material is forced into brittle overlying rocks. Geological process using e.g. Salt rises due to its lower density and plasticity and can break through layers of the overburden. In this way, the salt dome of the North German lowlands has emerged. This geological process is also observed in other rocks (for example granite).

Diffusive transport—see transport.

Dispersion—Hydromechanical dispersion describes the distribution or mixing of the same ingredients (or heat) in the moving pore water. This process is caused in porous media by the different flow velocities in the pore, the pore size distribution and the path length, which can traverse the individual particulate matter.

Dispersivity—Dispersivity is a characteristic that can be measured in the laboratory or in the field. It is a geometric measure of the permeability- and storage heterogeneity of the aquifer.

The order of magnitude of the dispersion coefficient depends on the scale as well as porosity, grain shape and grain size.

It is necessary to distinguish between longitudinal dispersivity [m], transversal-horizontal dispersivity [m], and transversal-vertical dispersivity [m]. Tracer tests are suitable as a detection method.

Dose—Dose is a measure of the energy that is transmitted by the radiation to a living being and the connected potential damage. See also Dose assessment.

– **Absorbed dose**—is a measure of the energy deposited in a medium by ionising radiation. It is equal to the energy deposited per unit mass of a medium, and so has the unit joules (J) per kilogram (kg), with the adopted name of gray (Gy) where $1 \text{ Gy} = 1 \text{ Jkg}^{-1}$. The absorbed dose is not a good indicator of the likely biological effect. 1 Gy of alpha radiation would be much more biologically damaging than 1 Gy of photon radiation for example. Appropriate weighting factors can be applied reflecting the different relative biological effects to find the equivalent dose.

– **Effective dose**—Effective dose is a dose quantity in the International Commission on Radiological Protection (ICRP) system of radiological protection.

It is the tissue-weighted sum of the equivalent doses in all specified tissues and organs of the human body and represents the stochastic health risk to the whole body, which is the probability of cancer induction and genetic effects, of low levels of ionising radiation. It takes into account the type of radiation and the nature of each organ or tissue being irradiated, and enables summation of organ doses due to varying levels and types of radiation, both internal and external, to produce an overall calculated effective dose.

The SI unit for effective dose is the sievert (Sv). 1 sievert represents a 5.5% risk of causing cancer.

– **Equivalent dose**—The equivalent dose is a measure of the radiation dose to tissue, representing the stochastic health effects of low levels of ionizing radiation on the human body. It is derived from the physical quantity absorbed dose, but also takes into account the biological effectiveness of the radiation, which is dependent on the radiation type and energy. In the SI system of units, the unit of measure is the sievert (Sv).

Other units: Röntgen equivalent man (rem); In SI base units: $\text{Jkg}^{-1} = \text{m}^2\text{s}^{-2} = 1 \text{ Gy (Gray)}$.

The rem has been defined since 1976 as equal to 0.01 Sievert, which is the more commonly used SI unit outside the United States.

Dose rate—Dose per unit time. The unit habitually used to measure the dose equivalent is the Sievert or Milli-Sievert per hour (Sv/h or mSv/h).

Downcast ventilation shaft: Fresh air supply from the surface (intake air), see also upcast ventilation shaft.

Drift—Mining term for a underground mine structure, which unlike a tunnel (gallery) has no connection to the earth's surface and usually extending horizontally or with a slight inclination. General: A horizontal underground opening that follows along the length of a vein or rock formation as opposed to a crosscut which crosses the rock formation (Intake airway).

Drop (gravity) stowing—Backfill method, from a mine room above the backfill drift backfill material is dropped into this.

Effective Containment Zone (ECZ)—The geological area which in a permanent storage surrounds the storage chambers with radioactive waste. Effective Containment Zone (ECZ) should make sure in conjunction with technical and geotechnical barriers that the containment of the waste is achieved for the required period of time.

End (Working face)—Side rock wall of the drift in a mine.

Evaporite rocks (rock salt)—Chemical sediments formed by progressive water evaporation from seawater in the order of more rising solubility (first carbonates, then sulphates, halides and finally potassium and magnesium salts). There are two types of evaporite deposits: marine, which can also be described as ocean deposits, and non-marine, which are found in standing bodies of water such as lakes.

Event—In the context of the reporting and analysis of events, an event is any occurrence unintended by the operator, including operating error, equipment failure or other mishap, and deliberate action on the part of others, the consequences or potential consequences of which are not negligible from the point of view of protection and safety.

Exposure—see radiation exposure.

Favorable overall geological situation—Geological site characteristics which, taken as a whole, indicate that the requirements for final disposal are highly likely to be met.

Federal state collection facilities—facilities of the federal states in Germany for the interim storage or processing of radioactive waste generated in their territory.

FEP—"Features", "Events", and "Processes"; are terms used in the field of radioactive waste management to define relevant scenarios for safety assessment studies. For a radioactive waste repository, features would include the characteristics of the site, such as the type of soil (rock) or geological formation the repository is to be built on or under. Events would include things that may or will occur in the future, like, e.g., glaciations, droughts, earthquakes, or formation of faults. Processes are things that are ongoing, such as the erosion or subsidence of the landform where the site is located on, or near.

Final storage (Disposal)—Permanent storage of radioactive waste (in deep geological formations).

Flooding monitoring—Monitoring of all consequences on the environment of the flooding of underground mining plant.

Fresh mine air supply, also mine ventilation—Mining term for technical measures to supply mines with fresh air.

Fuel assembly—Common fuel elements consist of several tens to several hundred fuel rods, which are bundled by components. They form together with the other internals the reactor core. The hermetically sealed fuel rods contain the nuclear fuel. Burnt fuel elements additionally contain the highly radioactive products of fission and newly formed actinides.

GAU—worst possible or foreseeable accident. The worst possible accident considered for controlling, engineering, or planning when designing something such as a nuclear or industrial plant.

Geological barrier—Geological layers between the storage area and the biosphere which, due to their properties and dimensions, hinder or prevent pollutant dispersion (contamination plume).

Geotechnical barrier—Closure measures for shafts, galleries, tunnels and/or boreholes after completion of the storage of waste in a repository. These closure measures will be used to achieve a secure long-term seal.

Governance—comprises all of the processes of governing—whether undertaken by the government of a state, by a market or by a network—over a social system (family, tribe, formal or informal organization, a territory or across territories) and whether through the laws, norms, power or language of an organized society. It relates to "the processes of interaction and decision-making among the actors involved in a collective problem that lead to the creation, reinforcement, or reproduction of social norms and institutions".

GWd—Giga Watt-days; Electricity companies express power output from their plants in terms of gigawatt-days rather than kilowatt-hours. One gigawatt-day equals 24 million kilowatt-hours, i.e. 1 million kilowatts during 24 h.

Half-lives ($T_{1/2}$)—The time in which radioactivity of a nuclide is halved by radioactive decay.

HAW-coquilles—Stainless steel container filled with vitrified High Active Waste (HAW) from reprocessing.

HLW/HAW—HLW (abbreviation for High Level Waste)/HAW (abbreviation for High Active Waste), both means highly radioactive waste.

Host rock—Suitable geological formation for receiving a repository. The nature, composition and properties of the rocks surrounding of repository sites primarily determine how well these natural barriers protect the environment from radiation. Internationally, rock salt, clay, claystone and magmatic rocks—crystalline rocks (e.g. granite) are considered as possible host rocks.

IAEA—International Atomic Energy Agency.

Immobilisat—An active ingredient that achieves immobilization, with which one of the effects can be achieved: Physical encapsulation; chemical integration; precipitation; sorption.

Immobilisation—Immobilisation is intended to influence the contaminated material in such a way, that emissions of the contained pollutants are prevented in the long term or are below specified target values. Physical encapsulation; chemical integration; precipitation; sorption. As a rule, there is no elimination or destruction of the pollutants. **Example**: In-situ immobilisation: In-situ immobilization of arsenic and uranium by injectable amounts of iron particles and stimulated autotrophic sulfate reduction.

Infiltration water—the part of the precipitation seeping into the ground.

Integrity—The term describes the preservation of the properties of the containment capacity of the host rock and the ECZ and all other barriers of a repository in the period under review.

Ionization—is the process by which an atom or a molecule acquires a negative or positive charge by gaining or losing electrons to form ions, often in conjunction with other chemical changes.

Ionising radiation—Any radiation that is able to cause ionization processes in atoms and molecules in the matter it passes through.

Isolation—in the context of final disposal means that the release and transport of radionuclides are impeded to such an extent that no or only small quantities of radioactive substances enter the biosphere.

Isolation period—is the period of time for which technical and geotechnical barriers have to work together so, that the isolation of waste and residues is ensured.

Isotopes—Atoms of the same chemical element with the same number of protons and electrons, but different number of neutrons. Isotopes therefore have the same chemical properties, but different nuclear physical characteristics.

Interim storage (temporary)—Storage of radioactive waste in an interim storage facility until it is transported to the repository or conditioned for final storage.

Level—A series of structures in a mine all at the same horizontal level or elevation.

Liquid waste—Liquid wastes from nuclear power plants, industry, medicine and research include oils, sludges, evaporator concentrates, ion exchange resins and filter aids.

LLW/LAW—LLW (abbreviation for Low Level Waste)/ LAW (abbreviation for Low Active Waste), both describe low level radioactive waste.

Local dose rate—The radiation dose of gamma radiation absorbed per time unit, measured at a fixed location, expressed in sieverts per hour (Sv/h).

Long term prognosis—A prognosis is generally defined as the prediction of future developments with the aid of a probability conclusion based on experience or/and calculations. The long-term prognosis is the scientifically based derivation of possible developments of the repository system with its components relevant for safety and their respective probability.

Long-term safety—Long-term safety is ensured by a repository system that separates stored radioactive waste materials from the material cycles of the biosphere for the required period of time.

Long-term safety analysis—describes the analysis of the long-term behaviour of the repository after decommissioning. A central aspect is the analysis of the containment capacity of the repository system and its reliability. It is a prerequisite for long-term safety assessment (proof).

Long-term safety assessment—Long-term safety assessment is the assessment of the long-term safe containment of radioactive waste in geotechnical environmental structures, in this case repositories in geological formations. By combining all analyses and arguments, the long-term safety assessment is founded to justify the safety for the post-operational phase of the repository. The long-term safety assessment may be divided into a set of radiological and non-radiological safety assessments to be carried out for the operation and post-closure operation of a repository.

One element of the long-term safety assessment is the long-term safety analysis. In this analysis, the future behaviour of the repository is investigated and the possible future radiation exposure at the repository site is determined. In addition, the long-term safety assessment also contains further justifications for the robustness and reliability of the safety evaluation. The long-term safety assessment is not identical to the approval of the repository. The long-term safety assessment is submitted to the competent authority. The competent authorities have to check the proofs submitted by the applicant. You can request changes or additions and confirm the assessments as guided or refuse their confirmation.

Mine air—mining term for the air or the gas mixture in a mine.

Mining claim—The area of land which has been set aside to be developed by mining activities.

Mine ventilation—Targeted control of the quality and quantity of the mine air through the mine.

MLW/MAW—MLW (abbreviation for Medium Level Waste)/MAW (abbreviation for Middle Active Waste), both mean medium level radioactive waste.

MOX—Abbreviation for Mixed oxide fuel, commonly referred to as MOX fuel, is nuclear fuel that contains more than one oxide of fissile material, usually consisting of plutonium blended with natural uranium, reprocessed uranium, or depleted uranium.

NEA—Nuclear Energy Agency, the OECD's nuclear energy department (Organisation for Economic Co-operation and Development).

(German) **Nuclear Waste Management Commission (ESK)**—advises the Federal Environment Ministry on all aspects of nuclear waste management (Conditioning, temporary storage and transport of radioactive substances and waste, decommissioning and dismantling of nuclear facilities, final disposal in deep geological formations). The members work voluntarily and are appointed on a temporary basis.

Nucleon—the main components of the atomic nucleus (protons and neutrons).

Nuclide—An atomic nucleus defined by its mass- and nuclear charge number.

Optimization—Optimization is the term used to describe the process that aims to optimize the risks arising from final deposited waste to make them as low as reasonably possible, taking into account the state of the art in science and technology and taking into account all the circumstances of the individual case, including social and economic factors.

Overburden—This refers to the geological strata above the effective containment zone (ECZ); generally not of economic mineral value.

Partitioning—Separation of radionuclides with long half-lives (e.g. plutonium, americium, curium) from a waste matrix.

Passive safety—Safety, in particular long-term safety, of a repository system derived from the particularly favorable permanent inclusion properties of geological and geotechnical barriers (in Germany according to German Federal Ministry for Environment, Protection of Nature and Reactor Safety 2010: safety requirements for the disposal of heat-generating radioactive waste, in particular the effective containment zone) and the technical barriers of a repository system, and the measures to ensure their protective function for humans and the environment (monitoring, maintenance, repair) demonstrably not required. The security functions of the barriers, which are relevant for security, are based on natural laws, the validity and applicability of which, however, has to be proven for every repository system.

Permeability—Permeability in fluid mechanics and the earth sciences (commonly symbolized as k) is a measure of the ability of a porous material (often, a rock or an unconsolidated material) to allow fluids to pass through it. The permeability of a medium is related to the porosity, but also to the shapes of the pores in the medium and their level of connectedness.

Period under review (Detection period)—the relevant (required) period in a long-term safety assessment for the radioactive waste stored in a repository that one can almost certainly be assumed that these remain separated of material cycles of biosphere.

Pilot storage facility—Temporary storage facility for monitoring a representative portion of waste, underground geoscientific structures, infrastructure facilities and access tunnels or shafts. The pilot storage facility is integrated into the repository concept. A representative part of the planned total inventory is stored and transferred to the pilot storage facility at a relatively early stage. Systematic monitoring is prepared for this. The technical and geotechnical barriers are monitored beyond the operating time of the main storage facility. Its storage sections and the final storage containers will be surrounded by an extensive and complex measuring and monitoring system (wired probes, sensors, etc.). This measuring and observation system should be designed in such a way that the integrity of the chambers and or sections is not violated. In this way, the behaviour of the safety barriers after closure could be analysed and results could be obtained in an observation phase that is currently not specified in more detail. These could allow conclusions to be drawn about the functioning of the main bearing and thus provide early optimisation and empirical evidence for its safety, as well as signal emergency measures (such as retrievals) which would be necessary if central conceptual weaknesses were already to become apparent in the pilot storage facility.

Plan approval procedure—Special approval procedures with citizen participation in major projects, regulated in the German Administrative Procedure Act (VwVfG).

Pollux Container—containers possibly intended for final storage/disposal of waste.

Pore diffusion—Pollutants are transported through water-filled rock pores. This must be distinguished from solid-state diffusion, which is considerably slower goes on than pore diffusion.

Pore volume—is the volume fraction, the sum of air and water volumes in a porous medium, which consisting of a three-phase system of air/water/soil.

Pore water—Water or aqueous solutions in a porous medium, here in soil or rock pores. The understanding of such reactions, for example pore diffusion, which determine the solubility of the corresponding substances in pore water and thus also the diffusion-controlled leaching, is of great importance.

Porosity—Porosity or void fraction is a measure of the void (i.e. "empty") spaces in a material, and is a fraction of the volume of voids over the total volume, between 0 and 1, or as a percentage between 0 and 100%.

Post closure phase—Period after the safe closure of the repository; Period after all radioactive waste has been stored and the repository has been sealed.

Proliferation—Proliferation of nuclear weapons from states that possess such technologies to other states that do not yet have them, or the expansion and further development of their own stocks.

Protection objective (goal)—Since the radioactivity is detectable throughout the biosphere, a standard of comparison must be used, e.g. which deviation by introduced radioactive waste and/or residues is tolerable in comparison to the natural environmental radioactivity.

Pump stowing—Pumping a viscous thick matter into an open mine room to fill it.

Qualified permission issued under water law—Issued Permission (under water law) for a project that has not yet been used and will later be activated (qualified). Also "conditional approval".

Radiation exposure—Influence of radiation on the human body. In the case of external radiation exposure, the radiation acts on the body from the outside. In the case of internal radiation exposure, radionuclides were taken in with the inhaled air and the food (ingestion). Until the elimination of radionuclides, their radiation from the inside affects organs and tissue. The measure of exposure to radiation by ionizing radiation is the effective dose.

Radiation Protection Commission [German Radiation Protection Commission (SSK)]—The German Radiation Protection Commission (SSK) advises the Federal Ministry for Environment, Protection of Nature and Reactor Safety in all matters of protection against ionizing and non-ionizing radiation. The results of the deliberations are addressed to the Federal Ministry as recommendations or statements and among others posted on a website. Its members are appointed on a temporary basis and work voluntarily.

The International Commission on Radiological Protection (ICRP) is an independent, international, non-governmental organization, with the mission to provide recommendations and guidance on radiation protection. The ICRP is a not-for-profit organization registered as a charity in the United Kingdom and has its scientific secretariat in Ottawa, Ontario, Canada.

Radioactivity—Decay of unstable atomic nuclei with the emission of ionizing radiation.

Radiolysis—Radiolysis is the disassembling of a chemical bond under the influence of ionizing.

Radionuclides—Are unstable atomic nuclei, which spontaneously decay with the emission of high-energy (ionizing) radiation and thereby transform themselves into other types of atoms by releasing core components. Characteristic of each radionuclide is its half-life. There are both natural and artificial radionuclides.

Radiotoxicity—Harmfulness of incorporated radionuclides due to their ionizing radiation. Radiotoxicity depends on the length of time the radionuclide is in the body, the type of radiation (alpha rays, beta rays, gamma rays, relative biological activity), the energy, the half-life (biological half-life) of the nuclide and its chemical behavior in the body. Most dangerous are radionuclides with a large half-life that emit alpha or beta rays. In addition to tissue damage and genetic damage (mutagens) may occur as late damage i.a. Carcinomas and sarcomas (cancer).

Recoverability/Recovery—The term recovery is not included in the NEA concept. It is a term from the German Repository Site Selection Act.

By "recoverability" is meant a retrieval of containers with high level radioactive waste when the repository mine is completely closed (covert, sealed). For the recovery of high radioactive waste, no special precautions have been taken or no special precautions are known. The recovery of waste from the already closed repository or repository area, if necessary for a certain period of time after closure of the repository, if necessary by excavating it again. For this purpose, the already stored waste containers must remain intact for a fixed period of time (in Germany 500 years). Recovery is the unplanned retrieval of radioactive waste from a repository as an emergency measure.

Redissolution (dissolving and reprecipitating)—Interaction of unsaturated solutions with rocks by solution and crystallization of individual components.

Reference level—In an emergency exposure situation or an existing exposure situation, the level of dose, risk or activity concentration above which it is not appropriate to plan to allow exposures to occur and below which optimization of protection and safety would continue to be implemented.

Reference measuring—Representative measuring point which can be used to assess a change in the environment during environmental monitoring.

Reference scenario—A hypothetical but possible evolution of a disposal facility and its surroundings on the basis of activities, such as construction work, mining or drilling, that have a high probability of being undertaken by people in the future and that could cause a human intrusion into the disposal facility, and which can be evaluated.

Regional and Local authorities—Authorities whose jurisdiction and membership is determined territorially.

Release—The release is a formal act to release waste for disposal whose radionuclide concentration no longer poses a danger. A distinction is made between conditional and unconditional release with regard to the further use of the material.

Remaining opencast—after completion of the mining activity remaining part of an opencast mine (also void or pit).

Repository system—A repository system consists of the repository mine, the effective containment zone (ECZ) and the surrounding or overlapping geological strata up to the earth's surface, insofar as they are important in terms of safety.

Legacy mine—A legacy mine consists of various components such as shafts, drifts and pits as well as waste rock dumps which no longer has a responsible owner and/or operator" and is unrestored.

Reprocessing—A method of separating unused fissile material from the remaining spent fuel components with the goal these are used again for energy generating. Not allowed in Germany since 2005. Therefore, spent fuel since then is sent for direct disposal.

Retrieval: retrieval of the deposited waste some time after closure of repository.

Retrievability—"Retrievability is the ability in principle to recover waste or entire waste packages once they have been emplaced in a repository; retrieval is the concrete action of removal of the waste. Retrievability implies making provisions in order to allow retrieval should it be required."

Retrievability: Legally prescribed in different countries so in Germany: Guaranteed recovery of the radioactive waste over a period of 500 years after closure of the repository. In the legislation of the USA, France, GB, Sweden, Finland, Switzerland, etc., the retrievability anchored (The Nuclear Waste Policy Act- 1987- The U.S. Nuclear Regulatory Commission requirements are given in Title 10 of the Code of Federal Regulations (CFR), Part 60 and draft Part 63).

Reversibility—"Reversibility describes the ability in principle to reverse or reconsider decisions taken during the progressive implementation of a disposal system; reversal is the concrete action of overturning a decision and moving back to a previous situation." A new concept: is needed to allow mistakes to be corrected, to leave scope for action for future generations—for example to take account of new scientific knowledge—and can help to build trust in the procedure. Concepts for the retrievability or recovery of the waste that enable the reversibility of decisions are central here.

Risk—The International Atomic Energy Agency's (IAEA) Glossary for Nuclear Safety and radiation protection cites as a possible definition for the term risk: A multi-parameter variable that is a measure of any accidental or causal, actual or potential hazard that results in actual or potential health hazard or harm. (Risk: A multi-attribute quantity expressing hazard, danger or chance of harmful or associated with actual or potential exposures.)

The risk indicator used here is the probability that a person living near the repository suffers serious health damage during its lifetime caused by radioactive substances released from the effective containment zone (ECZ) of the repository.

As a measure of risk is often the expected loss, i. e. the sum over the amount of possible damage, multiplied by the respective probabilities of their occurrence used. Such a quantitative description of the overall risks posed by the repository is only limited.

Robustness—Robustness denotes the reliability and quality, and thus the insensitivity of, the security functions of the repository system and its barriers to internal and external influences and disturbances, as well as the insensitivity of the safety analysis results to deviations from the underlying assumptions.

Rock mass—rock which is loose from the deposit.

Roof (of a cavern)—Mining term for the false ceiling between two chambers in a mine.

Safety—The term safety is used in a technical sense. The repository is technically safe enough if the risks to humans and the environment from the final disposal of radioactive waste are less than the permissible risks.

 - **Active safety**—Safety, in particular long-term safety, of a warehouse which relies substantially or solely on the technical barriers of the storage system and measures to ensure its protective function for man and the environment (monitoring, maintenance, repair).

 - **Passive safety**—Safety, in particular long-term safety, of a repository system derived from the particularly favorable permanent inclusion properties of geological and geotechnical barriers (in Germany according to German Federal Ministry for Environment, Protection of Nature and Reactor Safety 2010: safety requirements for the disposal of heat-generating radioactive waste, in particular the effective containment zone) and the technical barriers of a repository system, and the measures to ensure their protective function for humans and the environment (monitoring, maintenance, repair) demonstrably not required. The security functions of the barriers, which are relevant for security, are based on natural laws, the validity and applicability of which, however, has to be proven for every repository system.

Safety analysis—The behavior of the repository system is analyzed under a variety of load situations, taking into account data uncertainties, malfunctions and future development opportunities with regard to the safety functions.

Safety case—A collection of arguments and evidence in support of the safety of a facility or activity. This will normally include the findings of a safety assessment and a statement of confidence in these findings. Safety assessment is the main component of the safety case. For a repository, the safety case may relate to a given stage of development. In such cases, the safety case should acknowledge the existence of any unresolved issues and should provide guidance for work to resolve these issues in future development stages. Includes the review and evaluation of data, measures, analysis and arguments that demonstrate the safety of the repository. It includes the pooling of all evidence to be provided. A distinction is made between evidence for the operational phase of the repository and the post-operational phase (long-term safety assessment).

Safety function—A property or process that ensures the safety-relevant requirements in a repository. The interaction of such functions ensures the fulfillment of all safety requirements, both in the operating phase and in the post-closure phase.

Safety margins (reserves)—Properties that lead to a degree of safety beyond the required safety, e.g. Safety reserves can be due to the sealing properties of the overburden.

Safety management—All activities of planning, organization, management and control with the aim of defining and continue developing strategies and processes that lead to the implementation of safety requirements and to the continuous improvement of the safety level.

Salt creep—property of salt rock to self close existing cavities.

Salt grit—Backfill material in the final disposal of radioactive waste in the host rock salt.

Scenario—describes an assumed development of the repository system and its security functions. For the long-term safety analysis, scenarios are to be selected that represent probable developments and a representative range of less probable developments.

Silts (ISO 14688 grades slits)—Uncemented fine soils and sedimentary rocks consisting of at least 95% components with a particle size of 0.002 to 0.063 mm.

Sinking—Mining term for creating vertical cavities, generally include shafts, boreholes or bunkers.

Solid waste—Solid wastes from nuclear power plants, industry, medicine and research include paper, plastics, textiles, metal parts, tools, insulation material, construction waste and structural parts.

Solution path—Transport path from radionuclides in dissolved form to the earth's surface. Pathway of radionuclides in dissolved form from the ECZ to the earth's surface.

Sorel concrete—Special salt concrete used for backfilling (pump stowing) of mine salt mines. Components are rock salt, sand, limestone, cement and water.

Sorption—To be understood here as adsorption of pollutants to the host rock. Disables and limits the transport of radionuclides into the biosphere.

Spent nuclear fuel—Fuel elements that have reached their intended service life and therefore need to be replaced, also reprocessed spent nuclear fuel.

Storage area—A storage area is part of a repository mine, where radioactive waste is stored in storage fields and finally sealed against the infrastructure area and possibly other against storage areas.

Storage field—A storage field is part of a storage area. It contains storage drifts and/or—boreholes and -chambers, including cross cuts, through which the storage field is connected to the transport galleries of the storage area.

Stratigraphy—Stratigraphy is a branch of geology concerned with the study of rock layers (strata) and layering (stratification). It is primarily used in the study of sedimentary and layered volcanic rocks. Science of the temporal succession of deposited rocks. Rocks that have been deposited over other rocks are younger.

Subcriticality—State in which by fission fewer neutrons are produced, than disappear by absorption and leakage, i.e. the number of nuclear fissions decreases continuously. This condition must be permanently ensured in a repository or in the repository containers.

Tailings—Mining residues, ground up rock residue from mineral processing and extraction; here stored residues from the uranium ore processing in tailings ponds (sedimentation ponds).

Tailings pond (sedimentation pond; tailings storage)—Sedimentation pond for stored tailings, here residues from the uranium ore processing.

Tectonics—Science of the structure of the earth's crust in its structure and its large-scale movement.

Temporary (interim) storage—Storage of radioactive waste in an interim storage facility until it is transported to the repository or conditioned for final storage.

Test value—The safety proof is about detection, calculation, estimation or measurement defined properties. For these properties, certain values have to be complied with in some cases—so-called test values—and are specified (e.g. the risk indicators or the criteria for evaluating the integrity of the ECZ).

If all the required properties are met and if all test values have a high probability to be met, the safety assessment should assume that the protection goals are achieved and the protection criteria are met.

Tons of heavy metal (tHM)—Tons of heavy metal as a unit of measure in the reprocessing or disposal of waste from nuclear technology. Corresponds to megagram (mg) heavy metal, usual unit of measurement for nuclear fuel; Mass of uranium and plutonium in the fuel element before use in the reactor.

Transmutation—Conversion of long lived nuclides into shorter-lived radionuclides to reduce radiotoxicity.

Transport—Here the movement of substances in another medium. Advective transport refers to the movement of a substance dissolved or suspended in water with the flow of water. Diffusive transport refers to the movement of a substance through Brownian motion and concentration differences. The latter takes place with very slow speed also in solid matter.

Transuranium elements—Elements with a higher atomic number than uranium. They arise during nuclear fission.

Tunnel (gallery)—Mine structure, almost horizontal connection of a mine to the earth's surface.

Upcast air—from underground used air; exhaust air from mining installations.

Upcast ventilation shaft—Shaft through which used air and harmful gases from the mine structures are discharged to the surface; Often the suction is amplified by fans. See Mine ventilation.

Unexploited ground (deposit)—Mountain massif or deposit without mining improvement, here without excavation for final disposal, underground laboratories or sampling by drilling etc.

Vitrified, high-level radioactive fission products—In a borosilicate glass matrix cemented high-level radioactive waste, which arise during the reprocessing of spent fuel.

Waste rock (Tail) dump—Filling of waste rock, which arise in excavation of access to the underground and extraction to ore.

Water handling—That means in mining, a part of the mine infrastructure, underground rooms and technical facilities that serve to keep the mine clear of mine water. This includes the removal of mine water from the underground area; the natural draining from a tunnel; well systems for removing and lowering groundwater around the open pit area (keeping dry). May also involve water treatment to achieve suitable quality for discharge to the environment.

Bibliography

[1] IAEA SAFETY GLOSSARY—TERMINOLOGY USED IN NUCLEAR SAFETY AND RADIATION PROTECTION; 2018 EDITION; © IAEA, 2019; ISBN 978–92–0–104718–2 or: https://www-pub.iaea.org/MTCD/Publications/PDF/PUB1830_web.pdf

[2] IAEA Safety Glossary; Terminology used in Nuclear Safety and Radiation Protection; IAEA; Vienna 2007, ISBN 92-0-100-707-8

[3] Features, Events and Processes (FEPs) for Geologic Disposal of Radioactive Waste - An International Database; OECD PUBLICATIONS, 2, rue André-Pascal, 75775 PARIS CEDEX 16 PRINTED IN FRANCE (66 2000 14 1 P) ISBN 92-64-18514-3 – No. 51449 2000

[4] Radioactive waste management glossary; Vienna: International Atomic Energy Agency, 2003; STI/PUB/1155; ISBN 92-0-105303-7

[5] Home/Nuclear Basics/Glossary; World-Nuclear-Association; (Updated March 2014) http://www.world-nuclear.org/nuclear-basics/glossary.aspx

[6] ISO 12749-3:2015(en); Nuclear energy, nuclear technologies, and radiological protection — Vocabulary—Part 3: Nuclear fuel cycle; https://www.iso.org/obp/ui/#iso:std:iso:12749:-3:ed-1:v1:en

[7] US Nuclear Regulatory Commission (NRC); Full-Text Glossary; https://www.nrc.gov/reading-rm/basic-ref/glossary/full-text.html

Printed in the United States
By Bookmasters